Methods in Molecular Biology

For further volumes:
http://www.springer.com/series/7651

For over 35 years, biological scientists have come to rely on the research protocols and methodologies in the critically acclaimed *Methods in Molecular Biology* series. The series was the first to introduce the step-by-step protocols approach that has become the standard in all biomedical protocol publishing. Each protocol is provided in readily-reproducible step-by-step fashion, opening with an introductory overview, a list of the materials and reagents needed to complete the experiment, and followed by a detailed procedure that is supported with a helpful notes section offering tips and tricks of the trade as well as troubleshooting advice. These hallmark features were introduced by series editor Dr. John Walker and constitute the key ingredient in each and every volume of the *Methods in Molecular Biology* series. Tested and trusted, comprehensive and reliable, all protocols from the series are indexed in PubMed.

Skin Tissue Engineering

Methods and Protocols

Second Edition

Edited by

Thomas Biedermann and Sophie Böttcher-Haberzeth

University Children's Hospital Zurich, Zurich, Switzerland

Humana Press

Editors
Thomas Biedermann
University Children's Hospital Zurich
Zurich, Switzerland

Sophie Böttcher-Haberzeth
University Children's Hospital Zurich
Zurich, Switzerland

ISSN 1064-3745 ISSN 1940-6029 (electronic)
Methods in Molecular Biology
ISBN 978-1-0716-4509-3 ISBN 978-1-0716-4510-9 (eBook)
https://doi.org/10.1007/978-1-0716-4510-9

This Humana imprint is published by the registered company Springer Science+Business Media, LLC, part of Springer Nature.
The registered company address is: 1 New York Plaza, New York, NY 10004, U.S.A.

If disposing of this product, please recycle the paper.

Preface

We are delighted to introduce the second edition of *Skin Tissue Engineering Methods and Protocols*, a volume in the esteemed Methods in Molecular Biology series of Springer Protocols. This new edition comes at a crucial time in the field of regenerative medicine, where skin tissue engineering has emerged as a transformative approach for addressing complex skin injuries, including extensive wounds caused by trauma, burns, and chronic skin diseases. The progress made in this field continues to push the boundaries of both clinical treatments and laboratory research, offering new hope for patients with otherwise untreatable conditions.

This edition builds on the foundation of the first, reflecting the latest advancements in the field while expanding on the protocols and methods that have proven most valuable to researchers. With a comprehensive array of protocols, this volume is specifically designed to assist researchers in isolating and culturing a wide range of skin-derived cell types and in applying these cells to develop constructs that support wound healing, bioengineering, and translational medicine. The scope of the book has been broadened to provide a deeper understanding of the complexities involved in skin regeneration and to address emerging challenges in applying these techniques in clinical practice.

The volume is organized into four distinct sections, offering a clear and structured approach to the various stages of skin tissue engineering. The first section focuses on the isolation and culture of diverse skin cell types, offering detailed methodologies for working with keratinocytes, fibroblasts, melanocytes, and other critical cell types that form the foundation of skin regeneration. The second section explores the integration and management of these cells within co-culture systems and advanced in vitro skin models, a necessary step toward replicating the intricate architecture and function of human skin. The third section shifts focus to the application and analysis of these skin models in experimental settings in vivo, providing insights into how engineered skin constructs perform under physiological conditions, including in wound healing and other translational contexts. Finally, the fourth section covers the translational applications of these techniques, emphasizing the clinical potential of skin tissue engineering and the steps required to bridge the gap between laboratory research and real-world medical applications.

This volume is intended to serve as a valuable resource for a wide audience of basic and clinical researchers, including biologists, physicians, and biomedical engineers, all of whom are engaged in advancing the science and clinical application of translational regenerative medicine. The methodologies presented are particularly focused on skin tissue engineering, but the principles and techniques are applicable to various aspects of tissue engineering and regenerative medicine. Each chapter offers a detailed list of materials, step-by-step methodological descriptions, and valuable personal insights from the authors to ensure the successful execution of the techniques described.

Thomas Biedermann
Sophie Böttcher-Haberzeth

Contents

Contributors

MIGUEL ALAMINOS • *Tissue Engineering Group, Department of Histology, University of Granada, Granada, Spain; Instituto de Investigacion Biosanitaria ibs.GRANADA, Granada, Spain*

MARGARETE ARRAS • *Biological Central Laboratory & Division of Surgical Research, University Hospital Zurich, University of Zurich, Zurich, Switzerland*

FRANÇOIS A. AUGER • *Centre de recherche en organogénèse expérimentale de l'Université Laval/LOEX, Québec, QC, Canada; CHU de Québec – Université Laval Research Center, Division of Regenerative Medicine, Québec, QC, Canada; Department of Surgery, Faculty of Medicine, Université Laval, Québec, QC, Canada*

CÉLINE AUXENFANS • *Banque de tissus et cellules, Plateforme de Biotherapies, Hospices Civils de Lyon, Lyon, France*

GERY BARMETTLER • *Center for Microscopy and Image Analysis, University of Zurich, Zurich, Switzerland*

FABIOLA BERMEJO-CASARES • *Tissue Engineering Group, Department of Histology, University of Granada, Granada, Spain*

THOMAS BIEDERMANN • *Tissue Biology Research Unit, Department of Pediatric Surgery, University Children's Hospital Zurich, Zurich, Switzerland; Children's Research Center, University Children's Hospital Zurich, Zurich, Switzerland; University of Zurich, Zurich, Switzerland*

THOMAS BIRNGRUBER • *HEALTH – Institute for Biomedical Research and Technologies, JOANNEUM RESEARCH Forschungsgesellschaft mbH, Graz, Austria*

SOPHIE BÖTTCHER-HABERZETH • *Pediatric Burn Center, Children's Skin Center, Department of Surgery, University Children's Hospital Zurich, University of Zurich, Zurich, Switzerland; Department of Surgery, University Children's Hospital Zurich, University of Zurich, Zurich, Switzerland; Children's Research Center (CRC), University Children's Hospital Zurich, University of Zurich, Zurich, Switzerland; University of Zurich (UZH), Zurich, Switzerland*

JENNIFER BOURLAND • *Centre de recherche en organogénèse expérimentale de l'Université Laval/LOEX, Québec, QC, Canada; CHU de Québec – Université Laval Research Center, Division of Regenerative Medicine, Québec, QC, Canada; Department of Surgery, Faculty of Medicine, Université Laval, Québec, QC, Canada*

LUDWIK K. BRANSKI • *Department of Surgery, University of Texas Medical Branch and Shriners Hospitals for Children® – Galveston, Galveston, TX, USA; Division of Hand, Plastic and Reconstructive Surgery, Department of Surgery, Medical University of Graz, Graz, Austria*

JEAN-FRANÇOIS BRUNET • *Cell Production Center, Service of Pharmacy, Lausanne University Hospital, Epalinges, Switzerland*

FERNANDO CAMPOS • *Tissue Engineering Group, Department of Histology, University of Granada, Granada, Spain; Instituto de Investigacion Biosanitaria ibs.GRANADA, Granada, Spain*

MURIEL CARIO • *INSERM 1312, University of Bordeaux, Bordeaux Cedex, France*

GLORIA CARMONA SÁNCHEZ • *Andalusian Network for the Design and Translation of Advanced Therapies (AND&TAT) – Fundación Pública Andaluza Progreso y Salud, Sevilla, Spain*

VÍCTOR CARRIEL • *Tissue Engineering Group, Department of Histology, University of Granada, Granada, Spain; Instituto de Investigacion Biosanitaria ibs.GRANADA, Granada, Spain*

BETTINA CATTIER • *Centre de recherche en organogénèse expérimentale de l'Université Laval / LOEX, Centre de recherche du CHU de Québec-Université Laval, Axe médecine régéné ratrice, and Département de Chirurgie, Faculté de Médecine, Université Laval, Québec, QC, Canada*

NIKOLA CESAROVIC • *Biological Central Laboratory & Division of Surgical Research, University Hospital Zurich, University of Zurich, Zurich, Switzerland*

JESÚS CHATO-ASTRAIN • *Tissue Engineering Group, Department of Histology, University of Granada, Granada, Spain; Instituto de Investigacion Biosanitaria ibs.GRANADA, Granada, Spain*

ALVIN WEN CHOONG CHUA • *Department of Plastic, Reconstructive and Aesthetic Surgery, Singapore General Hospital, Singapore, Singapore*

SERGIO CORTEZ GHIO • *Centre de recherche en organogénèse expérimentale de l'Université Laval / LOEX, Centre de recherche du CHU de Québec-Université Laval, Axe médecine ré génératrice, and Département de Chirurgie, Faculté de Médecine, Université Laval, Qué bec, QC, Canada; In Silico Data Science, Research Support Department, Québec, QC, Canada*

HENRI DE KONINCK • *Centre de recherche en organogénèse expérimentale de l'Université Laval / LOEX, Centre de recherche du CHU de Québec-Université Laval, Axe médecine ré génératrice, and Département de Chirurgie, Faculté de Médecine, Université Laval, Qué bec, QC, Canada*

ROCIO GARCÍA DE LA CRUZ VALENCIA • *Pediatric Burn Center, Children's Skin Center, Department of Surgery, University Children's Hospital Zurich, University of Zurich, Zurich, Switzerland; Department of Surgery, University Children's Hospital Zurich, University of Zurich, Zurich, Switzerland; Children's Research Center (CRC), University Children's Hospital Zurich, University of Zurich, Zurich, Switzerland*

PAOLA DE LOS SANTOS GOMEZ • *Department of Biosciences, Durham University, Durham, UK*

JÖRN DENGJEL • *Department of Biology, University of Fribourg, Fribourg, Switzerland*

JORIS A. VAN DONGEN • *Department of Plastic, Reconstructive and Hand Surgery, University Medical Center Utrecht, Utrecht University, Utrecht, The Netherlands*

MELINDA FARKAS • *Department of Surgery, Pediatric Burn Center, University Children's Hospital Zurich, University of Zurich, Zurich, Switzerland; Children's Research Center, University Children's Hospital Zurich, University of Zurich, Zurich, Switzerland; Faculty of Medicine, University of Zurich (UZH), Zurich, Switzerland*

KAREL FERLAND • *Centre de recherche en organogénèse expérimentale de l'Université Laval / LOEX, Centre de recherche du CHU de Québec-Université Laval, Axe médecine régéné ratrice, and Département de Chirurgie, Faculté de Médecine, Université Laval, Québec, QC, Canada*

THEA FLEISCHMANN • *Biological Central Laboratory & Division of Surgical Research, University Hospital Zurich, University of Zurich, Zurich, Switzerland*

INES FOESSL • *COREMED-Centre for Regenerative and Precision Medicine, JOANNEUM RESEARCH Forschungsgesellschaft mbH, Graz, Austria; Division of Endocrinology and Diabetology, Department of Internal Medicine, Medical University of Graz, Graz, Austria*

SASKIA M. FOGG • *School of Chemical Engineering, University of Birmingham, Edgbaston, UK*

JULIE FRADETTE • *Centre de recherche en organogénèse expérimentale de l'Université Laval/ LOEX, Québec, QC, Canada; CHU de Québec – Université Laval Research Center, Division of Regenerative Medicine, Québec, QC, Canada; Department of Surgery, Faculty of Medicine, Université Laval, Québec, QC, Canada*

LUCIE GERMAIN • *Centre de recherche en organogénèse expérimentale de l'Université Laval / LOEX, Centre de recherche du CHU de Québec-Université Laval, Axe médecine régéné ratrice, and Département de Chirurgie, Faculté de Médecine, Université Laval, Québec, QC, Canada*

KIRSTY GONCALVES • *Department of Biosciences, Durham University, Durham, UK*

ANA FERNÁNDEZ GONZÁLEZ • *Unidad de Producción Celular e Ingeniería Tisular, Hospital Universitario Virgen de las Nieves, Granada. Red Andaluza de diseño y traslación de Terapias Avanzadas de la Fundación Progreso y Salud, Sevilla, Spain*

LIAM M. GROVER • *School of Chemical Engineering, University of Birmingham, Edgbaston, UK*

ALFRED GUGERELL • *Laboratory for Cardiac and Thoracic Diagnosis and Regeneration, Department of Thoracic Surgery, Medical University of Vienna, Vienna, Austria*

MARTIN C. HARMSEN • *Department of Pathology & Medical Biology, University Medical Center Groningen, University of Groningen, Groningen, The Netherlands*

CRISTINA HAS • *Department of Dermatology, Medical Center-University of Freiburg, Faculty of Medicine, University of Freiburg, Freiburg, Germany*

YINGHONG HE • *Department of Dermatology, Medical Center-University of Freiburg, Faculty of Medicine, University of Freiburg, Freiburg, Germany*

ELISABETH HOFMANN • *COREMED-Centre for Regenerative and Precision Medicine, JOANNEUM RESEARCH Forschungsgesellschaft mbH, Graz, Austria; Division of Plastic, Aesthetic and Reconstructive Surgery, Department of Surgery, Medical University of Graz, Graz, Austria; Research Unit for Tissue Regeneration, Repair and Reconstruction, Division of Plastic, Aesthetic and Reconstructive Surgery, Department of Surgery, Medical University of Graz, Graz, Austria*

ALEXANDRA KARSTRÖM • *Department of Plastic and Maxillofacial Surgery, Uppsala University Hospital, Uppsala, Sweden*

MAIKE KECK • *Agaplesion Diakonieklinikum Hamburg, Hamburg, Germany; Clinic for Plastic Surgery, University of Lübeck, Lübeck, Germany*

FREDERIQUE M. KEMME • *Alliance of Dutch Burn Care (ADBC), Burn Centre, Red Cross Hospital, Beverwijk, The Netherlands; Department of Plastic, Reconstructive & Hand Surgery, Amsterdam UMC location VUmc, Amsterdam, The Netherlands; Amsterdam Movement Sciences, Tissue Function and Regeneration, Amsterdam UMC, Amsterdam, The Netherlands*

AGNES S. KLAR • *Tissue Biology Research Unit, University Children's Hospital Zurich, University of Zurich, Zurich, Switzerland; Children's Research Center, University Children's Hospital Zurich, Zurich, Switzerland*

JOHANNA KOBER • *Department of Plastic and Reconstructive Surgery, Medical University of Vienna, Vienna, Austria*

ANNE M. KONG • *O'Brien Institute Department, St Vincent's Institute of Medical Research, Fitzroy, VIC, Australia*

PETRA KOTZBECK • *COREMED-Centre for Regenerative and Precision Medicine, JOANNEUM RESEARCH Forschungsgesellschaft mbH, Graz, Austria; Division of Plastic, Aesthetic and Reconstructive Surgery, Department of Surgery, Medical University of Graz, Graz, Austria; Research Unit for Tissue Regeneration, Repair and Reconstruction, Division of Plastic, Aesthetic and Reconstructive Surgery, Department of Surgery, Medical University of Graz, Graz, Austria*

DANIELLE LAROUCHE • *Centre de recherche en organogénèse expérimentale de l'Université Laval / LOEX, Centre de recherche du CHU de Québec-Université Laval, Axe médecine régénératrice, and Département de Chirurgie, Faculté de Médecine, Université Laval, Québec, QC, Canada*

SENG-TEIK LEE • *Department of Plastic, Reconstructive and Aesthetic Surgery, Singapore General Hospital, Singapore, Singapore*

ALEXANDRE LEYTENS • *Department of Biology, University of Fribourg, Fribourg, Switzerland*

SHIANG Y. LIM • *O'Brien Institute Department, St Vincent's Institute of Medical Research, Fitzroy, VIC, Australia; Department of Surgery and Medicine, University of Melbourne, East Melbourne, VIC, Australia; Drug Discovery Biology, Monash Institute of Pharmaceutical Sciences, Monash University, Parkville, VIC, Australia; National Heart Research Institute Singapore, National Heart Centre, Singapore, Singapore*

MIRIAM LIPISKI • *Biological Central Laboratory & Division of Surgical Research, University Hospital Zurich, University of Zurich, Zurich, Switzerland; Center for Preclinical Development, University of Zurich, Zurich, Switzerland*

GUEI-SHEUNG LIU • *Ophthalmology, Department of Surgery, University of Melbourne, East Melbourne, VIC, Australia; Center for Eye Research Australia, Royal Victorian Eye and Ear Hospital, East Melbourne, VIC, Australia; Menzies Institute for Medical Research, University of Tasmania, Hobart, TAS, Australia*

DONGRUI MA • *Department of Plastic, Reconstructive and Aesthetic Surgery, Singapore General Hospital, Singapore, Singapore*

VICTORIA MALTMAN • *Department of Biosciences, Durham University, Durham, UK*

DOMINIQUE MAYRAND • *Centre de recherche en organogénèse expérimentale de l'Université Laval/LOEX, Québec, QC, Canada; CHU de Québec – Université Laval Research Center, Division of Regenerative Medicine, Québec, QC, Canada; Department of Surgery, Faculty of Medicine, Université Laval, Québec, QC, Canada*

ANNEBETH MEIJ-DE VRIES • *Alliance of Dutch Burn Care (ADBC), Burn Centre, Red Cross Hospital, Beverwijk, The Netherlands; Pediatric Surgical Centre, Emma Children's Hospital, Amsterdam UMC, Amsterdam, The Netherlands; Department of Surgery, Red Cross Hospital, Beverwijk, The Netherlands*

HILAL METE GUNAYDIN • *School of Chemical Engineering, University of Birmingham, Edgbaston, UK*

KATARZYNA MICHALAK-MICKA • *Tissue Biology Research Unit, University Children's Hospital Zurich, University of Zurich, Zurich, Switzerland; Children's Research Center, University Children's Hospital Zurich, Zurich, Switzerland*

ESTHER MIDDELKOOP • *Alliance of Dutch Burn Care (ADBC), Burn Centre, Red Cross Hospital, Beverwijk, The Netherlands; Department of Plastic, Reconstructive & Hand Surgery, Amsterdam UMC location VUmc, Amsterdam, The Netherlands; Amsterdam Movement Sciences, Tissue Function and Regeneration, Amsterdam UMC, Amsterdam, The Netherlands*

GERALDINE M. MITCHELL • *O'Brien Institute Department, St Vincent's Institute of Medical Research, Fitzroy, VIC, Australia; Department of Surgery and Medicine, University of Melbourne, East Melbourne, VIC, Australia; Faculty of Health Sciences, Australian Catholic University, Melbourne, VIC, Australia*
RICHARD J. A. MOAKES • *School of Chemical Engineering, University of Birmingham, Edgbaston, UK*
UELI MOEHRLEN • *Tissue Biology Research Unit, Department of Pediatric Surgery, University Children's Hospital Zurich, Zurich, Switzerland; Children's Research Center, University Children's Hospital Zurich, Zurich, Switzerland; University of Zurich, Zurich, Switzerland; Spina Bifida Center, University Children's Hospital Zurich, Zurich, Switzerland; Zurich Center for Fetal Diagnosis and Treatment, Zurich, Switzerland*
VÉRONIQUE J. MOULIN • *Centre de recherche en organogénèse expérimentale de l'Université Laval/LOEX, Québec, QC, Canada; CHU de Québec – Université Laval Research Center, Division of Regenerative Medicine, Québec, QC, Canada; Department of Surgery, Faculty of Medicine, Université Laval, Québec, QC, Canada*
FLORA NICHOLLS • *Biological Central Laboratory & Division of Surgical Research, University Hospital Zurich, University of Zurich, Zurich, Switzerland; Center for Preclinical Development, University of Zurich, Zurich, Switzerland*
ANOUK PIJPE • *Alliance of Dutch Burn Care (ADBC), Burn Centre, Red Cross Hospital, Beverwijk, The Netherlands; Department of Plastic, Reconstructive & Hand Surgery, Amsterdam UMC location VUmc, Amsterdam, The Netherlands; Amsterdam Movement Sciences, Tissue Function and Regeneration, Amsterdam UMC, Amsterdam, The Netherlands*
FILIPA PINTO • *Sapphiros Ai Bio., Boston, MA, USA*
LUCA PONTIGGIA • *Tissue Biology Research Unit, Department of Pediatric Surgery, University Children's Hospital Zurich, Zurich, Switzerland; Children's Research Center, University Children's Hospital Zurich, Zurich, Switzerland*
STEFAN PRZYBORSKI • *Department of Biosciences, Durham University, Durham, UK; Reprocell Europe Ltd, Glasgow, UK*
STEFAN PRZYBORSKI • *Department of Biosciences, Durham University, Durham, UK; Reprocell Europe Ltd, Glasgow, UK*
SERJOSHA ROBMANN • *Citus AG, c/o EMPA, Dübendorf, Switzerland*
DAVID SACHS • *Citus AG, c/o EMPA, Dübendorf, Switzerland*
SIBILLA SANDER • *Department of Biology, University of Fribourg, Fribourg, Switzerland*
NATHALIE A. SENN • *Tissue Biology Research Unit, Department of Surgery, University Children's Hospital Zurich, Zurich, Switzerland; Children's Research Centre, University Children's Hospital Zurich, Zurich, Switzerland; University of Zurich, Zurich, Switzerland*
MAKOTO SENCO • *FRACORA Co. Ltd., Shinjuku-ku, Tokyo, Japan; BIRTH Co. Ltd., Shinjuku-ku, Tokyo, Japan*
LUCY SMITH • *Department of Biosciences, Durham University, Durham, UK*
HIERONYMUS P. STEVENS • *PRSkliniek, Vlaardingen, The Netherlands; Department of Aesthetic Surgery, Velthuis Clinics, Rotterdam, The Netherlands*
MATTHEA M. STOOP • *Alliance of Dutch Burn Care (ADBC), Burn Centre, Red Cross Hospital, Beverwijk, The Netherlands; Department of Plastic, Reconstructive & Hand Surgery, Amsterdam UMC location VUmc, Amsterdam, The Netherlands*
ALBERTO STRACUZZI • *Citus AG, c/o EMPA, Dübendorf, Switzerland*
DAISUKE SUZUKI • *Department of Physiology and Biochemistry, Faculty of Nutrition, Kobe Gakuin University, Kobe, Japan*

ALAIN TAIEB • *INSERM 1312, University of Bordeaux, Bordeaux Cedex, France*

CHRISTIAN TAPKING • *Department of Hand, Plastic and Reconstructive Surgery, Burn Trauma Center, BG Trauma Center Ludwigshafen, University of Heidelberg, Heidelberg, Germany*

BETTINA THUMM • *Citus AG, c/o EMPA, Dübendorf, Switzerland; Institute for Mechanical Systems, Department of Mechanical and Process Engineering, ETH Zurich, Zurich, Switzerland*

NATHALIE TREMBLAY • *Centre de recherche en organogénèse expérimentale de l'Université Laval/LOEX, Québec, QC, Canada; CHU de Québec – Université Laval Research Center, Division of Regenerative Medicine, Québec, QC, Canada; Department of Surgery, Faculty of Medicine, Université Laval, Québec, QC, Canada*

MARINA TROUILLAS • *Institut de Recherche Biomédicale des Armées (IRBA), Unité de Thé rapie Cellulaire et Tissulaire (U.TCT), Antenne Centre de Transfusion Sanguine des Armées, 1 rue du Lieutenant Raoul BATANY, CLAMART Cedex, France*

MUSTAFA UGUTEN • *Department of Plastic, Reconstructive and Hand Surgery, University Medical Center Utrecht, Utrecht University, Utrecht, The Netherlands; Department of Plastic, Reconstructive and Hand Surgery, Medical Center Leeuwarden, ADLeeuwarden, The Netherlands*

JOERI VAN BOXTEL • *Department of Plastic, Reconstuctive and Hand Surgery, Catharina Hospital Eindhoven, Eindhoven, The Netherlands*

ANNA S. VAN DEN BOSCH • *Alliance of Dutch Burn Care (ADBC), Burn Centre, Red Cross Hospital, Beverwijk, The Netherlands; Department of Plastic, Reconstructive & Hand Surgery, Amsterdam UMC location VUmc, Amsterdam, The Netherlands; Amsterdam Movement Sciences, Tissue Function and Regeneration, Amsterdam UMC, Amsterdam, The Netherlands*

PAUL P. M. VAN ZUIJLEN • *Alliance of Dutch Burn Care (ADBC), Burn Centre, Red Cross Hospital, Beverwijk, The Netherlands; Department of Plastic, Reconstructive & Hand Surgery, Amsterdam UMC location VUmc, Amsterdam, The Netherlands; Amsterdam Movement Sciences, Tissue Function and Regeneration, Amsterdam UMC, Amsterdam, The Netherlands; Pediatric Surgical Centre, Emma Children's Hospital, Amsterdam UMC, Amsterdam, The Netherlands*

VIVIENNE WOODTLI • *Pediatric Burn Center, Children's Skin Center, Department of Surgery, University Children's Hospital Zurich, University of Zurich, Zurich, Switzerland; Department of Surgery, University Children's Hospital Zurich, University of Zurich, Zurich, Switzerland; Children's Research Center (CRC), University Children's Hospital Zurich, University of Zurich, Zurich, Switzerland; University of Zurich (UZH), Zurich, Switzerland*

HESHAM YOSEF • *microphotonX GmbH, Planegg, Germany*

URS ZIEGLER • *Center for Microscopy and Image Analysis, University of Zurich, Zurich, Switzerland*

FABIANA ZILIANI • *Tissue Biology Research Unit, University Children's Hospital Zurich, University of Zurich, Zurich, Switzerland; Children's Research Center, University Children's Hospital Zurich, Zurich, Switzerland*

Part I

Isolation and Culture of Skin Cells

Chapter 1

Expansion of Human Epidermal Keratinocytes in Co-culture with Human Feeder Cells

Daisuke Suzuki, Filipa Pinto, and Makoto Senoo

Abstract

The co-culture of human epidermal keratinocytes with mouse 3T3-J2 feeder cells, a method developed by Green and colleagues, has been widely used since the early 1980s to generate skin autografts. Additionally, co-culture with 3T3-J2 cells has been a crucial tool in skin stem cell biology, as it facilitates the evaluation of self-renewal capacity and differentiation of epidermal stem cells. This chapter presents a recent enhancement of the Green method, aimed at further promoting the expansion of human epidermal keratinocytes by utilizing a small molecule inhibitor of TGF-β signaling. This new protocol also enables the rapid expansion of human epidermal keratinocytes in co-culture with human feeder cells, including human dermal fibroblasts and human preadipocytes—two key alternatives to 3T3-J2 cells—with the long-term goal of developing customized Xeno-free skin autografts.

Key words Co-culture, Feeder cells, Human keratinocytes, Skin graft, TGF-β signaling

1 Introduction

Homeostasis of the epidermis is maintained by tissue-specific stem cells capable of self-renewal, proliferation, and differentiation [1, 2]. Culture of epidermal stem cells has contributed to the development of therapeutic strategies in regenerative medicine [3], including skin transplantation for patients with severe burn wounds. The Green method, developed for the expansion of autologous keratinocytes from a small skin biopsy, has enabled permanent wound closure with autografts in burn victims [3–6] and more recently with the transgenic autografts in patients with a lethal genetic disease termed Junctional Epidermolysis Bullosa (JEB) with recurrent severe blistering [7]. In addition, the Green method has been essential in studying skin stem cell biology as it allows the expansion of epithelial stem cells and the assessment of their self-renewal capacity [8–11].

Thomas Biedermann and Sophie Böttcher-Haberzeth (eds.), *Skin Tissue Engineering: Methods and Protocols*, Methods in Molecular Biology, vol. 2922, https://doi.org/10.1007/978-1-0716-4510-9_1,

The Green method utilizes 3 T3-J2 cells, a unique mouse mesenchymal cell line, as a feeder layer to support the proliferative potential of epidermal progenitor cells long term [3, 8]. The co-culture of human primary epidermal keratinocytes with 3T3-J2 cells produces cultured epidermal autografts (CEA). Since its inception in the 1980s, the CEA have been used worldwide for patients with severe burn wounds, often saving their lives [3–6]. However, patients with massive burn wounds can spare only limited skin donor sites and it requires several weeks of amplification of keratinocytes to prepare sufficient numbers of CEA. A methodology to promote the expansion of keratinocytes in Green's protocol would be beneficial as it can decrease hospitalization time for patients with massive burn injuries who undergo replacement of skin covering nearly the entire body surface area. Moreover, the use of mouse feeder cells has been a debate and its replacement with human feeder cells to generate Xeno-free grafts has been increasingly anticipated.

Transforming growth factor-β (TGF-β) signaling has been shown to suppress the growth of keratinocytes by inhibiting their cell cycle progression [12]. Effects of TGF-β signaling are mediated through a TGF-β receptor complex consisting of the Type I TGF-β receptor (ALK5) and the Type II TGF-β Receptor (TGFBRII), both of which possess intrinsic serine/threonine kinase activity [13]. Upon binding of TGF-β ligands, TGFBRII activates ALK5 kinase by phosphorylation which, in turn, activates Smad2/3-mediated gene transcription required for the growth arrest of keratinocytes [14]. In co-culture with 3 T3-J2 cells, there are three major sources of TGF-β ligands: keratinocytes [15], 3T3-J2 cells [16], and serum present in the co-culture media [17]. The presence of TGF-β ligands in co-culture could restrict the potential growth of human epidermal keratinocytes. Indeed, we have shown recently that TGF-β signaling is active in human keratinocytes in co-culture with 3T3-J2 cells as determined by the nuclear localization of Smad2/3 [18]. Accordingly, suppression of TGF-β signaling by RepSox, a cell-permeable small molecule inhibitor of ALK5 kinase, increases the proliferation of human epidermal keratinocytes in co-culture with 3T3-J2 cells [18]. In addition, the use of TGF-β signaling inhibitors also enhances significantly the expansion of human epidermal keratinocytes in co-culture with other feeder cell types including human dermal fibroblasts and human preadipocytes, two major cell types utilized as alternatives to 3T3-J2 cells, with a long-term goal of developing customized Xeno-free skin autografts [19, 20].

In this chapter, we present an enhanced version of Green's method, where the expansion of human epidermal keratinocytes is significantly boosted through pharmacological inhibition of TGF-β signaling. Notably, this improved protocol also permits the use of human feeder cells, eliminating the need for mouse-derived 3 T3-J2 feeder cells, aiming at generating Xeno-free skin grafts.

2 Materials

2.1 Cells

1. Human primary epidermal keratinocytes (Cellntec) (*see* **Note 1**).
2. 3T3-J2 cells, derived from mouse embryonic fibroblasts [8].
3. PrimaPure human dermal fibroblasts (hDFs). hDFs used in this protocol were derived from the dermis of normal human neonatal foreskin and have been verified to express fibroblast surface proteins.
4. Human preadipocytes (hPAs). hPAs used in this protocol were isolated from adipose tissues of patients who underwent plastic surgery in the abdominal region. The subjects were free of metabolic or endocrine diseases as assessed by routine clinical examination and laboratory tests.

2.2 Cell Culture Reagents

1. Dulbecco's Modified Eagle's Medium (DMEM): 4.5 g/L glucose, 584 mg/L L-glutamine, and 110 mg/L sodium pyruvate in DMEM.
2. Ham's F-12: 1.8 g/L glucose, 146 mg/L L-glutamine, and 110 mg/L sodium pyruvate in Ham's F-12.
3. Adenine (minimum 99%): 24 mg/mL in 0.2 N HCl, sterilize with a 0.2 μm syringe filter. Store in 0.5 mL aliquots at −20 °C.
4. T_3 (3,3′,5-triiodo-L-thyronine sodium salt) (minimum 95% HPLC): 10 μM in PBS (Phosphate-buffered saline without calcium and magnesium), sterilize with a 0.2 μm syringe filter. Store in 0.5 mL aliquots at −20 °C.
5. Hydrocortisone (>98%): 5 mg/mL in 95% EtOH. Store in 0.2 mL aliquots at −20 °C.
6. Cholera toxin (Approx. 95%): 0.1 mg/mL in dH_2O, sterilize with a 0.2 μm syringe filter. Store in 0.4 mL aliquots at −20 °C.
7. Insulin from bovine pancreas: 0.1 mg/mL in 0.005 N HCl, sterilize with a 0.2 μm syringe filter. Store in 0.5 mL aliquots at −80 °C.
8. Human epidermal growth factor (EGF): 20 μg/mL in 1% BSA (Bovine serum albumin) aqueous solution, sterilize with a 0.2 μm syringe filter. Store in 0.5 mL aliquots at −80 °C.
9. Basic fibroblast growth factor (bFGF): 10 μg/mL in 1% BSA aqueous solution. Store in 10 μL aliquots at −80 °C.

10. Biotin: 1 mg/mL in 0.01 N NaOH, sterilize with a 0.2 μm syringe filter. Store in 0.1 mL aliquots at −80 °C.
11. D-Pantothenic acid hemicalcium salt: 4 mg/mL in dH_2O, sterilize with a 0.2 μm syringe filter. Store in 0.1 mL aliquots at −80 °C.
12. 2-[3-(6-Methylpyridine-2-yl)-1*H*-pyrazol-4-yl]-1,5-naphthyridine (RepSox): 25 mM in DMSO (Dimethyl Sulfoxide). Store in 50 μL aliquots at −80 °C.
13. Trypsin (0.25%) -EDTA (1 mM) solution.

2.3 Other Reagents and Chemicals

1. Antibodies
   (a) Rabbit anti-cytokeratin 5 (CK5).
   (b) Mouse anti-p63 (clone 4A4).
   (c) Alexa 488-goat anti-mouse IgG.
   (d) Alexa 488-goat anti-rabbit IgG.
   (e) Alexa 594-goat anti-rabbit IgG.
2. 0.4% Trypan blue solution: Used to determine the viability of cells in cell counting.
3. 1% Triton X-100: 1% (v/v) Triton X-100 in PBS, store at room temperature.
4. 0.1% Tween-20 (PBS-T): 0.1% (v/v) Tween-20 in PBS, store at room temperature.
5. Blocking solution: 10% (v/v) FBS (Fetal bovine serum) in PBS.
6. IntraSure kit (BD Biosciences): Used for intracellular staining.
7. 1% Rhodamine B: 1% (w/v) Rhodamine B in dH_2O, filter through Whatman filter papers, store at room temperature.
8. Hoechst 33342: 1 mg/mL in DMSO (Dimethyl Sulfoxide) as a stock solution. Store in 0.1 mL aliquots at −20 °C. To prepare working solution, dilute the stock solution to 1 μg/mL in PBS.

2.4 Instruments and Supplies

1. Humidified CO_2 incubators.
2. CO_2 gas cylinders.
3. 37 °C water bath.
4. Hemocytometer.
5. Inverted microscope.
6. Fluorescence microscope.
7. Flow cytometer.
8. γ-irradiator with a Cesium-137 source.
9. Tissue culture plates (6-well, 35 mm, 60 mm, 100 mm).
10. 15 mL and 50 mL polypropylene conical tubes.

11. 1.5 mL Eppendorf tubes.
12. Cryovials.
13. Syringe Filter unit (0.2 μm).
14. Parafilm.

2.5 Cell Culture Media and Freezing Media

1. J2 medium: 100 U/mL penicillin, 100 μg/mL streptomycin, 10% (v/v) CS (Bovine calf serum) in DMEM [8].
2. hPA medium: Mix DMEM and Ham's F-12 at a 1:1 ratio (v/v) and supplement with 10% FBS (Fetal bovine serum) (v/v), 10 ng/mL EGF, 1 ng/mL bFGF, 8 μg/mL Biotin, 4 μg/mL D-Pantothenic acid, 8.7 μM insulin, 100 U/mL penicillin, and 100 μg/mL streptomycin [21].
3. FAD medium: Mix DMEM and Ham's F-12 at a 3:1 ratio (v/v) and supplement with 10% FBS (v/v), 100 U/mL penicillin, and 100 μg/mL streptomycin [8].
4. Complete FAD (cFAD) medium: 10 ng/mL EGF, 5 μg/mL insulin, 2×10^{-9} M 3,3′,5- triiodo-L-thyronine, 0.4 μg/mL hydrocortisone, 24 μg/mL adenine, and 1×10^{-10} M cholera toxin in FAD medium [8].
5. CnT-PR medium: 100 U/mL penicillin, 100 μg/mL streptomycin in CnT-PR media (CnT-Prime epithelial proliferation medium).
6. Freezing medium: Mix DMEM, serum (CS or FBS), and DMSO (Dimethyl Sulfoxide) at a 4:5:1 ratio (v/v/v). To freeze 3 T3-J2 cells and hDFs, use CS while FBS is used for human epidermal keratinocytes and hPAs.

3 Methods

3.1 Preparation of Feeder Cells

In this protocol, both 3 T3-J2 cells and hDFs are expanded in J2 medium under 10% CO_2 atmosphere while hPAs are propagated in hPA medium under 5% CO_2 atmosphere. Keep in mind that the conditions of feeder cells, such as the viability, density, and passage number (*see* **Note 2**), are critical for successful expansion of human epidermal keratinocytes in feeder cell co-culture.

3.1.1 Culture, Passage, and Storage of 3T3-J2 Cells and hDFs

1. To expand 3T3-J2 cells and hDFs, one cryovial containing 1×10^6 cells is taken out of a liquid nitrogen tank and thawed in a 37 °C water bath for 3 min.
2. Spin down the cells at 200 × g for 5 min.
3. Remove freezing media and suspend the cells in 1 mL J2 medium.

4. Plate the resuspended cells in two 100 mm tissue culture plates that contain 10 mL J2 medium (5×10^5 cells per plate).
5. Place tissue culture plates in a humidified incubator with 10% CO_2 at 37 °C.
6. After overnight incubation (>16 h), replace media with 10 mL of fresh J2 medium. Culture medium should be changed every 2–3 days thereafter until the cells become subconfluent (*see* **Note 3**).
7. When the cells become subconfluent, remove culture media and wash the cells twice in 5 mL PBS.
8. Add 2 mL 0.25% trypsin-EDTA to the cells and incubate the plates in a humidified incubator with 10% CO_2 at 37 °C for 5 min.
9. Dissociate the cells by gentle pipetting, collect them in a 15 mL polypropylene tube, and neutralize the enzymatic reaction by adding the same volume (2 mL) of J2 medium.
10. Centrifuge the cells at $200 \times g$ for 5 min.
11. Remove supernatant and resuspend the cells in 1 mL J2 medium.
12. Count cell numbers. Approximately 3×10^6 cells are collected from one subconfluent 100 mm tissue culture plate.
13. Replate 3T3-J2 cells or hDFs at a 1:10–1:15 ratio at the density of 3.3×10^3 cells per cm^2 into 100 mm tissue culture plates that contain 10 mL J2 medium. These secondary cultures become subconfluent in 5–7 days.
14. To prepare frozen stock, resuspend the cell pellet in freezing medium at the density of 2×10^6 cells per mL and aliquot 0.5 mL per cryovial (1×10^6 cells per vial). Place the cryovials in −80 °C overnight and transfer them to a liquid nitrogen tank for longer storage.

3.1.2 Culture, Passage, and Storage of hPAs

1. To expand hPAs, one cryovial containing 1×10^6 hPAs is taken out of a liquid nitrogen tank and thawed in a 37 °C water bath for 3 min.
2. Spin down the cells at $200 \times g$ for 5 min.
3. Remove freezing media and suspend the cells in 1 mL hPA medium.
4. Plate 1×10^6 hPAs in one 100 mm tissue culture plate that contains 10 mL hPA medium.
5. Place the tissue culture plate in a humidified incubator with 5% CO_2 at 37 °C.

6. After overnight incubation (>16 hrs), replace media with 10 mL of hPA medium. Culture medium should be changed every 2–3 days thereafter until the cells become subconfluent.
7. When the cells become subconfluent, remove culture media and wash the cells twice in 5 mL PBS.
8. Add 2 mL 0.25% trypsin-EDTA to the cells and incubate the plates in a humidified incubator with 5% CO_2 at 37 °C for 5 min.
9. Dissociate the cells by gentle pipetting, collect them in a 15 mL polypropylene tube, and neutralize the enzymatic reaction by adding the same volume (2 mL) of hPA medium.
10. Centrifuge the cells at 200 × g for 5 min.
11. Remove supernatant and resuspend the cells in 1 mL hPA medium.
12. Count cell numbers.
13. Replate hPAs at a 1:10 ratio at the density of 3.3×10^3 cells per cm^2 into 100 mm tissue culture plates that contain 10 mL hPA medium. These secondary cultures become subconfluent in 7–10 days.
14. To prepare frozen stock, resuspend the cell pellet in freezing medium at the density of 2×10^6 cells per mL and aliquot 0.5 mL per cryovial (1×10^6 cells per vial). Place the cryovials in −80 °C overnight and transfer them to a liquid nitrogen tank for longer storage.

3.2 Preparation of Human Epidermal Keratinocyte Stocks

Human epidermal keratinocytes used in this protocol were derived from neonatal foreskin and were purchased from Cellntec. The cells have been tested negative for Hepatitis B, Hepatitis C, and HIV-1, and are free of bacteria, fungi, and mycoplasma contamination. Cryovials containing 5×10^5 viable cells were supplied on dry ice. The cells can be used immediately or transferred to a liquid nitrogen tank for later use.

1. To grow human epidermal keratinocytes in plastic culture without feeder cells, one cryovial containing 5×10^5 cells is thawed in a 37 °C water bath for 3 min.
2. Spin down the cells at 200 × g for 5 min.
3. Remove freezing media and suspend the cells in 1 mL CnT-PR medium.
4. Plate 5×10^5 human epidermal keratinocytes in one 100 mm tissue culture plate that contains 10 mL CnT-PR medium.
5. Place the tissue culture plate in a humidified incubator with 5% CO_2 at 37 °C.

6. After overnight incubation (>16 h), replace media with 10 mL of fresh CnT-PR medium. Culture medium should be changed every 2–3 days thereafter until the cells become subconfluent (*see* **Note 4**).
7. When the cells become subconfluent, remove culture media and wash the cells twice in 5 mL PBS.
8. Add 2 mL of 0.25% trypsin-EDTA to the cells and incubate the plate in a humidified chamber with 5% CO_2 at 37 °C for 5 min (*see* **Note 5**).
9. Tap the tissue culture plate to dislodge the cells.
10. Dissociate the cells by gentle pipetting, collect them in a 15 mL polypropylene tube, and neutralize the enzymatic reaction by adding 0.2 mL FBS (equivalent to 10% the volume of trypsin used).
11. Centrifuge the cells at 200 × g for 5 min.
12. Remove supernatant, wash the cells twice in 5 mL PBS, and resuspend in 1 mL CnT-PR medium.
13. Count cell numbers. Approximately 3–5 × 10^6 cells are collected from one subconfluent 100 mm tissue culture plate.
14. Replate human keratinocytes at a 1:15–1:20 ratio at the density of 4 × 10^3 cells per cm^2 into 100 mm tissue culture plates that contain 10 mL CnT-PR medium. These secondary cultures become subconfluent in 5–7 days.
15. To prepare frozen stock, resuspend the cell pellet in keratinocyte freezing medium at the density of 2 × 10^6 cells per mL and aliquot 0.5 mL per cryovial (1 × 10^6 cells per vial). Place the cryovials in −80 °C overnight and transfer them to a liquid nitrogen tank for longer storage.

3.3 Promotion of Human Epidermal Keratinocyte Expansion in Co-culture with Feeder Cells Using a TGF-β Signaling Inhibitor

3.3.1 Preparation of Feeder Cell Layers

Prior to seeding keratinocytes, feeder cells will be mitotically arrested by either lethal γ-irradiation or mitomycin C treatment (*see* **Note 6**) to prevent propagation of the feeder cells in subsequent passages. In this protocol, we describe the use of lethally γ-irradiated feeder cells on the scale of a 100 mm tissue culture plate as an example. When other sizes of cell culture vessels are used, the number of feeder cells seeded and volume of the cell culture media used should be adjusted proportionally based on the surface area of the culture vessels (*see* **Note 7**).

1. Follow the procedures described in Subheading 3.1 and prepare desired numbers of 100 mm tissue culture plates with subconfluent feeder cells.
2. When feeder cells become subconfluent, seal the culture plates with Parafilm to maintain sterility and γ-irradiate the cells at a dose of 60 Gy (6000 rads).

3. After irradiation, replace media with 10 mL of fresh feeder cell culture medium and place the plates in a humidified incubator until use (*see* **Note 8**).

3.3.2 Setting Up Co-culture

1. Prepare a minimum amount of cFAD medium that can be consumed in 1 week as it contains proteins and chemicals that may degrade once diluted in media.
2. Prior to seeding human keratinocytes, replace feeder cell culture media with 10 mL cFAD medium and place the plates in a humidified incubator with 5% CO_2 at 37 °C for 1–2 h.
3. By following the steps in Subheading 3.2, prepare human epidermal keratinocytes in suspension. Use cFAD instead of CnT-PR medium to suspend epidermal keratinocytes.
4. Count viable keratinocyte numbers with the aid of Trypan blue solution.
5. Adjust the density of keratinocytes to $10^3 \sim 10^5$ viable cells per mL by diluting in cFAD medium.
6. Seed the desired number of human keratinocytes onto feeder cell layers. When rapid confluency of keratinocytes is planned, $10^4 \sim 10^5$ cells per 100 mm tissue culture plate can be seeded (*see* **Note 9**). Seeding at a higher density than 10^5 cells per 100 mm plate should not be used as it does not allow immature keratinocytes to fully expand and accumulate. When the determination of the proliferative capacity of keratinocytes at a clonal level is required, $10^2 \sim 10^4$ cells per 100 mm plate should be seeded to permit clonal growth of individual keratinocytes (*see* **Note 10**).
7. To promote the expansion of human epidermal keratinocytes in feeder cell co-culture, add a predetermined dose (*see* Subheading 3.4.1 below) of RepSox in cFAD medium in **step 6** above. A range between 0.1 and 1 μM RepSox treatment significantly boosts the growth of human epidermal keratinocytes with high proliferative potential [18] (*see* **Note 11**).
8. cFAD medium containing RepSox should be replaced every 2–3 days until desired confluency of keratinocytes is achieved.

3.4 Evaluation of the Effects by RepSox Treatment

In this section, we describe the methods to evaluate the effects by RepSox in feeder cell co-culture on (a) the growth of human keratinocyte progenitor cells as determined by intracellular staining of cytokeratin 5 (CK5), a marker for basal epidermal cells [22]; (b) clonal expansion of human epidermal keratinocytes as determined by Rhodamine B staining; and (c) protein expression of p63, an essential transcription factor for the maintenance of the proliferative potential of epidermal stem cells [11].

3.4.1 Determining the Effective Doses of RepSox in Promoting the Growth of Human Epidermal Keratinocytes in Feeder Cell Co-culture

Optimum doses of RepSox that promote human epidermal keratinocyte growth in feeder cell co-culture vary from one feeder cell type to another. The optimum dose of RepSox or any other TGF-β signaling inhibitors should be determined for each feeder cell type prior to performing co-culture of keratinocytes.

1. Inoculate 5×10^3 human epidermal keratinocytes (*see* Subheading 3.2) onto γ-irradiated feeder cells in 60 mm tissue culture plates that contain 5 mL cFAD medium.
2. Add increasing concentration of RepSox to the culture media. Cell culture media are changed every 2–3 days.
3. At day 10, remove cell culture media from the plates and wash the cells twice in 5 mL PBS.
4. Add 2 mL 0.25% trypsin-EDTA to the cells and incubate the plates in a humidified incubator with 5% CO_2 at 37 °C for 10 min.
5. Dissociate the cells by gentle pipetting, collect them in 15 mL polypropylene tubes, and neutralize the enzymatic reaction by adding 0.2 mL FBS (10% the volume of trypsin used).
6. Centrifuge the cells at 200 × g for 5 min.
7. Remove supernatant and gently suspend the cells in 1 mL cFAD medium.
8. Count total cell numbers (*see* **Note 12**).
9. Resuspend $5–20 \times 10^5$ cells in 50 μL PBS in 1.5 mL Eppendorf tubes.
10. Add 100 μL Reagent A (supplied as a component of IntraSure kit) to each tube, vortex the tubes, and incubate for 5 min at room temperature.
11. Add 1 mL PBS to each tube and centrifuge the cells at 800 × g for 5 min.
12. Remove supernatant and gently suspend the cells in 50 μL Reagent B (IntraSure kit).
13. Add rabbit anti-CK5 polyclonal antibody (1:500 dilution) to the cells and incubate for 30 min at room temperature.
14. Add 1 mL PBS to each tube and centrifuge the cells at 800 × g for 5 min.
15. Remove supernatant and gently suspend the cells in 50 μL PBS.
16. Add Alexa 488-goat anti-rabbit IgG antibody (1:250 dilution) and incubate the cells for 30 min in the dark at room temperature.
17. Add 1 mL PBS to each tube and spin down the cells at 800 × g for 5 min.

18. Remove supernatant and gently suspend the cells in1 mL PBS and centrifuge the cells again.
19. Remove supernatant and gently suspend the cells in 0.3 mL PBS.
20. Determine the frequency of $CK5^+$ cells by a flow cytometer.
21. Calculate absolute $CK5^+$ cell numbers using the total cell numbers obtained in **step 8** and the $CK5^+$ cell frequency determined in **step 20**. Plot the results for each concentration of RepSox as shown in Fig. 1 (*see* also **Note 13**).

3.4.2 Visualization of Human Epidermal Keratinocyte Clones in Feeder Cell Co-culture by Rhodamine B Staining

1. Inoculate 1×10^3 human keratinocytes (*see* Subheading 3.2) onto γ-irradiated feeder cells in 60 mm tissue culture plates in 5 mL cFAD medium in the presence or absence of pre-determined doses of RepSox (*see* Subheading 3.4.1). Change cell culture media every 2–3 days.
2. At day 14, wash the cells twice in 2 mL PBS and fix them in 10% buffered formalin for 15 min at room temperature.
3. Remove formalin solution and wash the cells twice in 2 mL PBS.
4. Add 2 mL of 1% Rhodamine B solution to the cells and stain for more than 15 min at room temperature.
5. Pour off Rhodamine B solution into a separate container (*see* **Note 14**).
6. Gently rinse the cells with tap water.
7. Invert the plates to air-dry.
8. Take photographs of the stained plates (*see* Fig. 2) and determine the sizes of individual clones using the NIH Image J or Adobe Photoshop software [18].

3.4.3 Immunofluorescence Staining of p63 and CK5 in Human Epidermal Keratinocyte Clones in Feeder Cell Co-culture

1. Inoculate 1×10^3 human epidermal keratinocytes (*see* Subheading 3.2) onto γ-irradiated feeder cells in 60 mm tissue culture plates in 5 mL cFAD medium in the presence or absence of pre-determined doses of RepSox (*see* Subheading 3.4.1). Change cell culture medium every 2–3 days.
2. At day 14, wash the cells twice in 2 mL PBS and fix them in 10% buffered formalin for 10 min at room temperature.
3. Remove formalin solution and add 2 mL of 1% Triton X-100 solution to permeabilize the cells for 5 min at room temperature.
4. Remove Triton X-100 solution and wash the cells three times in 2 mL PBS-T.
5. Add 2 mL blocking solution to the cells and incubate the plates for 30 min at room temperature.

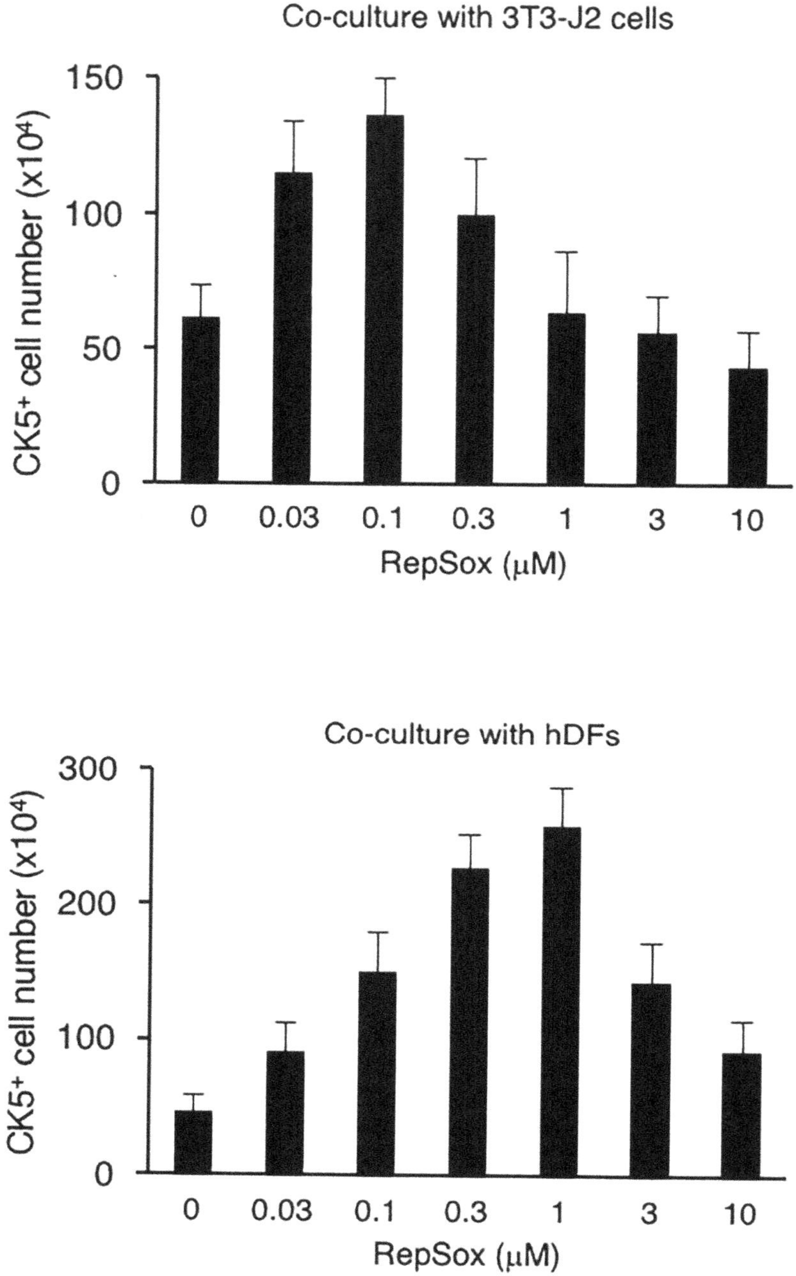

Fig. 1 Growth response of human epidermal keratinocytes to RepSox treatment in co-culture with feeder cells. Human epidermal keratinocytes were grown in co-culture with 3 T3-J2 cells (upper) and hDFs (lower) in the absence or presence of increasing concentration of TGF-β signaling inhibitor RepSox as indicated. At day 10, the cells were harvested and the cell numbers were counted, followed by flow cytometric analysis of intracellular staining of cytokeratin 5 (CK5). $CK5^+$ epidermal cell numbers were plotted. Data shown are mean $\pm$ s.e.m. ($n = 3$)

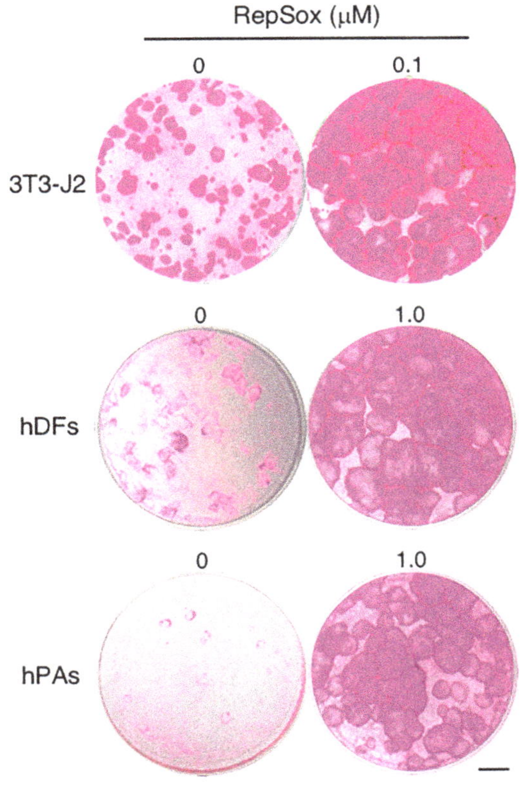

Fig. 2 Inhibition of TGF-β signaling promotes the expansion of human epidermal keratinocytes in co-culture with feeder cells. Data shown are representative images of Rhodamine B staining of human epidermal keratinocyte clones at day 14 of co-culture with 3 T3-J2 cells (upper), hDFs (middle), and hPAs (lower) in the absence (left) or presence (right) of RepSox. Note that while 0.1 μM RepSox was used in co-culture with 3 T3-J2 cells, 1.0 μM RepSox was used in co-culture with hDFs and hPAs. Equal numbers of human epidermal keratinocytes (1×10^3 cells) were plated in each well. Bar = 5 mm

6. Remove blocking solution and rinse the cells once in 2 mL PBS-T.
7. Add primary antibodies to the cells and incubate for 1 h at room temperature. Use a combination of rabbit anti-CK5 polyclonal antibody (1:2000) and mouse anti-p63 monoclonal antibody (1:1000), diluted in 2 mL PBS-T.
8. Remove the primary antibodies and wash the cells twice in 2 mL PBS-T for 5 min each.
9. Add secondary antibodies diluted in 2 mL PBS-T to the cells and incubate for 1 h in the dark at room temperature. Use a combination of Alexa 488-goat anti-mouse IgG and Alexa 594-goat anti-rabbit IgG (both 1:1000 dilution).

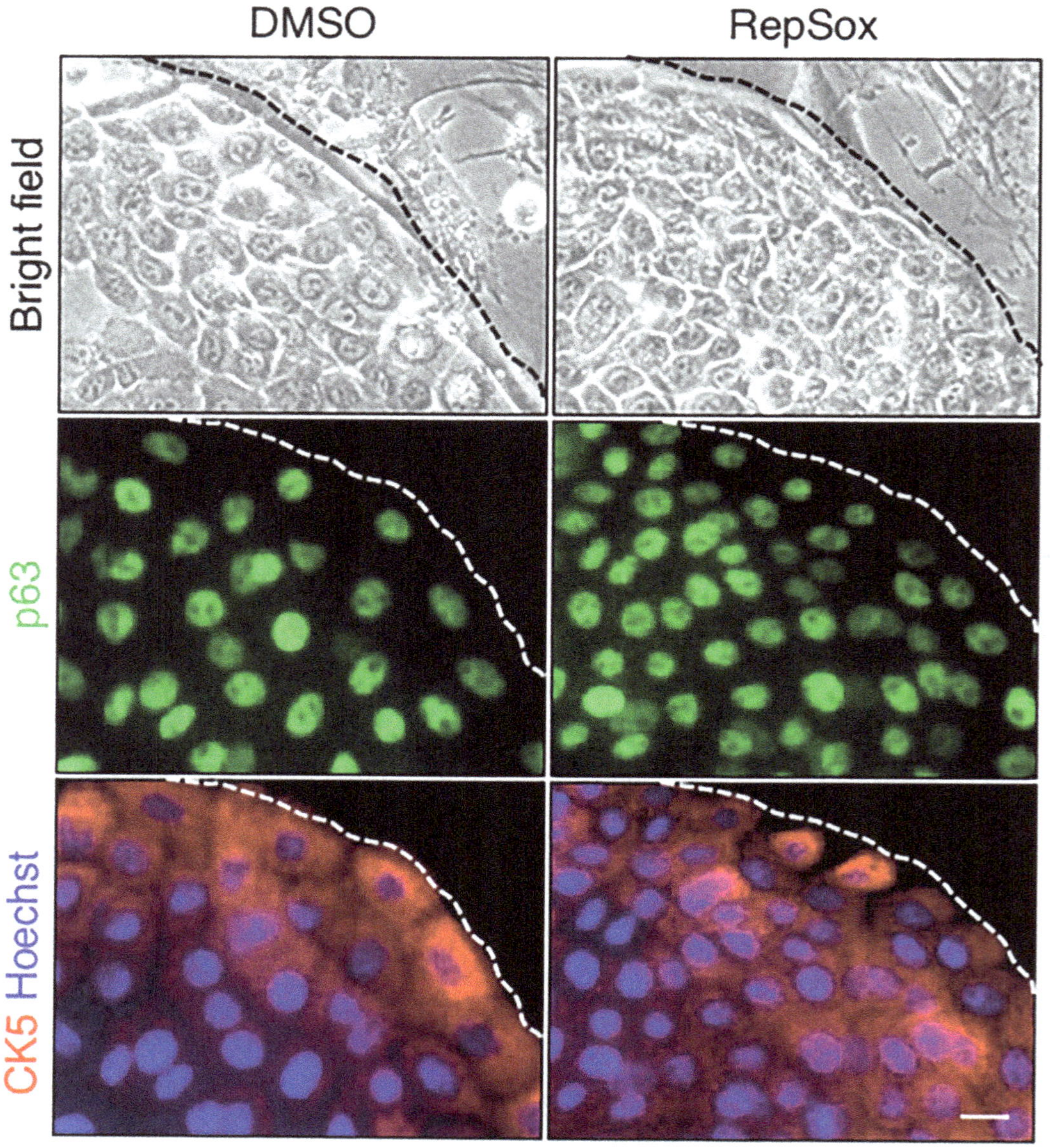

Fig. 3 Expression of p63 and CK5 in epidermal keratinocytes co-cultured with 3 T3-J2 cells in the absence or presence of RepSox. Human epidermal keratinocytes were grown in co-culture with 3 T3-J2 cells in the absence (left) or presence (right) of 0.1 μM RepSox for 14 days. Data shown are representative images of bright field (upper) and immunofluorescence staining with anti-p63 antibodies (middle) and anti-CK5 antibodies and nuclear counterstaining with Hoechst 33342 (lower). Dotted lines indicate the clone borders. Bar = 10 μm

10. Remove the secondary antibodies and wash the cells three times in 2 mL PBS-T for 10 min each.
11. Add 2 mL of Hoechst 33342 solution to the cells and incubate for 5 min in the dark at room temperature.

12. Remove Hoechst 33342 and rinse the cells twice in 2 mL PBS-T for 5 min each.
13. Cover the cells with 2 mL PBS-T and analyze the signals using a fluorescence microscope as shown in Fig. 3.

4 Notes

1. Isolation of primary human epidermal keratinocytes has been described elsewhere, see, for example, reference [23].
2. We routinely expand 3 T3-J2 cells at passage 4–7 and prepare a large number of frozen stocks before initiating epidermal cell co-culture. We typically use 3 T3-J2 cells in co-culture at passage 6–12.
3. Do not allow feeder cells to become confluent as the morphology of the cells is altered and the potential of feeder cells to support keratinocyte growth may decline.
4. Do not allow human keratinocytes to become confluent because confluent keratinocytes are difficult to dissociate by enzymatic digestion and they tend to differentiate prematurely in subsequent passages.
5. Accutase or TrypLE select can be used to more gently dissociate the cells than does trypsin-EDTA.
6. Instead of γ-irradiation, mitomycin C treatment can be used to induce growth arrest in feeder cells as follows.
   (a) Add mitomycin C to feeder cell culture media in a 50 mL tube at a final concentration of 10 μg/mL.
   (b) Aspirate cell culture media from the plates and add 6 mL of freshly prepared mitomycin C-containing feeder cell media in each 100 mm plate.
   (c) Place the plates in a humidified incubator with 10% CO_2 at 37 °C for 2 h.
   (d) Wash the cells three times in 5 mL PBS.
   (e) Add 10 mL feeder cell culture media and return the plates to a humidified incubator with 10% CO_2 at 37 °C until use.
7. See Table 1 for expected numbers of subconfluent feeder cells at the time of keratinocyte inoculation and appropriate amounts of cell culture media used for feeder cell co-culture. When feeder cells are seeded at the density of 3.3×10^3 cells per cm^2, it will take 5–7 days for 3T3-J2 cells and hDFs to become subconfluent while it will require additional 2–3 days for hPAs to become subconfluent. Higher numbers of feeder cells can be seeded to obtain subconfluent feeder cell layers sooner. In such

Table 1
Number of feeder cells and the volume of cell culture media

Culture vessel	Diameter	Area	Volume of medium	Number of feeder cells
6-well plate	35 mm	9.6 cm^2	3 mL	5.0×10^5 cells
60 mm plate	52 mm	21 cm^2	5 mL	1.1×10^6 cells
100 mm plate	84 mm	55 cm^2	10 mL	3.0×10^6 cells

a circumstance, up to 2×10^6 viable cells per 100 mm plate can be seeded and the cells can be γ-irradiated in 1–2 days when the feeder cells become subconfluent and adhere tightly to the culture vessels.

8. Although feeder cells treated with γ-irradiation or mitomycin C can remain viable for a few weeks, we typically use them in 5–7 days after the treatment. Feeder cell media should be replaced every 2–3 days until use.
9. Keratinocytes grow faster when they are plated at higher densities, possibly due to an autocrine effect on cell proliferation [24].
10. At this plating density, small colonies comprised of 4–10 cells begin to form at day 4–5. By day 7, epidermal clones become macroscopically visible and they reach 5–15 mm in diameter by day 14. Cell culture vessels smaller than 6-well plates are not recommended for clonogenic analysis as keratinocyte colonies may fuse before they reach the size suitable for clonal evaluation.
11. Human epidermal keratinocytes expanded by RepSox treatment in feeder cell co-culture maintain the original high proliferative capacity as determined by serial passages in clonogenic culture with 3 T3-J2 cells [18]. In addition, human epidermal keratinocytes expanded by RepSox treatment are capable of differentiation in response to Ca^{2+} stimulation [18].
12. Keratinocytes and feeder cells can be distinguished morphologically upon trypsin digestion: while keratinocytes are smaller in size with a smooth periphery, feeder cells are larger in size with an irregular shape. However, flow cytometric analysis with keratinocyte-specific markers is needed for more accurate quantification of keratinocyte numbers (*see* Fig. 1, for example).
13. Other widely used TGFBRI/ALK5 inhibitors, such as LY364947 and SB525334, also enhance the growth of human epidermal keratinocytes in co-culture with 3 T3-J2

cells in a dose-dependent manner [18]. However, the effect of RepSox was superior to any other TGFBRI/ALK5 inhibitors tested [18]. In the case of co-culture with hPAs, we investigated the effects of RepSox at 0.1 and 1 μM and found that the dose-dependent stimulation of human keratinocyte growth was similar to that in co-culture with hDFs.

14. Rhodamine B solution can be reused multiple times. Used solution should be collected in a separate container and filtered through Whatman filter papers to remove debris before use.

Acknowledgements

The authors thank Jeff Holcombe for proofreading of the manuscript. This study was supported by an R01AR066755 grant from the National Institute of Arthritis and Musculoskeletal and Skin Diseases of the National Institute of Health to M.S.

References

1. Blanpain C, Fuchs E (2014) Stem cell plasticity. Plasticity of epithelial stem cells in tissue regeneration. Science 344(1–10):1242281
2. Donati G, Watt FM (2015) Stem cell heterogeneity and plasticity in epithelia. Cell Stem Cell 16:465–476
3. Green H (2008) The birth of therapy with cultured cells. BioEssays 30:897–903
4. Chua AW, Khoo YC, Tan BK et al (2016) Skin tissue engineering advances in severe burns: review and therapeutic applications. Burns Trauma 4:3
5. Mcheik JN, Barrault C, Levard G et al (2014) Epidermal healing in burns: autologous keratinocyte transplantation as a standard procedure: update and perspective. Plast Reconstr Surg Glob Open 2:e218
6. Sun BK, Siprashvili Z, Khavari PA (2014) Advances in skin grafting and treatment of cutaneous wounds. Science 346:941–945
7. Hirsch T, Rothoeft T, Teig N et al (2017) Regeneration of the entire human epidermis using transgenic stem cells. Nature 551:327–332
8. Rheinwald JG, Green H (1975) Serial cultivation of strains of human epidermal keratinocytes: the formation of keratinizing colonies from single cells. Cell 6:331–343
9. Barrandon Y, Green H (1987) Three clonal types of keratinocyte with different capacities for multiplication. Proc Natl Acad Sci USA 84: 2302–2306
10. Oshima H, Rochat A, Kedzia C et al (2001) Morphogenesis and renewal of hair follicles from adult multipotent stem cells. Cell 104: 233–245
11. Senoo M, Pinto F, Crum CP et al (2007) p63 is essential for the proliferative potential of stem cells in stratified epithelia. Cell 129:523–536
12. Shipley GD, Pittelkow MR, Wille JJ Jr et al (1986) Reversible inhibition of normal human prokeratinocyte proliferation by type beta transforming growth factor-growth inhibitor in serum-free medium. Cancer Res 46:2068–2071
13. Schmierer B, Hill CS (2007) TGF-β-SMAD signal transduction: molecular specificity and functional flexibility. Nat Rev Mol Cell Biol 8: 970–982
14. Ikushima H, Miyazono K (2010) TGF-β signalling: a complex web in cancer progression. Nat Rev Cancer 10:415–424
15. Ghahary A, Marcoux Y, Karimi-Busheri F et al (2001) Keratinocyte differentiation inversely regulates the expression of involucrin and transforming growth factor β1. J Cell Biochem 83:239–248
16. Suzuki D, Senoo M (2015) Dact1 regulates the ability of 3T3-J2 cells to support proliferation of human epidermal keratinocytes. J Invest Dermatol 135:2894–2897
17. Robertson IB, Rifkin DB (2013) Unchaining the beast; insights from structural and evolutionary studies on TGF-β secretion,

sequestration, and activation. Cytokine Growth Factor Rev 24:355–372

18. Suzuki D, Pinto F, Senoo M (2017) Inhibition of TGF-β signaling promotes expansion of human epidermal keratinocytes in feeder cell co-culture. Wound Rep Reg. https://doi.org/10.1111/wrr.12541

19. Bullock AJ, Higham MC, MacNeil S (2006) Use of human fibroblasts in the development of a xenobiotic-free culture and delivery system for human keratinocytes. Tissue Eng 12:245–255

20. Sugiyama H, Maeda K, Yamato M et al (2008) Human adipose tissue-derived mesenchymal stem cells as a novel feeder layer for epithelial cells. J Tissue Eng Regen Med 2:445–449

21. Skurk T, Ecklebe S, Hauner H (2007) A novel technique to propagate primary human preadipocytes without loss of differentiation capacity. Obesity 15:2925–2931

22. Freedberg IM, Tomic-Canic M, Komine M et al (2001) Keratins and the keratinocyte activation cycle. J Invest Dermatol 116:633–640

23. Rasmussen C, Thomas-Virnig C, Allen-Hoffmann BL (2013) Classical human epidermal keratinocyte cell culture. Methods Mol Biol 945:161–175

24. Lim X, Tan SH, Koh WL et al (2013) Interfollicular epidermal stem cells self-renew via autocrine Wnt signaling. Science 342:1226–1230

Chapter 2

Isolation and Culture of Epidermal Melanocytes

Muriel Cario and Alain Taieb

Abstract

Melanocytes which represent around 5% of epidermal cells are located in the basal layer. To culture melanocytes, we used trypsin digestion instead of dispase to obtain a cell suspension containing only basal keratinocytes and melanocytes. Melanocytes are cells which needs a great attention. Indeed they dedifferentiate easily in culture as soon as they are in pure culture. Factors secreted by contaminating keratinocytes allow melanocytes to stay dendritic but by regulating their number avoid their growth. In order to, age, phototype and other individual dependent factors regulate the behavior of melanocytes in vitro. Thus, microscopic examination of melanocytes has to be performed each day to adapt conditions of culture to each primary cell culture. This is the secret to have nice melanocytes culture.

Key words Melanocytes, Epidermis, Isolation, Culture, Human skin

1 Introduction

During embryonic development in mammals, the neural crest-derived melanoblasts migrate along the dorsolateral axis, cross the basal membrane separating the dermis from the epidermis to reach their final location in the interfollicular epidermis and epidermal hair follicles [1]. In interfollicular epidermis, melanocytes are located in the basal layer (Fig. 1) and represent around 10% of epidermal basal cells [2] whereas melanocyte stem cells are located in the hair follicle bulge-subbulge area in mammals [3, 4]. The density of melanocytes in unexposed lower back skin is similar whatever the skin type (Asian, black, white) and is ranging from 12.2 to 12.8 melanocytes by mm [5]. However, according to skin localization, density of melanocytes differs. There are approximately 2000 melanocytes per square millimeter in the skin of the scrotum or in the foreskin and 1000 to 1500 melanocytes per square millimeter on the rest of nonpalmoplantar skin [6]. In order to, palmoplantar skin contains less melanocytes than non-palmoplantar skin [7, 8]. Melanocytes are dendritic cells which produce and transfer melanin via their dendrites to around

Thomas Biedermann and Sophie Böttcher-Haberzeth (eds.), *Skin Tissue Engineering: Methods and Protocols*, Methods in Molecular Biology, vol. 2922, https://doi.org/10.1007/978-1-0716-4510-9_2,

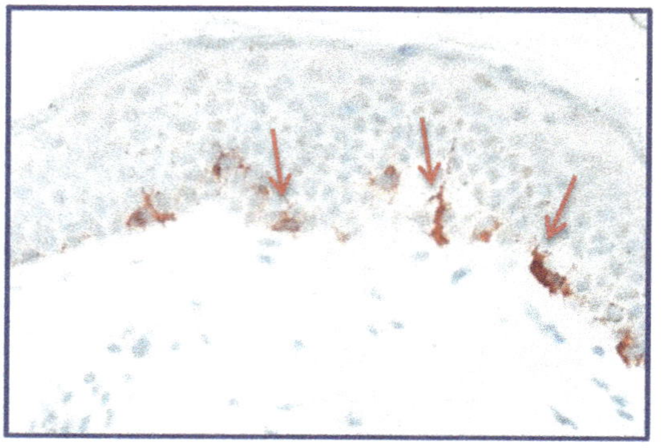

Fig. 1 Microscopic images of tissue showing melanocytes location in human epidermis. Immunohistochemistry of human skin using anti-melanA antibody and envision system (Dako) allowed visualization of melanocytes in red (arrow). Counterstaining of nucleus was done with hematoxylin (violet)

36 surrounding keratinocytes (Fig. 1) [9]. This complex is named epidermal melanin unit. Melanocytes produce pheomelanin (orange melanin) and eumelanins (brown, black melanins) which concentration depends of the phototype [10–12]. Melanin production is under the control of specific enzymes, tyrosinase which is common for the two melanogenic pathways whereas TRP-1 and TRP-2 are specific of the eumelanogenic pathway. Under stimulation by sunlight, melanocytes from darker skin have a greater capacity to produce melanin [13].

2 Materials

For human skin cells culture, all melanocytes are made in a biosafety level 2 culture laboratory, and all solutions are sterilized by heating or 0.22 μm filtration.

2.1 Human Skin Samples

Human skin samples are surgical wastes obtained after informed consent of donor. Common surgical wastes are mammary skin, foreskin, and abdominal skin. Samples from patients are often biopsy of 4 mm diameter.

2.2 Cell Culture

1. Hank's Balanced Salt Solution (HBSS).
2. Trypsin solution: 0.25% Trypsin, 0.1% EDTA in HBSS (*see* **Note 1**).
3. FBS solution: 10% Fetal bovine serum (FBS) diluted in HBSS.
4. PBS (phosphate buffered saline): 0.137 M NaCl, 0.05 M NaH_2PO_4, in deionized H_2O, adjust to pH 7.4.
5. Melanocyte medium: (a) Commercial MCDB153 medium: supplemented with 20 μg/mL insulin, 140 μg/mL bovine

pituitary extract, 1.75 μM hydrocortisone, 3% FBS and 1% penicillin-streptomycin. (b) Commercial M254 medium: supplemented with 3 mg/mL Heparin, 10 ng/mL PMA, 3 ng/mL FGF-2, 5 mg/mL insulin, 0.2% bovine pituitary extract, 0.18 mg/mL hydrocortisone, and with 1% penicillin-streptomycin. Some authors also add transferrin and FBS. (c) Commercial Melanocyte growth medium M2 or Melanocyte basal medium (MBM): supplemented with corresponding commercial supplement Mix (C42300 + C39420 or C24210 + C39410, Promocell) and with 1% penicillin-streptomycin. d) Commercial MBM: could also be supplemented with animal free 0.1 mg/mL GM-CSF, animal free 0.1 mg/mL SCF, 0.2 mg/mL animal free FGF-2 and with 1% penicillin-streptomycin to have clinical grade medium (*see* **Notes 2** and **3**).

6. Melanocytes are isolated from skin sample (*see* Subheading 3.1). Alternatively, one can purchase commercially available melanocytes (e.g., human skin-derived melanocytes, PromoCell, ATCC).
7. 0.4% trypan blue solution in PBS pH 7.4.
8. 3,7% paraformaldehyde in PBS pH 7.4.
9. 40 μm filter strainer.
10. Cell counter.
11. Chamber slide or cell culture wells (96-well plate).
12. Horseradish peroxidase (HRP).
13. 3% hydrogen peroxide in PBS pH 7.4.
14. Dapi (4′,6-Diamidin-2-phenylindol).

3 Methods

3.1 Isolation of Epidermal Melanocytes

1. Degrease, thin the samples and cut into fragments with a maximum size of 0.5 × 0.5 cm (*see* **Note 4**) (Fig. 2).

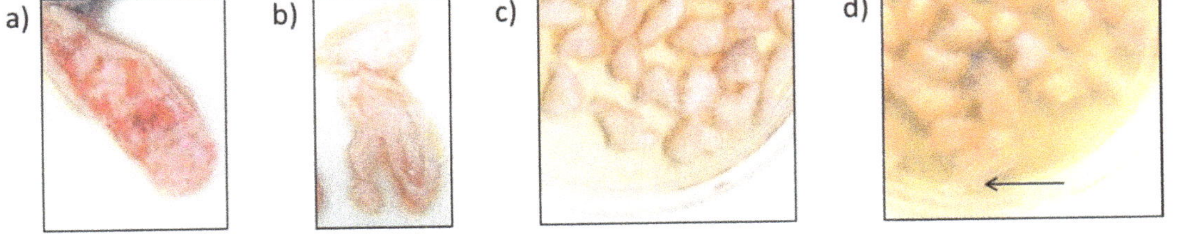

Fig. 2 Four images show stages of skin sample preparation. Appearance of skin before (**a**) and after (**b**) degreasing and thinning of skin. (**c**) Skin small pieces epidermal size up in trypsin-EDTA. (**d**) Epidermis peeling off (arrow)

Fig. 3 Microscopic image showing primo-culture of melanocyte. Depending on the medium used, more or less keratinocytes (arrow) can contaminate the culture making it difficult to purify melanocytes

2. Incubate in trypsin solution for around 3 h at 37 °C or overnight at 4 °C (*see* **Notes 5–7**).
3. Neutralize trypsin with same volume of FBS.
4. Remove the epidermis and scrape the basal layer not too hard to avoid fibroblasts in HBSS.
5. Filter the obtained suspension (40 μm filter strainer) and count cell.
6. Centrifuge at 250 g for 5 min.
7. Dilute 50 μL of solution containing cells in 50 μL trypan blue and count non-blue cells.
8. Suspend the pellet in melanocyte medium.
9. Seed cells at a density of $2x10^5$ per cm^2 (*see* **Notes 8** and **9**).
10. Change medium 3 times a week (*see* **Note 10**) after rinsing with HBSS.
11. After about a week, the primo-culture will be confluence and melanocytes may be purified (*see* **Note 8**) and amplified (*see* **Notes 11** and **12**) (Fig. 3).

3.2 Isolation of Epidermal Melanocytes from a Punch Biopsy

1. Thin biopsy by removing hypodermis and reticular dermis and cut biopsy in 2.
2. Incubate in trypsin solution for around 1–2 h at 37 °C.
3. Neutralize trypsin with same volume of FBS.
4. Remove the epidermis and scrape the basal layer in melanocyte medium.
5. Put cell suspension in a well of a 24-well plate.
6. Change medium without rinsing 24–48 h after seeding.
7. Change medium twice a week until confluence (*see* **Note 13**) (Fig. 4).

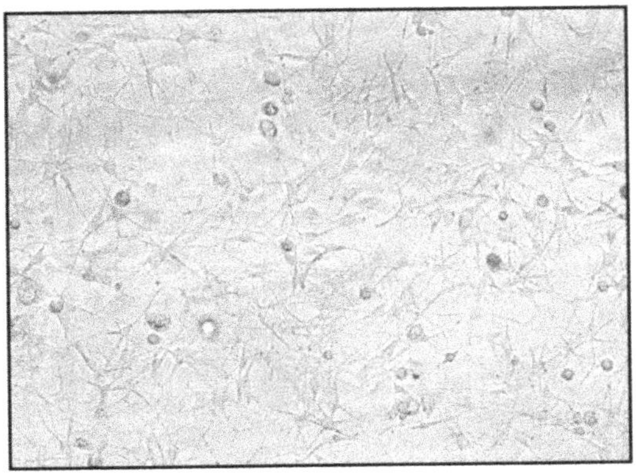

Fig. 4 Microscopic view of confluent low phototype melanocytes

3.3 Culture and Amplification of Melanocytes (See Notes 14 and 15)

1. Rinse primo-culture of melanocytes with HBSS to improve trypsin action when culture reach confluence or when too much keratinocytes contaminate the culture.
2. Incubate melanocytes in the trypsin working solution (0.025% trypsin, 0.01% EDTA) less than 1 min. Follow detachment under microscope (*see* **Notes 16** and **17**).
3. Neutralize trypsin with equal volume of 10% FBS (1.5 mL for 25 cm^2).
4. Collect the solution containing cells (note the volume) and before centrifugation take 50 μL to count cells.
5. Centrifuge the solution at 250 g at room temperature 5 min.
6. Dilute 50 μL of solution containing cells in 50 μL trypan blue and count non-blue cells.
7. Seed cells in culture medium usually by doubling the surface (*see* **Note 18**).
8. Repeat **step 1**–7 until having enough melanocytes to perform desired study (*see* **Note 20**) (Figs. 5 and 6).

3.4 Transduction of Melanocytes

Melanocytes from patient are often difficult to obtain thus you can model disease using gene transfer. Transduction efficiency either for overexpression or silencing lentiviral vectors is high in melanocyte.

1. Rinse pure culture of melanocytes passage 1 or 2 with HBSS to improve trypsin action.
2. Incubate melanocytes in the trypsin working solution (0.025% trypsin, 0.01% EDTA) less than 1 min. Follow detachment under microscope.
3. Neutralize trypsin with equal volume of 10% FBS (1.5 mL for 25 cm^2).

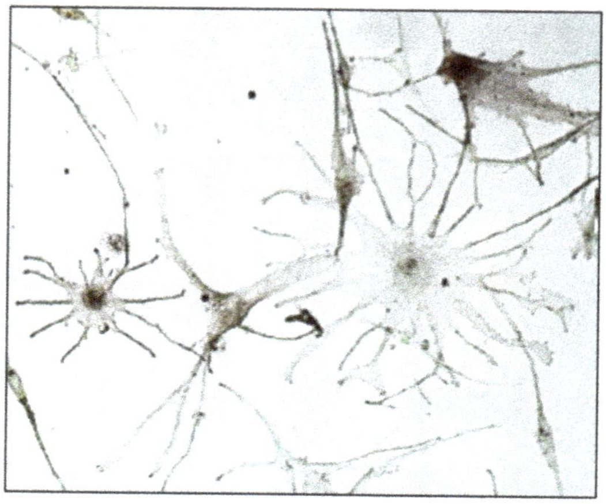

Fig. 5 Microscopic observation of melanocytes from phototype VI skin without any staining. Two populations of high dendritic melanocytes can be distinguished according to their intrinsic pigmentation

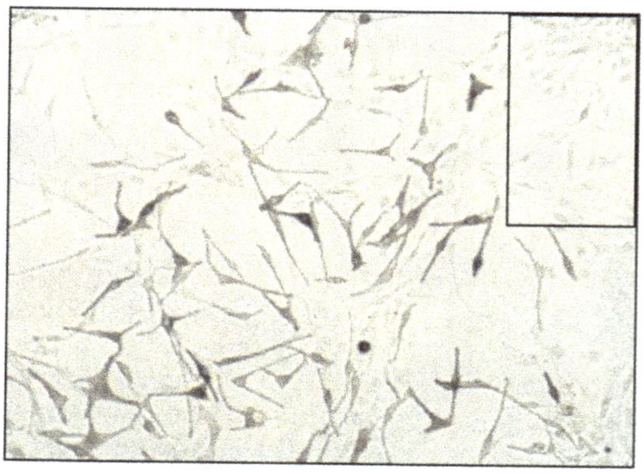

Fig. 6 Micrscopic observation of a culture of melanocytes and melanoblasts without any staining. Two populations can be distinguished, according to their intrinsic pigmentation, shape, and organization. Melanocytes are pigmented bi or tripolar cells whereas melanoblasts (rectangle) are unpigmented bipolar cells. Melanoblasts are more in close contact than melanocytes

4. Collect the solution containing cells (note the volume) and before centrifugation take 50 μL to count cells.
5. Centrifuge the solution at 250 g at room temperature 5 min.
6. Dilute 50 μL of solution containing cells in 50 μL trypan blue and count non-blue cells.
7. Seed melanocytes at a density of 18,000–20,000 by cm^2.
8. 6–18 hours after plating add virus at a multiplicity of infection (MOI) of 1 to 10, to determine the efficient MOI, in a small

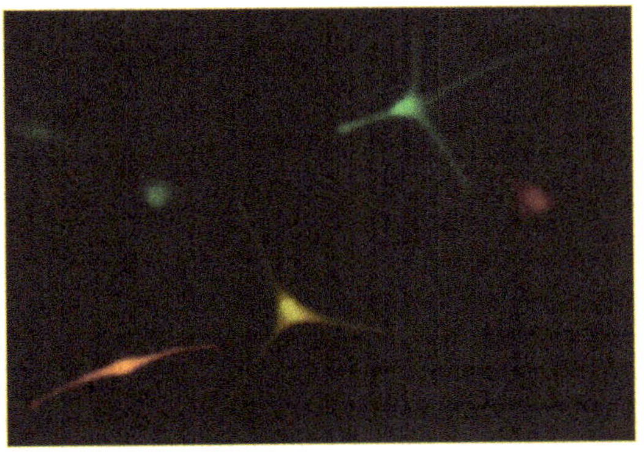

Fig. 7 Aspect of culture of melanocyte after double transduction MOI 4. Melanocytes were transduced with a lentivector containing gene of interest Y and GFP coding sequence and a second lentivector containing another gene of interest Y'and tdTomato coding sequence. Each transduction was at MOI 2. Double transduced melanocytes appear in yellow and single transduced melanocytes appeared in red (TdTomato) or green (GFP)

volume of medium without FBS and incubate at 37 °C, 5% CO_2 (*see* **Note 21**).

9. 6–18 hours after infection add complete melanocyte medium.
10. 48 hours after infection rinse two times the culture and add complete medium.
11. Transduction is efficient 72 h after infection.
12. 100% transduction can be achieved by fluorescent cell sorting if lentivector contains fluorescent protein, or by incubation with antibiotics (such as 10 μg/mL puromycin) if lentivector contains an antibiotic-resistant gene (*see* **Note 22**) (Fig. 7).

3.5 Characterization of Melanocytes by Immunohistochemistry

1. Culture cells in chamber slide or cell culture wells (96-well plate) for 6–24 h.
2. Rinse cells three times with cold PBS pH 7.4.
3. Fix cells with 3,7% paraformaldehyde for 10 min at 4 °C.
4. Rinse cells three times with cold PBS pH 7.4.
5. Permeabilize cells with 0.2% triton x-100 in PBS pH 7.4 (*see* **Note 23**).
6. Wash three times with PBS pH 7.4.
7. If HRP (horseradish peroxidase) staining is used, block endogenous peroxidase with 3% hydrogen peroxide during 10 min and wash three times with PBS pH 7.4.
8. Incubate with 2.5–10% FBS 20 min at 4 °C (*see* **Note 24**).

Table 1
List of useful antibodies to characterize melanocyte

Antibody	Reference	Dilution
HMB45	Dako M0634	1/20°
C-kit (CD117)	Dako A4502	1/1000°
NKI/beteb	Monosan Mon7006	1/20°
TRP-1	Abcam ab83774	1/100°
MITF	Abcam ab20663	1/100°
TRP-2	Abcam ab74073	1/500°
Tyrosinase	Novusbio NBP1	1/100°
MelanA (Mart-1)	Dako M7196	1/100°

9. Incubate with primary antibody (Table 1) diluted according to manufacturer in PBS pH 7.4 1 h at room temperature to overnight 4 °C in a moisture chamber.
10. Collect diluted antibody and keep this dilution at the antibody storage temperature (*see* **Note 25**).
11. Wash with PBS pH 7.4 three times.
12. Incubate with secondary fluorescent antibody raised against the host species used to generate the primary antibody diluted in PBS pH 7.4 according to manufacturer, 1 h at room temperature in the dark or with HRP-coupled secondary antibody.
13. Collect diluted antibody and keep this dilution at the antibody storage temperature (*see* **Note 25**).
14. Wash with PBS pH 7.4 three times.
15. Incubate with chromogen according to manufacturer's instructions for HRP staining.
16. Incubate with Dapi (4′,6-Diamidin-2-phenylindol) 10 min for fluorescent staining.
17. Wash with PBS pH 7.4 three times.
18. Mount with adapted mounting medium (Fig. 8 and Table 2).

3.6 Evaluation of Tyrosinase Activity in Melanocytes

Melanoblasts are dopa negative; thus this technique allows differentiating melanoblasts from melanocytes by detection of tyrosinase activity [14–16].

1. Culture cells in Lab-Tek® chamber slide or cell culture wells (96-well plate) for 6–24 h.
2. Rinse cells three times with cold PBS pH 7.4.
3. Fix cells with 3.7% paraformaldehyde for 15 min at 4 °C.

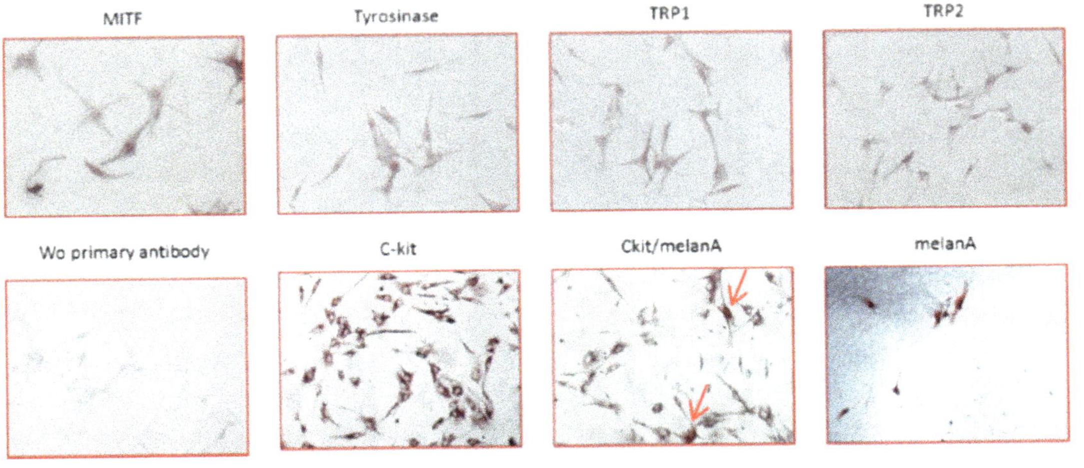

Fig. 8 A grid of eight microscopic images showing expression of various markers in melanocyte derived cells. Except for c-kit which chromogen was vector VIP (violet), chromogen was AEC (rouge). In the double staining c-Kit/melanA, melanA positive cells were identified with an arrow. MelanA staining revealed melanocytes

4. Rinse cells three times with cold PBS pH 7.4.
5. Incubate with L-DOPA (0.1% in PBS pH 7.4) 1 h at 37 °C.
6. Renew L-DOPA solution and incubate for 3 h at 37 °C.
7. Wash three times with PBS pH 7.4.
8. Mount in aqueous mounting medium (Fig. 9).

4 Notes

1. Make sure to be at 37 °C no more to avoid inactivation of trypsin.
2. Melanin produced by high phototype melanocytes hampers cell growth. Thus, for high phototype melanocyte M2 medium is more adapted or MCDB153 has to be supplemented with 300 μM PTU (phenyl thiourea), an inhibitor of tyrosinase. We commonly use MBM for phototype I-IV melanocytes and M2 medium for phototype V and VI.
3. Phorbol 12-myristate 13-acetate (PMA), is a skin tumor promotor, present in the supplement kit for MBM. According to experiment, it could be better to use PMA-free medium.
4. After degreasing and thinning, skin must almost seem transparent.
5. Pieces of skin must be placed in trypsin-EDTA with epidermal size up.
6. Trypsin cuts above basal layer at the difference of dispase which cuts under the basal layer at the dermo-epidermal junction.

Table 2
Reactivity of melanocytes and melanoblasts against antibodies

Marker	Melanocyte / melanoblast (dedifferentiated melanocyte)	Melanocyte/ melanoblast	Melanocyte/ dedifferentiated melanocyte	Melanocyte	Mouse melanoblast	Mouse melanoblast	Human melanoblats
Reference	[17]	[14]	[18]	[20, 21]	[17, 22]	[23]	[24]
HMB45	+/−	+/+	ND	ND	ND	ND	–
c-Kit (CD117)	−/+[a]	ND	+[a]/−	+	ND	ND	ND
TRP-1	ND	+/+	+/−	+	−/+ (according to medium	ND	ND
TRP-2	+/+	ND	ND	ND	—/+ (according to medium)	+	ND
Tyrosinase	ND	+/+$	ND	+	–	–[a]	ND
MelanA (Mart-1)	ND	+/+[a]	ND	ND	ND	ND	ND
S-100	+/−	ND	ND	ND	ND	ND	ND

$ detection of tyrosinase by RT-PCR or immunocytochemistry does not predict its activity. Indeed a great difference between melanoblasts and melanocytes is the localization of markers [14]

[a]Mixed cells of positive and negative

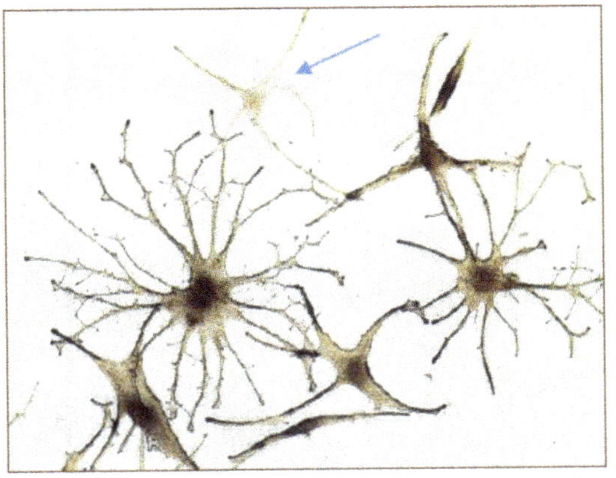

Fig. 9 Microscopic appearance of melanocytes after Dopa staining. According to Hirobe, melanobasts are dopa negative cells which have not to be restricted to bipolar and tripolar cells [15]. The dopa- negative cell with five dendrites may be a melanoblast (blue arrow)

7. Epidermis must peel off by itself. If not, incubation has to be continued. This may due to the thickness or the size of the samples.
8. Melanocytes growth is better in 25 cm^2 than 75 cm^2, so for example it is better to seed 15 millions melanocytes in 3 flasks of 25 cm^2 than 1 flask of 75 cm^2.
9. Melanocytes from patients suffering of pigment disorders such as vitiligo must be seeded at a greater density of 2.105 cells by cm^2.
10. If no melanocytes were seen after 2–3 days of culture, the scratching was not strong enough.
11. If too many keratinocytes contaminate the primo-culture, melanocytes must be purified before 70% confluence otherwise purification may be difficult [16].
12. If melanocytes have difficulty to grow, instead of amplifying melanocytes, you can concentrate them in a flask of smaller size.
13. If melanocytes have difficulties to maintain in culture and are for experiments such as genetic analysis (sequencing), PMA could be added in the medium.
14. Dendricity of melanocytes varies according to phototype. Phototype I–III are less dendritic than phototype IV–VI.
15. Melanocytes from phototype VI have often two populations of melanocytes, one pigmented and one unpigmented.
16. If primo-cultures are not pure, you must follow carefully cells' detachment under microscope, to avoid keratinocytes contamination.

17. If melanocytes detach too quickly and if it is difficult to purified culture, use a less concentrated solution of trypsin-EDTA as 0.0125% trypsin, 0.005% EDTA.
18. If melanocytes grow faster, you have to increase the size of seeding to avoid more than 1 passage by week.
19. Along culture melanocytes become bigger and less dendritic. However, the behavior of melanocytes in culture is individual dependent.
20. Shape of melanocytes has to be managed carefully since in culture, especially without PMA, melanocytes dedifferentiate quickly in melanoblasts [17, 18].
21. A MOI range has to be tested prior to do the experiment since modifying gene is not harmless and may induce death or detachment of melanocytes due to genotoxicity. Indeed, increasing MOI is associated with increase in vector copy number (VCN). It is admitted that there is only one copy of gene by cells when the percentage of transduction is around 30% [19].
22. Sorting or treatment with antibiotics may not be done too early since in the beginning transgene expression is due to pseudo-transduction (circular DNA) and not to integrated gene.
23. Permeabilization step can be avoided if the protein is membranous and that antibody target extracellular domain.
24. Manufacturer recommends using serum from species of the secondary antibody, but FBS is efficient for most staining. Incubation with 2.5–10% FBS avoids a specific fixation; the percentage has to be adjusted according to the background. Start with 2.5%. Otherwise, commercial blocking solutions are available.
25. Some diluted antibodies can be used up to three times.

References

1. Bonaventure J, Domingues MJ, Larue L (2013) Cellular and molecular mechanisms controlling the migration of melanocytes and melanoma cells. Pigment Cell Melanoma Res 26(3):316–325
2. Haass NK, Herlyn M (2005) Normal human melanocyte homeostasis as a paradigm for understanding melanoma. J Investig Dermatol Symp Proc 10(2):153–163
3. Jimbow K, Quevedo WC, Fitzpatrick TB et al (1976) Some aspects of melanin biology: 1950-1975. J Invest Dermatol 67(1):72–89
4. Nishimura EK (2011) Melanocyte stem cells: a melanocyte reservoir in hair follicles for hair and skin pigmentation. Pigment Cell Melanoma Res 24(3):401–410
5. Tadokoro T, Yamaguchi Y, Batzer J et al (2005) Mechanisms of skin tanning in different racial/ethnic groups in response to ultraviolet radiation. J Invest Dermatol 124(6): 1326–1332
6. Démarchez M (2011) Melanocyte and pigmentation. Available from: https://biologiedelapeau.fr/spip.php?article7
7. Szabo G (1967) The regional anatomy of the human integument with special reference to the distribution of hair follicles, sweat glands and melanocytes. Philos Trans R Soc Lond Ser B Biol Sci 252(779):447–485

8. Yamaguchi Y, Itami S, Watabe H et al (2004) Mesenchymal-epithelial interactions in the skin: increased expression of dickkopf1 by palmoplantar fibroblasts inhibits melanocyte growth and differentiation. J Cell Biol 165(2):275–285
9. Fitzpatrick TB, Breathnach AS (1963) The epidermal melanin unit system. Dermatol Wochenschr 147:481–489
10. Hennessy A, Oh C, Diffey B et al (2005) Eumelanin and pheomelanin concentrations in human epidermis before and after UVB irradiation. Pigment Cell Res 18(3):220–223
11. Hunt G, Kyne S, Ito S et al (1995) Eumelanin and phaeomelanin contents of human epidermis and cultured melanocytes. Pigment Cell Res 8(4):202–208
12. Tobin D, Quinn AG, Ito S et al (1994) The presence of tyrosinase and related proteins in human epidermis and their relationship to melanin type. Pigment Cell Res 7(4):204–209
13. Quevedo WC, Fitzpatrick TB, Pathak MA et al (1975) Role of light in human skin color variation. Am J Phys Anthropol 43(3):393–408
14. Cook AL, Donatien PD, Smith AG et al (2003) Human melanoblasts in culture: expression of BRN2 and synergistic regulation by fibroblast growth Factor-2, stem cell factor, and Endothelin-3. J Invest Dermatol 121(5): 1150–1159
15. Hirobe T (1992) Melanocyte stimulating hormone induces the differentiation of mouse epidermal melanocytes in serum-free culture. J Cell Physiol 152(2):337–345
16. Henrot P, Laurent P, Levionnois E et al (2020) A method for isolating and culturing skin cells: application to endothelial cells, fibroblasts, keratinocytes, and melanocytes from punch biopsies in systemic sclerosis skin. Front Immunol 11:566607
17. Zhao Z, Jin C, Ding K et al (2012) Dedifferentiation of human epidermal melanocytes into melanoblasts in vitro. Exp Dermatol 21(7): 504–508
18. Kormos B, Belső N, Bebes A et al (2011) In vitro dedifferentiation of melanocytes from adult epidermis. PLoS One 6(2):e17197
19. Kustikova OS, Wahlers A, Kühlcke K et al (2003) Dose finding with retroviral vectors: correlation of retroviral vector copy numbers in single cells with gene transfer efficiency in a cell population. Blood 102(12):3934–3937
20. Hachiya A, Kobayashi A, Ohuchi A et al (2001) The paracrine role of stem cell factor/c-kit signaling in the activation of human melanocytes in ultraviolet-B-induced pigmentation. J Invest Dermatol 116(4):578–586
21. Luo D, Chen H, Searles G et al (1995) Coordinated mRNA expression of c-Kit with tyrosinase and TRP-1 in melanin pigmentation of normal and malignant human melanocytes and transient activation of tyrosinase by kit/ SCF-R. Melanoma Res 5(5):303–309
22. Hirobe T, Osawa M, Nishikawa S-I (2004) Hepatocyte growth factor controls the proliferation of cultured epidermal melanoblasts and melanocytes from newborn mice. Pigment Cell Res 17(1):51–61
23. Sviderskaya EV, Wakeling WF, Bennett DC (1995) A cloned, immortal line of murine melanoblasts inducible to differentiate to melanocytes. Dev Camb Engl 121(5):1547–1557
24. Holbrook KA, Underwood RA, Vogel AM et al (1989) The appearance, density and distribution of melanocytes in human embryonic and fetal skin revealed by the anti-melanoma monoclonal antibody, HMB-45. Anat Embryol (Berl) 180(5):443–455

Chapter 3

Initiating Separate Cultures of Keratinocytes and Melanocytes from a Single Biopsy

Karel Ferland, Henri De Koninck, Bettina Cattier, Sergio Cortez Ghio, Danielle Larouche, and Lucie Germain

Abstract

A significant stride in enhancing the care for extensively burned patients has been made through the cultivation of autologous keratinocytes, resulting in the formation of cohesive epithelial tissue sheets. Beyond clinical applications, both keratinocyte and melanocyte cultures play vital roles in investigating diverse cellular and molecular mechanisms underlying various skin pathologies. In epithelial cell culture, preserving stem cells is crucial, necessitating dependable methods and optimal conditions. Well-cultured keratinocytes typically exhibit consistent cuboid morphology and maintain proliferative capacity over numerous passages. In contrast, melanocyte culture requires specialized additives to stimulate growth, sustain differentiation, and ensure purity. This chapter provides a comprehensive guide to the materials and techniques essential for efficient keratinocyte culture for clinical purposes and melanocyte culture for research endeavors. Covered aspects include tissue sampling and transportation, isolation, routine culture, subculture, and cryopreservation.

Key words Cell therapy, Cultured epithelial autograft, Epidermis, Epithelium, Regenerative medicine, Tissue engineering, Wound healing, Keratinocyte, Melanocyte

1 Introduction

Since 1957, researchers have been cultivating human epidermal cells (keratinocytes) [1]. In 1975, Rheinwald and Green pioneered the extensive expansion of keratinocytes in vitro using feeder layers [2], and shortly after in 1982, Eisinger and Marko established the first effective method for proliferative melanocyte culture [3]. The use of dispase in generating epidermal sheets [4] eventually led to the development of cultured autologous epidermal sheets for direct application onto burn wounds or donor sites after autograft harvesting, promoting skin regeneration [5, 6].

Karel Ferland and Henri De Koninck contributed equally with all other contributors.

Thomas Biedermann and Sophie Böttcher-Haberzeth (eds.), *Skin Tissue Engineering: Methods and Protocols*, Methods in Molecular Biology, vol. 2922, https://doi.org/10.1007/978-1-0716-4510-9_3,

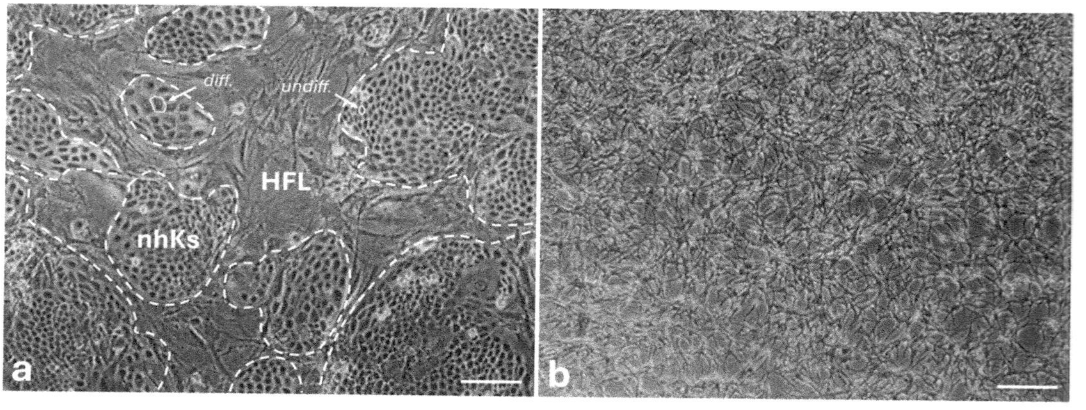

Fig. 1 The image consists of two panels showing microscopic views of cell cultures. Panel (**a**) displays second passage of normal human keratinocytes (nhKs) cultured on a human feeder layer (HFL) at 10× under a phase-contrast microscope. Dashed white lines delineate nhK colonies, composed of smaller cuboidal cells, from the larger HFL cells. The nhK confluence is estimated at 60%. Within the nhK population, undifferentiated cells ("undiff.") exhibit smaller size and a higher nucleus/cytoplasm ratio compared to differentiated nhKs ("diff."). Each phenotype's individual cell is outlined in white, with arrows for clarity. (**b**) Fourth passage of a human melanocyte (hMels) culture at 10× under a phase-contrast microscope. hMels are dendritic, the confluence is estimated at 90%. Both panels include a scale for reference, bar is 200 μm

Keratinocytes, being the primary constituents of the epidermis and responsible for its essential functions, are also valuable in exploring cellular and molecular mechanisms underlying various pathologies like skin cancer [7], epidermolysis bullosa [8], psoriasis [9], and amyotrophic lateral sclerosis [10]. Similarly, melanocytes, originating from the neural crest and accountable for skin pigmentation via melanin synthesis and transfer to keratinocytes, are found in the epidermal basal layer at a ratio of approximately 1:35 to 40 epidermal keratinocytes [11]. Culturing melanocytes proves beneficial for studying pigmentary disorders such as vitiligo [12].

The therapeutic efficacy of cultured epithelial autologous grafts hinges on the cells' ability to maintain their proliferative potential post-implantation. An effective method for assessing the quality of keratinocyte culture over time involves observing cell morphology through phase-contrastmicroscopy [13]. Properly cultured keratinocytes demonstrate a cuboid shape and form sizable colonies comprising small, undifferentiated cells with a high nucleus/cytoplasm ratio, referred to as holoclones. As passages progress and cells undergo differentiation, their proliferative capacity diminishes. Consequently, a greater proportion of keratinocytes give rise to small colonies comprising large cells with a low nucleus/cytoplasm ratio, known as paraclones (abortive colonies). Fully differentiated keratinocytes exhibit severely impaired proliferation and adopt an irregular morphology (refer to Fig. 1a). In contrast, melanocytes showcase a dendritic morphology and spread across the Petri dish (refer to Fig. 1b).

In our laboratory, we employ culture techniques rooted in those initially outlined by Rheinwald and Green to amplify keratinocytes for both research and clinical purposes [2]. While these methods have improved over time, we have transitioned to using human non-proliferative skin fibroblast feeder layers instead of murine embryonic immortalized fibroblast layers [14–16]. Additionally, we have replaced cholera toxin with isoproterenol, a readily available synthetic cyclic AMP inducer that serves the same function in epithelial cells [17, 18]. For the isolation of melanocytes, we utilize approaches adapted from those delineated by Halaban and Alfano to inhibit rapidly dividing cells such as keratinocytes and fibroblasts, employing geneticin, an antibiotic, for the purification of melanocytes [19]. Following isolation, melanocytes are cultured in commercial media containing essential factors to support their growth and preserve their differentiated state.

This chapter provides comprehensive details on the reliable methods and materials required for all aspects of keratinocyte and melanocyte culture, including tissue sampling and transport, isolation, routine culture, subculture, and cryopreservation. All described methods for keratinocytes have undergone rigorous testing and are utilized in the clinical treatment of patients [20–23]. The selection of products, reagents, and materials recommended for keratinocyte culture in this chapter is based on criteria such as availability, efficiency, quality, and safety, adhering to Good Manufacturing Practice (GMP) guidelines and utilizing United States Pharmacopeia (USP) or GMP grade materials whenever possible. However, it's important to note that the methods and protocols outlined for melanocytes are solely intended for research purposes due to the carcinogenic nature of some products used, such as phorbol 12-myristate 13-acetate, rendering them unsuitable for clinical studies [24].

2 Materials

2.1 Base Keratinocyte Medium

DME-Ham Combine 3 parts Dulbecco's modified Eagle's medium (DMEM) with 1 part Ham's F-12 medium in apyrogenic ultrapure water. Add 3.07 g/L of $NaHCO_3$ (36.54 mM), 24.3 mg/L of adenine (0.18 mM) solubilized in 312.5 μL/L of 0.2 N HCL. Adjust pH to 7.1. Sterilize by filtration through a 0.22 μm low binding disposable filter and store in the dark at 4 °C.

2.2 Sera

1. *Bovine growth serum.*
2. *FetalClone II serum.*

Thaw sera at 4 °C or in cold water (*see* **Note 1**). Gently swirl it to resuspend its components. Inactivate in a 56 °C water bath for

30 min. To avoid repeated thawing and freezing cycles, distribute in single use aliquots. Store at 4 °C (1 month) or at −80 °C for long term storage.

2.3 Additives for Keratinocyte Media

1. *Recombinant human insulin.* Dissolve at 5 mg/mL in 5 mM HCl. Yields a 1000× stock solution. Can be stored at 4 °C for up to a year (according to manufacturer's recommendations).
2. *Epidermal growth factor (EGF).* Dissolve at 333.3 μg/mL in PBS. Dilute (1:33.3) the solution with DME-Ham containing 10% v/v FetalClone II serum. Yields a 1000× stock solution.
3. *Hydrocortisone.* Dissolve at 7.5 mg/mL in 95% ethanol. Dilute (1:37.5) the solution with DME-Ham. Yields a 500× stock solution.
4. *Penicillin G/Gentamicin.* Dissolve penicillin G at 50,000 IU/mL and active gentamicin at 12.5 mg/mL in apyrogenic ultrapure water. Yields a 500× stock solution.
5. *Isoproterenol hydrochloride.* Store at 4 °C. Vials are at 0.2 mg/mL and are single use. Leftover isoproterenol should not be stored.
6. *Fungizone.* Dissolve amphotericin B at 0.25 mg/mL in apyrogenic ultrapure water to make a 500× stock solution.

For all additives, except isoproterenol, sterilize by filtration through a 0.22 μm low binding disposable filter. To avoid repeated thawing and freezing cycles, distribute in single use aliquots. Store at −80 °C.

2.4 Complete Keratinocyte Media

1. *Tissue transport medium (tDMEM).* Add 10% v/v of bovine growth serum to high (4.5 g/L) glucose DMEM (containing sodium pyruvate and L-glutamine). Then add fungizone, and penicillin G/gentamicin (dilute to 1×). Store in the dark at −20 °C to −80 °C for up to 6 months or at 4 °C for up to 10 days.
2. *Complete keratinocyte culture medium (ckDME-Ham).* Add 5% v/v of FetalClone II serum to DME-Ham. Then add 1.06 mL/L of isoproterenol. Then, in this order, add insulin, epidermal growth factor, hydrocortisone, and penicillin G/gentamicin (dilute to 1×). Store in the dark at 4 °C for up to 10 days.
3. *Cryopreservation medium.* Add 10% v/v of dimethyl sulfoxide (DMSO) to bovine growth serum. Keep on ice or store at 4 °C and use within the day (*see* **Note 2**).

All frozen components can be thawed at 4 °C (*see* **Note 1**).

2.5 Melanocyte Medium

1. *Commercial Melanocyte Growth Medium-4 BulletKit (MGM-4)*. Reconstitute MGM-4 medium with MBM™-4 Basal Medium with MGM-4 SingleQuots™ supplements: 1 mL $CaCl_2$, 0.5 mL phorbol 12-myristate-13-acetate (PMA), 1 mL recombinant human Fibroblast Growth Factor-B (hFGF-B), 1 mL recombinant human insulin, 1 mL bovine pituitary extract (BPE), 0.5 mL hydrocortisone, 2.5 mL fetal bovine serum (FBS) and 0.5 mL Gentamicin Sulfate-Amphotericin (GA-1000). Use melanocyte medium within 6 months.

2.6 Other Solutions

1. *Phosphate buffered saline (PBS)*. Dissolve NaCl at 127 mM, KCl at 2.7 mM, Na_2HPO_4 at 6.5 mM and KH_2PO_4 at 1.5 mM in apyrogenic ultrapure water. Verify pH is between 7.30 and 7.45. Yields a 10× stock solution. Store at room temperature.
2. *PBS—Penicillin G/Gentamicin/Fungizone (PBS-P/G/F)*. Dilute 10× PBS to 1× with apyrogenic ultrapure water. Sterilize by filtration through a 0.22 μm low binding disposable filter. Add Penicillin G/Gentamicin and Fungizone (dilute to 1×).
3. *HEPES/KCl/NaCl*. Dissolve 4-(2-hydroxyethyl)-1-piperazineethanesulfonic acid (HEPES) at 0.05 M, KCl at 33.5 mM and NaCl at 0.71 M in apyrogenic ultrapure water. Adjust pH to 7.3. Yields a 5× stock solution. Store in the dark at 4 °C (*see* **Note 3**).
4. *HEPES/KCl/NaCl– $CaCl_2$*. Dilute 5× HEPES/KCl/NaCl to 1× with apyrogenic ultrapure water. Add 1 mM of $CaCl_2$. Adjust pH to 7.45. Store in the dark at 4 °C.
5. *Thermolysin*. Dissolve at 500 μg/mL in HEPES/KCl/NaCl—$CaCl_2$. Sterilize by filtration through a 0.22 μm low binding disposable filter. Store at 4 °C and use within the day.
6. *Trypsin/EDTA diluent*. Dissolve D-glucose at 0.05% w/v and EDTA at 0.01% w/v in 1× PBS. Adjust pH to 7.45. Sterilize by filtration through a 0.22 μm low binding disposable filter. Store at 2–8 °C.
7. *Trypsin/EDTA concentrate*. Dissolve recombinant trypsin in trypsin/EDTA diluent. Adjust pH to 7.45. Sterilize by filtration through a 0.22 μm low binding disposable filter. Store at – 20 °C. Yields a 60X stock solution.
8. *Trypsin/EDTA*. Dilute 60× Trypsin/EDTA concentrate to 1× with trypsin/EDTA diluent. Store at 4 °C for up to 10 days.
9. *Geneticin*. Dissolve at 100 mg/mL in apyrogenic ultrapure water. Sterilize by filtration through 0.22 μm low binding disposable filter. Store at 2–8 °C.

2.7 Tissues and Cells

1. *Normal human keratinocytes (nhKs)*. Surgically removed 3 to 6 cm^2 skin biopsies.
2. *Human fibroblast layer (HFL)*. Non-proliferative irradiated fibroblasts (*see* **Note 4(1)**) extracted from normal skin biopsies (*see* **Note 4(2)**) are seeded at 150,000 to 200,000/25 cm^2 into a culture flask in 5 to 6 mL/25 cm^2 of ckDME-Ham. Medium must be changed every 7 days. HFL cultures are incubated in an 8% CO_2 and 100% humidity atmosphere at 37 °C. HFL cultures must be used within 30 days.
3. *Human melanocytes (hMels)*. Surgically removed 3 to 6 cm^2 skin biopsies.

2.8 Labware

1. Filters. For volumes inferior to 100 mL: 0.22 μm low binding disposable filter. For volumes superior to 100 mL: filtration unit mounted with a 47 mm diameter and 0.22 μm filter set.
2. Sterile specimen containers.
3. 50 mL tubes.
4. 100 × 15 mm cell culture Petri dishes.
5. Dissecting jewelers microforceps.
6. Size 4 scalpel and size 22 blade.
7. Celstir® 50 mL suspension culture flask.
8. Parafilm® M.
9. 25 or 75 cm^2 tissue culture flasks.
10. Sterile cryogenic vials.
11. Freezing container.

3 Methods

Manipulations must be performed under a sterile laminar flow cabinet.

3.1 Tissue Sampling and Transport

1. Immediately following aseptic (*see* **Note 5**) surgical removal, put the skin biopsy into a sterile container filled with cold (4 ° C) tDMEM.
2. Keep samples cold and transport to a cell culture facility without delay.

3.2 Isolation of nhKs

1. Wash the skin sample by transferring it into a 50 mL tube containing 30 mL of PBS-P/G/F and agitating vigorously. With sterile forceps, transfer the sample into another 50 mL tube containing 30 mL to 45 mL of fresh PBS-P/G/F. Repeat the transfer 8 additional times for a total of 10 washes.

2. Use curved forceps to gently spread out the skin sample (epidermis facing top) in a Petri dish.
3. Use a scalpel with a size 22 blade to cut the sample into as many 3 mm × 100 to 150 mm strips as necessary. If the skin is hairy (i.e., scalp tissue), 3 mm × 3 mm strips are preferable.
4. Add 10 mL of cold thermolysin into the Petri dish containing the small skin strips (*see* **Note 6**).
5. Seal the Petri dish with Parafilm® M and incubate overnight at 4 °C.
6. Use two microforceps to carefully separate the epidermis from the dermis (delicately pull hair if present to collect hair follicle keratinocytes). To dissociate epithelial cells, put the epithelial strips (epidermis and hairs) in a Celstir® suspension culture flask containing 20 mL warm (37 °C) trypsin/EDTA.
7. Incubate under agitation for 15–30 min at 37 °C.
8. Add 10 mL of ckDME-Ham into the flask. Transfer the 20 mL cell suspension into a 50 mL tube containing 10 mL of ckDME-Ham. Rinse the culture suspension flask with 10 mL ckDME-Ham and add it to the 50 mL tube.
9. Use an automated cell counter or a hemocytometer to count the epithelial cells. Count is expected to be approximately 3 to 4 million cells/cm^2 of skin biopsy and cell size should be between 6 and 14 μm.
10. Use trypan blue staining and a hemocytometer to estimate cell viability. Cell viability is expected to be greater than 80%.
11. Centrifuge (300 × g) the cell suspension for 10 min at room temperature.
12. Remove the supernatant and resuspend epithelial cells at the desired concentration in ckDME-Ham.
13. Seed epithelial cells between 1 and 2 million cells/25 cm^2 or more (*see* **Note 7**) into a culture flask pre-seeded with a HFL. Seed directly into the culture medium. Total medium volume should not exceed 7 mL/25 cm^2. If HFL medium change is scheduled the same day cells are seeded (7 days since last HFL medium change), replace half the HFL conditioned medium with warm (37 °C) ckDME-Ham.

3.3 Culture of nhKs

1. Incubate nhK-HFL co-cultures (*see* **Note 8**) in an 8% CO_2 and 100% humidity atmosphere at 37 °C.
2. Change the culture medium three times a week, every 2 to 3 days. Remove the medium from the culture flask. Replace it with warm (37 °C) ckDME-Ham.
3. Monitor nhK morphology and confluence (*see* **Note 9**) daily under a microscope.

4. When nhKs reach 75–95% confluence, subculture or cryopreserve them. Do not let nhKs reach 100% confluence.

3.4 Subculture (Passage) of nhKs

1. Remove the culture medium.
2. Depending on culture flask size, swiftly rinse nhKs with either 1 or 2 mL (for either 25 or 75 cm^2 culture flasks, respectively) trypsin/EDTA. Remove trypsin/EDTA.
3. Depending on culture flask size, add either 3 or 8 mL (for either 25 or 75 cm^2 culture flasks, respectively) of trypsin/EDTA into the culture flask.
4. Incubate at 37 °C until all nhKs are completely detached from the flask (verify cell detachment under a microscope). Time for complete detachment should be around 10 min. Do not incubate for more than 15 min (*see* **Note 10**).
5. Depending on culture flask size, neutralize trypsin activity by adding either 3 or 8 mL of ckDME-Ham and swirl the nhK suspension.
6. Vigorously pipette the nhK suspension up and down at least ten times to ensure suspension homogeneity.
7. Transfer the nhK suspension into a 50 mL tube.
8. Thoroughly rinse the culture flask with 4 mL ckDME-Ham and transfer the suspension to the 50 mL tube.
9. Use an automated cell counter or a hemocytometer to count nhKs.
10. Use trypan blue staining and a hemocytometer to estimate cell viability. Cell viability is expected to be greater than 95%.
11. Centrifuge (300 × g) the nhK suspension for 10 min at room temperature.
12. Remove the supernatant and resuspend nhKs at the desired concentration in ckDME-Ham.
13. Seed nhKs at 100,000/25 cm^2 or more (*see* **Note 7**) into a culture flask pre-seeded with a HFL. Seed directly into the culture medium. Total medium volume should not exceed 7 mL/25 cm^2. If HFL medium change is scheduled the same day nhKs are seeded (7 days since last HFL medium change), replace half the HFL conditioned medium with warm (37 °C) ckDME-Ham.

3.5 Cryopreservation of nhKs

1. Fill a freezing container with 100% isopropyl alcohol and store it at 4 °C until cool.
2. Follow **steps 1** through **11** from Subheading 3.4.
3. Remove the supernatant and resuspend nhKs at the desired concentration (max. 10 million/mL) in cryopreservation medium. Put the tube on ice.

4. Aliquot in cryogenic vials on ice.
5. Put the cryogenic vials in the freezing container.
6. Store the container overnight at −80 °C. In these conditions, cell temperature should drop 1 °C/min.
7. After 24 h, store cryogenic vials in liquid nitrogen for long term storage.

3.6 Thawing of nhKs

1. Thaw cryogenic vial in a 37 °C water bath completely without agitating. Ensure that there is no remaining ice in the nhK suspension, while keeping it cold.
2. Transfer the cell suspension from the cryovial(s) into an empty 50 mL tube.
3. Rinse the cryogenic vial with 1.5 mL of warm ckDME-Ham. If there are multiple cryovials, rinse all cryovials with the same volume of 1.5 mL.
4. Transfer the rinsing medium drop by drop into the tube containing the nhK suspension, while manually agitating the tube to homogenize.
5. Add the required volume of warm ckDME-HAM drop by drop to the nhK suspension tube, ensuring a tenfold dilution of the cell suspension to dilute DMSO.
6. Centrifuge (300 × g) the nhK suspension for 10 min at room temperature.
7. Remove the supernatant and gently resuspend nhKs pellets in 10 mL of warm (37 °C) ckDME-Ham.
8. Centrifuge a second time (300 × g) the nhK suspension for 10 min at room temperature, to eliminate DMSO (*see* **Note 11**).
9. Use an automated cell counter or a hemocytometer to count the nhKs.
10. Use trypan blue staining and a hemocytometer to estimate cell viability. Cell viability is expected to be greater than 80%.
11. Remove the supernatant and resuspend nhKs at the desired concentration in ckDME-Ham.
12. Seed nhKs at 100,000/25 cm^2 or more (*see* **Note 7**) into a culture flask pre-seeded with a HFL. Seed directly into the culture medium. Total medium volume should not exceed 7 mL/25 cm^2. If HFL medium change is scheduled the same day nhKs are seeded (7 days since last HFL medium change), replace half the HFL conditioned medium with warm (37 °C) ckDME-Ham.

3.7 Isolation of hMels

1. Perform **steps 1** through **11** of the nhK isolation protocol (*see* Subheading 3.2).
2. Remove the supernatant and resuspend epidermal cells at the desired concentration in MGM-4 medium.
3. Seed epidermal cells at 8.25 million cells/25 cm^2 into a culture flask in 5 mL/25 cm^2 of MGM-4 medium.
4. Add 0.1 mg/mL geneticin to the medium in the culture flask.

3.8 Culture of hMels

1. Incubate hMel cultures (*see* **Note 8**) in an 8% CO_2 and 100% humidity atmosphere at 37 °C.
2. Change the culture medium 3 times a week, every 2 to 3 days. Remove the medium from the culture flask. Replace it with warm (37 °C) MGM-4.
3. Monitor hMel morphology and confluence daily under a microscope.
4. When hMels reach 100% confluence, subculture or cryopreserve them.

3.9 Subculture (Passage) of hMels

1. Remove the culture medium.
2. Depending on culture flask size, swiftly rinse hMels with either 1 or 2 mL (for either 25 or 75 cm^2 culture flasks, respectively) trypsin/EDTA. Remove trypsin/EDTA.
3. Depending on culture flask size, add either 2 or 3 mL (for either 25 or 75 cm^2 culture flasks, respectively) of trypsin/EDTA into the culture flask.
4. Incubate at 37 °C until all hMels are completely detached from the flask (verify cell detachment under a microscope). Time for complete detachment should be around 2 min.
5. Depending on culture flask size, neutralize trypsin activity by adding either 2 or 3 mL of MGM-4 and swirl the hMel suspension.
6. Vigorously pipette the hMel suspension up and down at least ten times to ensure suspension homogeneity.
7. Transfer the hMel suspension into a 50 mL tube.
8. Thoroughly rinse the culture flask with 4 mL MGM-4 and transfer the suspension to the 50 mL tube.
9. Use an automated cell counter or a hemocytometer to count hMels.
10. Use trypan blue staining and a hemocytometer to estimate cell viability. Cell viability is expected to be greater than 95%.
11. Centrifuge (300 x g) the hMel suspension for 10 min at room temperature.

12. Remove the supernatant and resuspend hMels at the desired concentration in MGM-4.
13. Seed hMels at 625,000/25 cm^2 into a culture flask in 5 mL/ 25 cm^2 of MGM-4.

3.10 Cryo-preservation of hMels

1. Fill a freezing container with 100% isopropyl alcohol and store it at 4 °C until cool.
2. Follow steps **1** through **11** from Subheading 3.9.
3. Remove the supernatant and resuspend hMels at the desired concentration (max. 10 million/mL) in cryopreservation medium. Put the tube on ice.
4. Aliquot in cryogenic vials on ice.
5. Put the cryogenic vials in the freezing container.
6. Store the container overnight at −80 °C. In these conditions, cell temperature should drop 1 °C/min.
7. Store cryogenic vials in liquid nitrogen for long term storage.

3.11 Thawing of hMels

1. Thaw cryogenic vial in a 37 °C water bath completely without agitating. Ensure that there is no remaining ice in the hMel suspension, while keeping it cold.
2. Transfer the cell suspension from the cryovial(s) into an empty 50 mL tube.
3. Rinse the cryogenic vial with 1.5 mL of warm MGM-4. If there are multiple cryovials, rinse all cryovials with the same volume of 1.5 mL.
4. Transfer the rinsing medium drop by drop into the tube containing the hMel suspension, while manually agitating the tube to homogenize.
5. Add the required volume of warm MGM-4 drop by drop to the hMel suspension tube, ensuring a 10-fold dilution of the cell suspension to dilute DMSO.
6. Centrifuge (300 × g) the hMel suspension for 10 min at room temperature.
7. Remove the supernatant and gently resuspend hMels pellets in 10 mL of warm (37 °C) MGM-4.
8. Centrifuge a second time (300 × g) the hMel suspension for 10 min at room temperature, to eliminate DMSO (*see* **Note 11**).
9. Use an automated cell counter or a hemocytometer to count the hMels.
10. Use trypan blue staining and a hemocytometer to estimate cell viability. Cell viability is expected to be greater than 80%.

11. Remove the supernatant and resuspend hMels at the desired concentration in MGM-4.
12. Seed hMels at 625,000/25 cm^2 into a culture flask in 5 mL/25 cm^2 of MGM-4.

4 Notes

1. To expedite the thawing process of sera and keratinocyte media components, they can be thawed rapidly either at room temperature or in a 37 °C water bath. However, it is advised to dilute them into culture media immediately for use and avoid refreezing.
2. DMSO acts as a toxic oxidative agent above 10 °C. Therefore, when working with it, swift action is necessary, and it should be kept on ice.
3. Exposure to fluorescent light leads to the degradation of HEPES, rendering it cytotoxic [25].
4. (1) Various alternative methods are employed by other teams to restrict fibroblast proliferation. (2) After irradiation (or a preferred treatment), fibroblasts can be cryopreserved in liquid nitrogen.
5. Prior to culture, the skin biopsy should be disinfected with 0.5% v/v chlorhexidine gluconate in isopropyl alcohol. The use of any other disinfectant may compromise cell viability. After disinfection, thorough rinsing with cold tDMEM is essential to remove any residual disinfectant, as its presence may also compromise cell viability.
6. Thermolysin facilitates the separation of the epidermis from the dermis, thus minimizing fibroblast contamination in nhK and hMel cultures [26].
7. The seeding density of nhKs is chosen by the operator. Expect nhK density to double daily, though proliferation rates may vary between populations and decrease with successive passages. It is recommended to initially seed at a density between 1 and 2 million nhKs/25 cm^2 or higher for uncharacterized populations, aiming for 95% confluence within 4 to 6 days for the initial passages.
8. Keratinocytes represent approximately 90% of epidermal cells and thus take precedence in culture over other epithelial cells [27]. Therefore, the culture is considered exclusively nhKs, disregarding any surviving minority cell types. Conversely, after culturing with geneticin, surviving cells are considered hMel, as only slow-proliferating cells like melanocytes survive the treatment.

9. Confluence of nhKs refers to the approximate surface percentage occupied by nhK colonies, disregarding the HFL. It is assessed under a microscope by evaluating the coverage of nhK colonies in a given field of vision (*see* Fig. 1). Evaluation should involve technical replicates and multiple fields to enhance accuracy. Seeding distribution should be adjusted if confluence varies significantly between fields. However, caution should be exercised when trypsinizing nhKs below 75% confluence to avoid count errors due to overrepresented HFL.
10. To ensure optimal trypsinization efficiency, avoid stacking culture flasks in the incubator, as the temperature is typically higher on the flask surface directly in contact with the incubator shelf.
11. It is advisable to perform a double wash to effectively remove DMSO from the suspension (*see* **Note 2**).
12. Geneticin supplementation for hMel culture is necessary only until the first medium change. If living nhKs persist in the culture, geneticin may be added during the second medium change.

Acknowledgments

The authors would like to thank the current and former members of the LOEX who have contributed to develop these protocols. This work was supported by the Canadian Institutes for Health Research (CIHR) grant IC-132948 (L.G.), Fondation des Pompiers du Québec pour les Grands Brûlés (FPQGB), the Fonds de Recherche du Québec-Santé (FRQS), and the Réseau de thérapie cellulaire, tissulaire et génique du Québec (ThéCell), a network supported by the FRQS. L.G. is the holder of a Tier 1 Canadian Research Chair from CIHR on Stem Cells and Tissue Engineering, and a Research Chair from the Fondation de l'Université Laval on Tissue-Engineered Organs and Translational Medicine.

References

1. Prunieras M, Bonjean M, Chardonnet Y et al (1957) Etuce sur la trypsination de l'epiderme humain. Bull Soc Fr Dermatol Syphiligr 67: 58–65
2. Rheinwald JG, Green H (1975) Serial cultivation of strains of human epidermal keratinocytes: the formation of keratinizing colonies from single cells. Cell 6(3):331–343
3. Eisinger M, Marko O (1982) Selective proliferation of normal human melanocytes in vitro in the presence of phorbol ester and cholera toxin. Proc Natl Acad Sci USA 79(6): 2018–2022
4. Green H, Kehinde O, Thomas J (1979) Growth of cultured human epidermal cells into multiple epithelia suitable for grafting. Proc Natl Acad Sci USA 76(11):5665–5668
5. Gallico GG, O'Connor NE, Compton CC et al (1984) Permanent coverage of large burn wounds with autologous cultured human epithelium. N Engl J Med 311(7):448–451

6. Nicholas E, O'Connor JM-S (1981) Grafting of burns with cultured epithelium prepared from autologous epidermal cells. Lancet 1(8211):75–78
7. Purdie KJ, Lambert SR, Teh MT et al (2007) Allelic imbalances and microdeletions affecting the PTPRD gene in cutaneous squamous cell carcinomas detected using single nucleotide polymorphism microarray analysis. Genes Chromosomes Cancer 46(7):661–669
8. Bauer JW, Koller J, Murauer EM et al (2017) Closure of a large chronic wound through transplantation of gene-corrected epidermal stem cells. J Invest Dermatol 137(3):778–781
9. Piskin G, Sylva-Steenland RM, Bos JD et al (2006) In vitro and in situ expression of IL-23 by keratinocytes in healthy skin and psoriasis lesions: enhanced expression in psoriatic skin. J Immunol 176(3):1908–1915
10. Paré B, Touzel-Deschênes L, Lamontagne R et al (2015) Early detection of structural abnormalities and cytoplasmic accumulation of TDP-43 in tissue-engineered skins derived from ALS patients. Acta Neuropathol Commun 3:5
11. Fitzpatrick TB, Breathnach AS (1963) The epidermal melanin unit system. Dermatol Wochenschr 147:481–489
12. Mosenson JA, Flood K, Klarquist J et al (2014) Preferential secretion of inducible HSP70 by vitiligo melanocytes under stress. Pigment Cell Melanoma Res 27(2):209–220
13. Barrandon Y, Green H (1987) Three clonal types of keratinocyte with different capacities for multiplication. Proc Natl Acad Sci USA 84(8):2302–2306
14. Braye F, Dumortier R, Bertin-Maghit M et al (2001) Cultured epidermis for the treatment of severe burns. A 2-year study (18 patients). Ann Chir Plast Esthet 46(6):599–606
15. Black AF, Bouez C, Perrier E et al (2005) Optimization and characterization of an engineered human skin equivalent. Tissue Eng 11(5–6):723–733
16. Bisson F, Rochefort E, Lavoie A et al (2013) Irradiated human dermal fibroblasts are as efficient as mouse fibroblasts as a feeder layer to improve human epidermal cell culture lifespan. Int J Mol Sci 14(3):4684–4704
17. Green H (1978) Cyclic AMP in relation to proliferation of the epidermal cell: a new view. Cell 15(3):801–811
18. Ghoubay-Benallaoua D, Pécha F, Goldschmidt P et al (2012) Effects of isoproterenol and cholera toxin on human limbal epithelial cell cultures. Curr Eye Res 37(7):644–653
19. Halaban R, Alfano FD (1984) Selective elimination of fibroblasts from cultures of normal human melanocytes. In Vitro 20(5):447–450
20. Larouche D, Cantin-Warren L, Desgagné M et al (2016) Improved methods to produce tissue-engineered skin substitutes suitable for the permanent closure of full-thickness skin injuries. Biores Open Access 5(1):320–329
21. Germain L, Larouche D, Nedelec B et al (2018) Autologous bilayered self-assembled skin substitutes (SASSs) as permanent grafts: a case series of 14 severely burned patients indicating clinical effectiveness. Eur Cell Mater 36:128–141
22. Kelly C, Wallace D, Moulin V et al (2021) Surviving an extensive burn injury using advanced skin replacement technologies. J Burn Care Res 42(6):1288–1291
23. Cortez Ghio S, Larouche D, Doucet EJ et al (2021) The role of cultured autologous bilayered skin substitutes as epithelial stem cell niches after grafting: a systematic review of clinical studies. Burns Open 5(2):56–66
24. Emerit I, Cerutti PA (1981) Tumour promoter phorbol-12-myristate-13-acetate induces chromosomal damage via indirect action. Nature 293(5828):144–146
25. Zigler JS Jr, Lepe-Zuniga JL, Vistica B et al (1985) Analysis of the cytotoxic effects of light-exposed HEPES-containing culture medium. In Vitro Cell Dev Biol 21(5):282–287
26. Germain L, Rouabhia M, Guignard R et al (1993) Improvement of human keratinocyte isolation and culture using thermolysin. Burns 19(2):99–104
27. Goldsmith LA, Katz SI, Gilchrest BA et al (2012) Fitzpatrick's dermatology in general medicine, 8e. The McGraw-Hill Companies, New York

Chapter 4

Long-Term Expansion of Mouse Primary Epidermal Keratinocytes Using a TGF-β Signaling Inhibitor

Filipa Pinto, Daisuke Suzuki, and Makoto Senoo

Abstract

Mouse models have been used to study the physiology and pathogenesis of the skin. However, propagation of mouse primary epidermal keratinocytes remains challenging. In this chapter, we introduce a newly developed protocol that enables long-term expansion of $p63^+$ mouse epidermal keratinocytes in low Ca^{2+} media without the need of progenitor cell-purification steps or support by a feeder cell layer. Pharmacological inhibition of TGF-β signaling in crude preparations of mouse epidermis robustly increases proliferative capacity of $p63^+$ epidermal progenitor cells, while preserving their ability to differentiate. Suppression of TGF-β signaling also permits $p63^+$ epidermal keratinocytes to form macroscopically large clones in 3 T3-J2 feeder cell co-culture. This simple and efficient approach will facilitate the use of mouse models by providing $p63^+$ primary epidermal keratinocytes in quantity.

Key words Mouse models, Primary epidermal keratinocytes, Small molecule inhibitor, Transcription factor p63, TGF-β signaling

1 Introduction

Mouse models have been used in studies of normal and disease conditions of the skin. However, the growth of primary epidermal keratinocytes derived from mice rapidly declines in culture and the cells become terminally differentiated [1], a limit that restricts the use of mouse primary keratinocytes for functional analyses of the skin. As proliferation and differentiation of epithelial cells are tightly coupled through the induction of cyclin-dependent kinase (CDK) inhibitor genes [2–4], suppression of growth arrest and of premature differentiation are both potential approaches to improve the lifespan of mouse primary epidermal keratinocytes [2, 5–8].

Although the use of low calcium (Ca^{2+}) media has extended proliferation of mouse epidermal keratinocytes short term [1], the most effective protocols to propagate these primary cells rely on modified 3 T3-J2 feeder co-culture [5, 7] or the use of fluorescence

Thomas Biedermann and Sophie Böttcher-Haberzeth (eds.), *Skin Tissue Engineering: Methods and Protocols*, Methods in Molecular Biology, vol. 2922, https://doi.org/10.1007/978-1-0716-4510-9_4,

activated cell sorting (FACS)-purified progenitor populations with triple drug inhibitors [8]. Highly efficient protocols that eliminate the use of undefined factors (e.g., feeder cells), labor-intensive purification steps, and the potentially complex effects of multiple drugs will facilitate the use of primary cells of mice in studies of skin biology.

Transforming growth factor-β (TGF-β) signaling regulates proliferation and differentiation of many different epithelial progenitor cells [9], including those that are controlled by the transcription factor p63 (e.g., epidermal keratinocytes) [10]. TGF-β signaling is mediated through the receptor complex consisting of the Type I TGF-β receptor (TGFβR1/ALK5) and the Type 2 TGF-β receptor (TGFβR2) [11]. Upon binding of the TGF-β ligands, TGFβR2 phosphorylates and activates TGFβR1/ALK5, resulting in the phosphorylation and nuclear translocation of Smad2/3, downstream effectors of TGF-β signaling [12]. We have shown recently that inhibition of TGF-β signaling suppresses Smad2/3 nuclear localization and consequently increases the proliferative capacity of $p63^+$ mouse epidermal keratinocytes in vitro [13].

In this chapter, we introduce a newly developed protocol that enables long-term proliferation of $p63^+$ mouse primary epidermal keratinocytes utilizing a TGF-β signaling inhibitor without the need of a feeder layer or progenitor cell-purification steps. Suppression of TGF-β signaling also permits the expanded $p63^+$ mouse epidermal keratinocytes to form macroscopically large clones in 3 T3-J2 feeder cell co-culture for clonal evaluation. We anticipate that this simple and efficient approach will facilitate the use of mouse models for studying the physiology and pathogenesis of the epidermis.

2 Materials

2.1 Biological and Other Material

1. C57BL/6 mice.
2. 3 T3-J2 cells [14].
3. Isojin.

2.2 Instruments and Supplies

1. Humidified CO_2 incubators (5% and 10%).
2. CO_2 gas cylinders.
3. 37 °C water bath.
4. Hemocytometer.
5. Inverted microscope.
6. γ-irradiator with a Cesium-137 source.
7. Tissue culture plates (35 mm, 60 mm, 100 mm).
8. 15 mL and 50 mL polypropylene conical tubes.

9. 1.5 mL Eppendorf tubes.
10. Cryovials.
11. Syringe filter unit (0.2 μm).
12. 70 μm cell strainer.
13. Parafilm.
14. Whatman filter paper.

2.3 Cell Culture Media and Freezing Media

1. CnT-PR medium: 100 U/mL penicillin, 100 μg/mL streptomycin in CnT-PR media (CnT-Prime epithelial proliferation medium).
2. DMEM: 4.5 g/L glucose, 584 mg/L L-glutamine, 110 mg/mL sodium pyruvate in Dulbecco's Modified Eagle's Medium (DMEM).
3. Ham's F-12: 1.8 g/L glucose, 146 mg/L L-glutamine, and 110 mg/L sodium pyruvate in Ham's F-12.
4. J2 medium: 100 U/mL penicillin, 100 μg/mL streptomycin, 10% (v/v) CS (Bovine calf serum) in DMEM [14].
5. FAD medium: Mix DMEM and Ham's F-12 media at a 3:1 ratio (v/v) and supplement with 10% v/v FBS (Fetal bovine serum), 100 U/mL penicillin, and 100 μg/mL streptomycin [14].
6. Complete FAD (cFAD) medium: 10 ng/mL EGF, 5 μg/mL insulin, 2×10^{-9} M T_3, 0.4 μg/mL hydrocortisone, 24 μg/mL adenine, 1×10^{-10} M cholera toxin in FAD medium [14].
7. Freezing medium: Mix cell culture media above (CnT-PR medium, J2 medium, or FAD medium), serum (CS or FBS), and DMSO (Dimethyl sulfoxide) at a 4:5:1 ratio (v/v/v). To freeze 3 T3-J2 cells use a combination of J2 medium and CS. To freeze epidermal cells, use either CnT-PR or FAD medium along with FBS.

2.4 Cell Culture Reagents

1. Adenine stock: 24 mg/mL adenine in 0.2 N HCl. Sterilize with a 0.2 μm syringe filter. Store in 0.5 mL aliquots at −20 °C.
2. T_3 (3,3′,5-triiodo-L-thyronine sodium salt) stock: 10 μM T_3 in PBS (Phosphate-buffered saline without calcium and magnesium). Sterilize with a 0.2 μm syringe filter. Store in 0.5 mL aliquots at −20 °C.
3. Hydrocortisone stock: 5 mg/mL hydrocortisone in 95% EtOH. Store in 0.2 mL aliquots at −20 °C.
4. Cholera toxin stock: 0.1 mg/mL cholera toxin in dH_2O. Sterilize with a 0.2 μm syringe filter. Store in 0.4 mL aliquots at −80 °C.

5. Insulin stock: 0.1 mg/mL insulin in 0.005 N HCl. Sterilize with a 0.2 μm syringe filter. Store in 0.5 mL aliquots at −80 °C.
6. EGF stock: 20 μg/mL EGF (Human epidermal growth factor) in 1% BSA (Bovine serum albumin) aqueous solution. Sterilize with a 0.2 μm syringe filter. Store in 0.5 mL aliquots at −80 °C.
7. Inhibitor stock: 25 mM RepSox in DMSO (Dimethyl sulfoxide). Store in 25 μL aliquots at −80 °C.
8. 1% Rhodamine B: 1% (w/v) Rhodamine B in dH_2O, filter through Whatman filter papers, and store at room temperature (RT).

3 Methods

3.1 Collection of Crude Primary Epidermal Cells from Mouse Skin

1. Euthanize embryos to neonatal mice by decapitation. Postnatal mice should be euthanized by overdosing of Isoflurane or CO_2 inhalation, followed by cervical dislocation.
2. When mice older than 2 weeks old are used, hair should be clipped prior to isolation of the skin.
3. Harvest 1–2 cm^2 back skin per animal as a single sheet of tissue using sterilized surgical scissors.
4. Sterilize the isolated skin by immersing in Isojin solution for 10 s, followed by three washes in PBS.
5. Place the skin on a sterilized 2 × 2 cm Whatman filter paper with the dermis facing down in a 35 mm tissue culture plate. The dermis is a viscous tissue and adheres tightly to the dried filter paper.
6. Slowly add 2 mL fresh 0.25% trypsin-EDTA in the tissue culture plate above, enough to soak the filter paper, and incubate for 2–4 h at RT in a tissue culture biosafety cabinet.
7. Carefully separate the epidermis from the dermis using sterilized forceps with fine tips. While the epidermis easily peels off as a whitish tissue, the dermis, as a translucent tissue, remains attached to the filter paper.
8. Transfer the isolated epidermis to 1.5 mL Eppendorf tubes.
9. Add 1 mL 0.25% trypsin-EDTA to the tubes and incubate for 5 min at RT.
10. Add 100 μL FBS (equivalent to 10% the volume of trypsin used) to inactivate the enzymatic reaction, vigorously suspend the cells by pipetting, and spin down the cells at 400 × g for 5 min.
11. Resuspend the cells in 1 mL PBS and filter through a 70 μm cell strainer.
12. Centrifuge the cells at 400 × g for 5 min.

13. Wash the cells two additional times in PBS in order to completely remove FBS, which potentially inhibits epidermal cell growth in the subsequent culture.
14. Resuspend the cells in 100 μL CnT-PR media and count the cell number using a hemocytometer.

3.2 Enrichment and Expansion of Mouse Primary Epidermal Keratinocytes in Serum-Free Media

1. Plate 1×10^6 crude primary cells (*see* Subheading 3.1) in one 35 mm tissue culture plate (or one well in a 6-well plate) that contains 3 mL CnT-PR media.
2. Add RepSox to the culture media at a final concentration of 1 μg/mL (*see* **Note 1**).
3. Place the tissue culture plate in a humidified incubator with 5% CO_2 at 37 °C. Allow the cells to adhere to the plastic plate for 2–3 days (*see* **Note 2**).
4. After the incubation, replace media with fresh 3 mL CnT-PR media. Culture media should be changed every 2–3 days thereafter until the cells become subconfluent (*see* **Note 3**).
5. When the cells become subconfluent, remove culture media and wash the cells twice in 3 mL PBS.
6. Add 1 mL of 0.25% trypsin-EDTA to the cells and incubate the plate in a humidified chamber with 5% CO_2 at 37 °C for 5 min (*see* **Note 4**).
7. Tap the tissue culture plate to dislodge the cells.
8. Dissociate the cells by gentle pipetting, collect them in a 15 mL polypropylene tube, and neutralize the enzymatic reaction by adding 100 μL FBS (equivalent to 10% the volume of trypsin used).
9. Centrifuge the cells at $200 \times g$ for 5 min.
10. Remove supernatant, wash the cells twice in 5 mL PBS, and resuspend in 1 mL CnT-PR media.
11. Count cell numbers. Approximately 3–5 $\times 10^5$ cells are collected from one subconfluent 35 mm tissue culture plate.
12. Replate the cells at a 1:10 ratio in CnT-PR media containing 1 μg/mL RepSox. These secondary cultures become subconfluent in 5–7 days for subsequent experiments. By passage (P) 1 to P2, virtually all growing cells are cytokeratin $(CK)^+p63^+$ epidermal progenitor cells [13, 16] (*see* **Note 5**).
13. To prepare frozen stock, resuspend the cell pellet in CnT-PR-based freezing media at the density of 2×10^6 cells per mL and aliquot 0.5 mL per cryovial (1×10^6 cells per vial). Place the cryovials in −80 °C overnight and transfer them to a liquid nitrogen tank for longer storage.

3.3 Expansion of Mouse Epidermal Keratinocytes in Co-culture with 3 T3-J2 Feeder Cells for Clonal Analysis

3 T3-J2 cells are expanded in J2 media under 10% CO_2 atmosphere. Keep in mind that the conditions of feeder cells, such as the viability, density, and passage number (*see* **Note 6**), are critical for successful expansion of mouse epidermal keratinocytes in feeder cell co-culture.

3.3.1 Culture, Passage, and Storage of 3 T3-J2 Cells

1. To expand 3 T3-J2 cells, one cryovial containing 1×10^6 cells is taken out of a liquid nitrogen tank and thawed in a 37 °C water bath for 3 min.
2. Spin down the cells at 200 × g for 5 min.
3. Remove freezing media and suspend the cells in 1 mL J2 media.
4. Plate the resuspended cells in two 100 mm tissue culture plates that contain 10 mL J2 media (5×10^5 cells per plate).
5. Place tissue culture plates in a humidified incubator with 10% CO_2 at 37 °C.
6. After overnight incubation (>16 hrs), replace media with 10 mL of fresh J2 media. Culture media should be changed every 2–3 days thereafter until the cells become subconfluent (*see* **Note 7**).
7. When the cells become subconfluent, remove culture media and wash the cells twice in 5 mL PBS.
8. Add 2 mL 0.25% trypsin-EDTA to the cells and incubate the plates in a humidified incubator with 10% CO_2 at 37 °C for 5 min.
9. Dissociate the cells by gentle pipetting, collect them in a 15 mL polypropylene tube, and neutralize the enzymatic reaction by adding the same volume (2 mL) of J2 media.
10. Centrifuge the cells at 200 × g for 5 min.
11. Remove supernatant and resuspend the cells in 1 mL J2 media.
12. Count cell numbers. Approximately 3×10^6 cells are collected from one subconfluent 100 mm tissue culture plate.
13. Replate 3 T3-J2 cells at a 1:10–1:15 ratio at the density of 3.3×10^3 cells per cm^2 into 100 mm tissue culture plates that contain 10 mL J2 media. These secondary cultures become subconfluent in 5–7 days.
14. To prepare frozen stock, resuspend the cell pellet in freezing media at the density of 2×10^6 cells per mL and aliquot 0.5 mL per cryovial (1×10^6 cells per vial). Place the cryovials in −80 °C overnight and transfer them to a liquid nitrogen tank for longer storage.

Table 1
Number of 3 T3-J2 cells and the volume of cell culture media

Culture vessel	Diameter	Area	Volume of media	Number of 3 T3-J2 cells
35 mm plate	35 mm	9.6 cm^2	3 mL	5.0×10^5 cells
60 mm plate	52 mm	21 cm^2	5 mL	1.1×10^6 cells
100 mm plate	84 mm	55 cm^2	10 mL	3.0×10^6 cells

3.3.2 Preparation of Feeder Cell Layers

Prior to seeding mouse epidermal cells, feeder cells will be mitotically arrested by either lethal γ-irradiation or mitomycin C treatment (*see* **Note 8**) to prevent propagation of the feeder cells in subsequent passages. In this protocol, we describe the use of lethally γ-irradiated feeder cells on the scale of a 60 mm tissue culture plate as an example. When other sizes of cell culture vessels are used, the number of feeder cells seeded and volume of the cell culture media used should be adjusted proportionally based on the surface area of the culture vessels (*see* **Note 9**).

1. Follow the procedures described in Subheading 3.3.1 (also Table 1) and prepare desired numbers of 60 mm tissue culture plates with subconfluent 3 T3-J2 cells.
2. When 3 T3-J2 cells become subconfluent, seal the culture plates with Parafilm to maintain sterility and γ-irradiate the cells at a dose of 60 Gy (6000 rads).
3. After irradiation, replace media with 5 mL of fresh J2 media and place the plates in a humidified incubator until use (*see* **Note 10**).

3.3.3 Setting Up Co-culture

1. Prepare a minimum amount of cFAD media that can be consumed in 1 week as it contains proteins and chemicals that may degrade once diluted in media.
2. Prior to seeding mouse epidermal cells, replace J2 media with 5 mL cFAD media and place the plates in a humidified incubator with 5% CO_2 at 37 °C for 1–2 h.
3. By following the steps in Subheading 3.2, prepare mouse epidermal cells in suspension. Use cFAD instead of CnT-PR media to suspend mouse epidermal cells.
4. Count viable cell numbers with the aid of Trypan blue solution.
5. Adjust the density of epidermal cells to $10^3 \sim 10^5$ viable cells per mL by diluting in cFAD media.
6. Seed the desired number of epidermal cells onto feeder cell layers. When rapid confluency of keratinocytes is planned, $10^3 \sim 10^4$ cells per 60 mm tissue culture plate can be seeded (*see* **Note 11**). Seeding at a higher density than 10^4 cells per

60 mm plate is not recommended as it does not fully allow relatively immature epidermal keratinocytes to expand. When the determination of the proliferative capacity of epidermal cells at a clonal level is required, 10^2 ~ 10^3 cells per 60 mm plate should be seeded to permit clonal growth of individual epidermal cells (*see* **Note 12**).

7. Add 1 μM RepSox in cFAD media. Note that RepSox treatment significantly boosts the growth of mouse epidermal cells with high proliferative potential [13] (*see* **Note 13**).
8. cFAD media containing RepSox should be replaced every 2–3 days until epidermal cell clones reach desired clone sizes for evaluation.

3.3.4 Visualization of Mouse Epidermal Keratinocyte Clones in Feeder Cell Co-culture by Rhodamine B Staining

1. Wash the 60 mm co-culture plates above (*see* Subheading 3.3.3) twice in 5 mL PBS and fix them in 10% buffered formalin for 15 min at RT.
2. Remove formalin solution and wash the cells twice in 5 mL PBS.
3. Add 5 mL of 1% Rhodamine B solution to the cells and stain for more than 15 min at RT.
4. Pour off Rhodamine B solution into a separate container (*see* **Note 14**).
5. Gently rinse the cells with tap water.
6. Invert the plates to air-dry.
7. Take photographs of the stained plates (*see* Fig. 4) and determine the sizes of individual clones as needed using NIH Image J or Adobe Photoshop software [13].

4 Notes

1. RepSox treatment robustly increases the CK^+ epidermal cell number in a dose-dependent manner (*see* Fig. 1). Other widely used TGF-β signaling inhibitors also stimulate the growth of CK^+ mouse primary epidermal cells in a dose-dependent manner. However, we found that RepSox treatment showed the highest increase in epidermal cell number at all the doses we examined (Fig. 1). In addition, $p63^+$ mouse epidermal progenitor cells grow at a constant rate for at least 60 days in the presence of RepSox (Fig. 2a). Proliferation of mouse epidermal cells is RepSox-dependent, as removal of RepSox from the culture results in a rapid decline in the growth of epidermal cells (Fig. 2a), accompanied by elevated expression of phosphorylated Smad2/3 at both P5 and P20 stages (Fig. 2b, c).

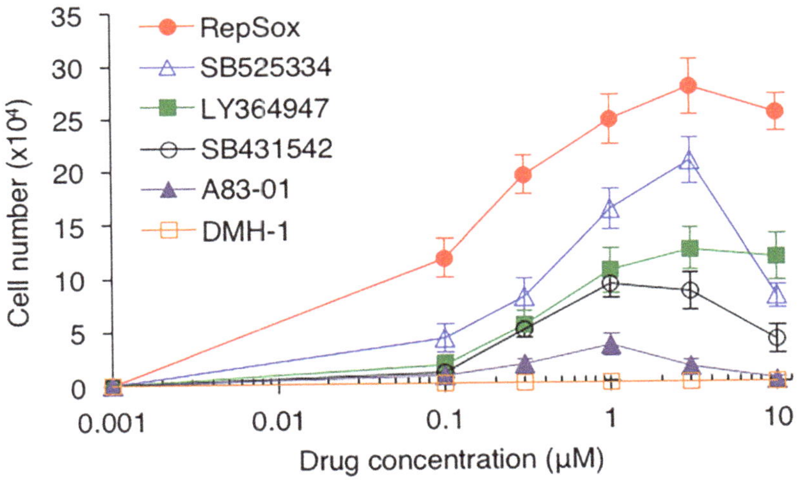

Fig. 1 Dose-dependent effects of TGF-β signaling inhibitors on cell proliferation of mouse primary epidermal keratinocytes. Newborn mouse-derived, CnT-PR-expanded cytokeratin-positive (CK^+) epidermal cells (2×10^4) were cultivated in CnT-PR media for 10 days in the presence of increasing concentration of TGF-β signaling inhibitors (RepSox, SB525334, LY364947, SB431542, and A83–01). A BMP signaling inhibitor DMH-1 was included as a negative control. Data shown are mean ± s.e.m. ($n = 3$). (Adopted from reference [13])

2. Although the majority of the cells are still floating in the culture media 3 days after the inoculation, approximately 10% of the cells adhere to the plates and initiate the growth.
3. Do not over grow mouse epidermal cells as confluent epidermal cells are difficult to dissociate by trypsinization and they tend to differentiate prematurely in the subsequent passage. The days for primary epidermal cells to become confluent in culture seem to depend on the age of mice (*see* Fig. 3).
4. Accutase or TrypLE select can be used to more gently dissociate the cells compared with trypsin-EDTA.
5. Inhibition of TGF-β signaling also enriches and expands other $p63^+$ epithelial progenitor cells in primary crude cultures of multiple epithelia, including the cornea, oral and lingual epithelia, salivary gland, esophagus, thymus, bladder [13], and tympanic membrane and middle ear mucosa [16].
6. We routinely expand 3 T3-J2 cells at passage 4–7 and prepare a large number of frozen stocks before initiating epidermal cell co-culture. We typically use 3 T3-J2 cells in co-culture at passage 6–10.
7. Do not allow 3 T3-J2 cells to become confluent as the morphology of the cells is altered and the potential of feeder cells to support keratinocyte growth may decline.

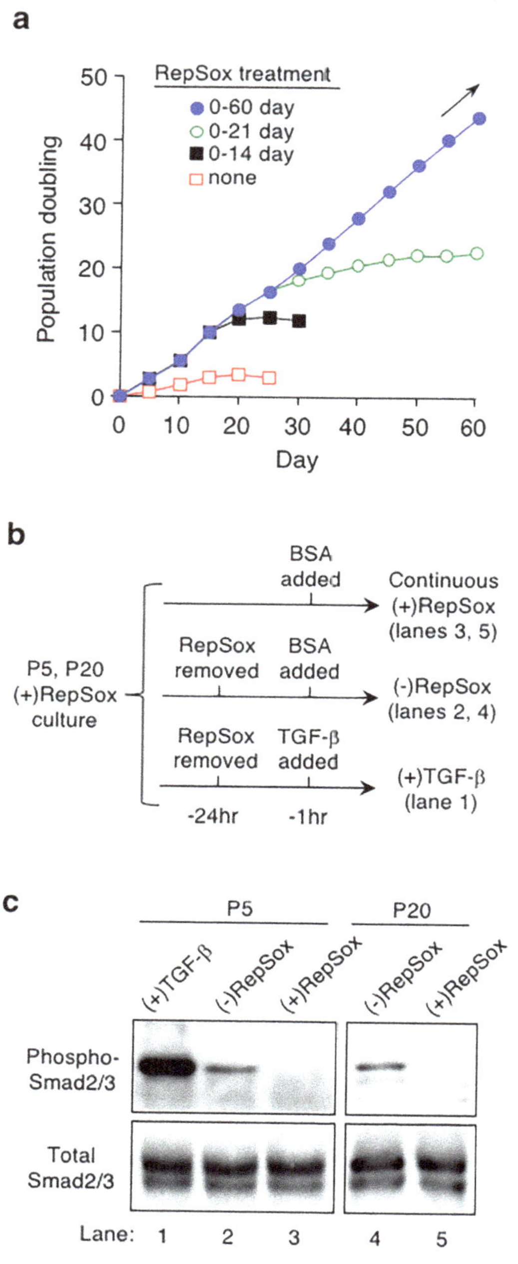

Fig. 2 TGF-β signaling inhibition-mediated enhancement of population doubling of p63$^+$ mouse primary epidermal keratinocytes. (**a**) Population doubling of newborn mouse-derived, RepSox-expanded p63$^+$ epidermal cells grown in CnT-PR media in the presence of 1 μM RepSox for 0, 14, 21, and 60 days. Data shown are representative of three independent experiments with similar results. Arrow indicates continuous cell growth. (**b**, **c**) Removal of RepSox associates with increased Smad2/3 phosphorylation. (**b**) Experimental design. Newborn mouse-derived, RepSox-expanded P5 and P20 epidermal keratinocytes were further grown in continuous presence (upper) or absence (middle and lower) of 1 μM RepSox for 24 hrs. Culture of P5 cells stimulated with 1 ng/ml TGF-β for 1 hr. prior to lysis (lower) was used as a positive control. (**c**)

8. Instead of γ-irradiation, mitomycin C treatment can be used to induce growth arrest in feeder cells as follows.
   (a) Add mitomycin C to 3 T3-J2 media in a 50 mL tube at a final concentration of 10 μg/mL.
   (b) Aspirate cell culture media from the plates and add 3 mL of freshly prepared mitomycin C-containing feeder cell media in each 60 mm plate.
   (c) Place the plates in a humidified incubator with 10% CO_2 at 37 °C for 2 h.
   (d) Wash the cells three times in 3 mL PBS.
   (e) Add 5 mL J2 media and return the plates to a humidified incubator with 10% CO_2 at 37 °C until use.
9. See Table 1 (Table 1) for expected numbers of subconfluent feeder cells at the time of epidermal cell inoculation and appropriate amounts of cell culture media used for feeder cell co-culture. When 3 T3-J2 cells are seeded at the density of 3.3×10^3 cells per cm^2, it will take 5–7 days for them to become subconfluent. Higher numbers of 3 T3-J2 cells can be seeded to obtain subconfluent feeder cell layers sooner. In such a circumstance, up to 6×10^5 viable cells per 60 mm plate can be seeded and the cells can be γ-irradiated in 1–2 days when the feeder cells become subconfluent and adhere tightly to the culture vessels.
10. Although 3 T3-J2 cells treated with γ-irradiation or mitomycin C can remain viable for a few weeks, we typically use them in 5–7 days after the treatment. Feeder cell media should be replaced every 2–3 days until use.
11. Epidermal cells grow faster when they are plated at higher densities, possibly due to an autocrine effect on cell proliferation [15].
12. At this plating density, small colonies comprised of 4–10 cells begin to form at day 4–5. By day 7, epidermal clones become macroscopically visible and they reach 5–15 mm in diameter by day 14 in the presence of 1 μM RepSox (see, for example, Fig. 4). Cell culture vessels smaller than 35 mm plates (or 6-well plates) are not recommended for clonogenic analysis

Fig. 2 (continued) Expression of total and phosphorylated Smad2/3 was determined by Western blot. Lane numbers correspond to those in (**b**). Antibodies used were rabbit anti-phosphorylated Smad2/3 (D27F4, Cell Signaling Technology), rabbit anti-Smad2/3 (D7G7, Cell Signaling Technology), and horseradish peroxidase (HRP)-conjugated goat anti-rabbit IgG (Cell Signaling Technology). (Adopted from reference [13])

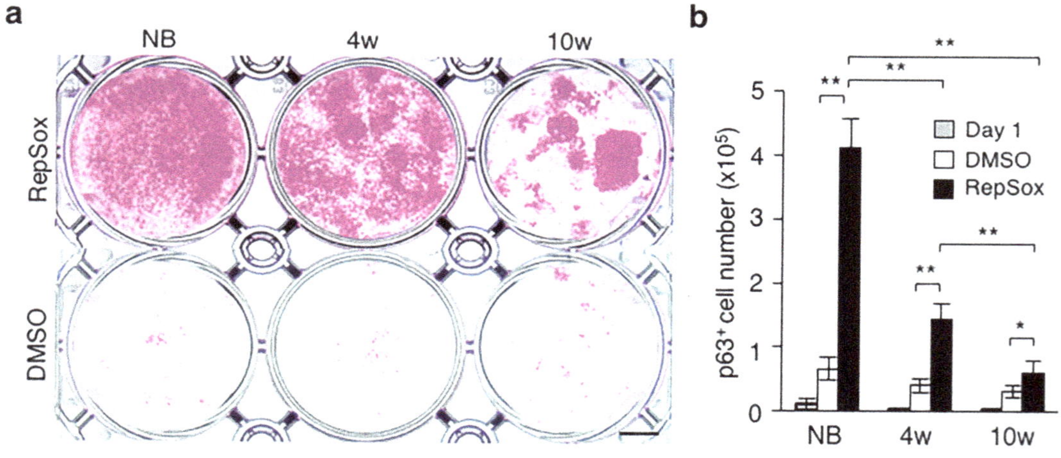

Fig. 3 Age-dependent decline in TGF-β signaling inhibition-mediated cell growth of mouse primary epidermal keratinocytes. (**a**) Representative images of Rhodamine B staining of crude culture of epidermal cells harvested from newborn (NB), 4-week-old (4w), and 10-week-old (10w) mice and grown in CnT-PR media in the presence (upper) or absence (lower) of 1 μM RepSox for 14 days. One million primary cells were seeded. Data shown are representative of two independent experiments with similar results. Bar = 5 mm. (**b**) Total number of $p63^+$ epidermal cells. One million primary cells were seeded and adherent cells were counted at day 1 in replicative wells (gray bars) as the majority of the cells remained in suspension. Solid bars and open bars represent total numbers of $p63^+$ epidermal cells at day 14 of culture in the subsequent presence of 1 μM RepSox or 0.1% DMSO, respectively ($n = 3$). $^*P < 0.05$; $^{**}P < 0.01$. (Adopted from reference [13])

as epidermal cell colonies may fuse before they reach the size suitable for clonal evaluation.

13. Mouse primary epidermal cells expanded by RepSox treatment in feeder cell co-culture maintain the original high proliferative capacity as determined by serial passages in clonogenic culture with 3 T3-J2 cells [13]. In addition, mouse primary epidermal cells expanded by RepSox treatment are capable of differentiation in response to Ca^{2+} stimulation [13].
14. Rhodamine B solution can be reused multiple times. Used solution should be collected in a separate container and filtered through Whatman filter papers to remove debris before use.

Acknowledgements

The authors thank Jeff Holcombe for proofreading of the manuscript. This study was supported by an R01AR066755 grant from the National Institute of Arthritis and Musculoskeletal and Skin Diseases of the National Institute of Health to M.S.

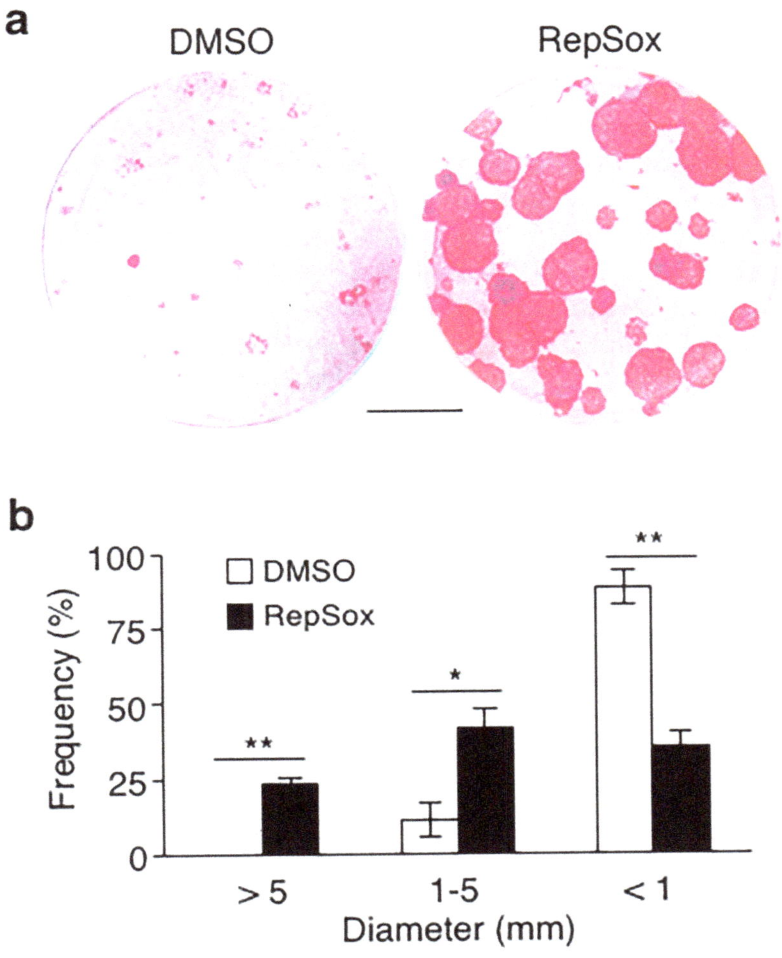

Fig. 4 Clonal expansion of mouse primary epidermal keratinocytes in co-culture with 3 T3-J2 feeder cells. (**a**) Rhodamine B staining of newborn mouse-derived, RepSox-expanded P1 epidermal keratinocytes grown in 3 T3-J2 co-culture in the presence or absence of 1 μM RepSox for 14 days. Bar = 10 mm. (**b**) Distribution of epidermal clone sizes at day 14. Data shown are mean ± s.e.m. ($n = 3$). *$P < 0.05$; **$P < 0.01$. (Adopted from reference [13])

References

1. Lichti U, Anders J, Yuspa SH (2008) Isolation and short-term culture of primary keratinocytes, hair follicle populations and dermal cells from newborn mice and keratinocytes from adult mice for in vitro analysis and for grafting to immunodeficient mice. Nat Protoc 3:799–810
2. Ha L, Ponnamperuma RM, Jay S et al (2011) Dysregulated ΔNp63α inhibits expression of Ink4a/arf, blocks senescence, and promotes malignant conversion of keratinocytes. PLoS One 6:e21877. https://doi.org/10.1371/journal.pone.0021877
3. Missero C, Di Cunto F, Kiyokawa H, l. (1996) The absence of p21$^{Cip/WAF1}$ alters keratinocyte growth and differentiation and promotes ras-tumor progression. Genes Dev 10:3065–3075
4. Paramio JM, Segrelles C, Ruiz S et al (2001) The ink4a/arf tumor suppressors cooperate with p21$^{cip1/waf}$ in the processes of mouse epidermal differentiation, senescence, and carcinogenesis. J Biol Chem 276:44203–44211
5. Chapman S, McDermott DH, Shen K et al (2014) The effect of Rho kinase inhibition on long-term keratinocyte proliferation is rapid and conditional. Stem Cell Res Ther 5:60. https://doi.org/10.1186/scrt449
6. King KE, Ponnamperuma RM, Yamashita T et al (2003) ΔNp63α functions as both a positive and a negative transcriptional regulator and blocks in vitro differentiation of murine keratinocytes. Oncogene 22:3635–3644
7. Liu X, Ory V, Chapman S et al (2012) ROCK inhibitor and feeder cells induce the conditional reprogramming of epithelial cells. Am J Pathol 180:599–607
8. Mou H, Vinarsky V, Tata PR et al (2016) Dual SMAD signaling inhibition enables long-term expansion of diverse epithelial basal cells. Cell Stem Cell 19:217–231
9. Watabe T, Miyazono K (2009) Roles of TGF-β family signaling in stem cell renewal and differentiation. Cell Res 19:103–115
10. Senoo M, Pinto F, Crum CP et al (2007) p63 is essential for the proliferative potential of stem cells in stratified epithelia. Cell 129:523–536
11. Schmierer B, Hill CS (2007) TGFβ-SMAD signal transduction: molecular specificity and functional flexibility. Nat Rev Mol Cell Biol 8:970–982
12. Ikushima H, Miyazono K (2010) TGFβ signalling: a complex web in cancer progression. Nat Rev Cancer 10:415–424
13. Suzuki D, Pinto F, Senoo M (2017) Inhibition of TGF-β signaling supports high proliferative potential of diverse p63$^+$ mouse epithelial progenitor cells in vitro. Sci Rep 7:6089. https://doi.org/10.1038/s41598-017-06470-y
14. Rheinwald JG, Green H (1975) Serial cultivation of strains of human epidermal keratinocytes: the formation of keratinizing colonies from single cells. Cell 6:331–343
15. Lim X, Tan SH, Koh WL et al (2013) Interfollicular epidermal stem cells self-renew via autocrine Wnt signaling. Science 342:1226–1230
16. Fukuda TY, Pinto F, Pitt K et al (2023) Inhibition of TGF-β signaling enables long-term proliferation of mouse primary epithelial stem/progenitor cells of the tympanic membrane and the middle ear mucosa. Sci Rep 13:4532. https://doi.org/10.1038/s41598-023-31246-y

Chapter 5

Isolation and Culture of Hair Follicle Dermal Sheath Mesenchymal Stromal Cells

Dongrui Ma, Seng-Teik Lee, and Alvin Wen Choong Chua

Abstract

Limited research has been published on the characterization and therapeutic potential of human mesenchymal stromal cells (MSCs) derived from hair follicle dermal sheath (DS). Here, we present protocols for isolating and culturing human DS-MSCs, beginning with the use of a dissecting microscope to separate dermal sheaths from hair follicles for trypsin digestion. Additionally, we outline protocols for adipogenic, osteogenic and chondrogenic differentiation of DS-MSCs, as we explore their potential applications in stem cell therapy and tissue engineering.

Key words Hair follicle, Dermal sheath, Mesenchymal stromal cells, MSCs, Adipogenic differentiation, Osteogenic differentiation, Chondrogenic differentiation, Stem cells, Tissue engineering

1 Introduction

Hair follicles contain discrete populations of mesenchymal and epithelial cells [1]. While research has demonstrated that dermal papilla and dermal sheath cells in the hair follicle possess strong hair-inducing capabilities, the stem cell characteristics of the hair follicle dermal sheath remain underexplored. Herein, we describe the isolation of fibroblast-like cells from human hair follicle dermal sheath [2] similar to other works reported [3, 4]. These cells which we termed as dermal sheath mesenchymal stromal cells (DS-MSCs) exhibited proliferative capacity and colony forming ability in in vitro culture. They also displayed immunophenotypic properties similar to that of bone marrow mesenchymal stromal cells (BM-MSCs), although with some variation. DS-MSCs have the ability to differentiate into adipocytes, chondrocytes, and osteoblasts and demonstrate high telomerase activity compared to BM-MSCs. Furthermore, in vitro experiments found that DS-MSCs exhibited immunomodulatory properties comparable or exceeding to those MSCs derived from adipose tissue. These

Thomas Biedermann and Sophie Böttcher-Haberzeth (eds.), *Skin Tissue Engineering: Methods and Protocols*, Methods in Molecular Biology, vol. 2922, https://doi.org/10.1007/978-1-0716-4510-9_5,

were based on comparison of their respective crosstalk data with cells of both the innate and adaptive immune systems [5]. Therefore, human hair follicle DS-MSCs area promising alternative to other sources of MSCs for applications in stem cell therapy and tissue engineering.

2 Materials

2.1 Solutions

1. 2.5 mg/mL Dispase II: Dissolve 500 mg Dispase II powder in 200 mL Dulbecco's modified Eagle's Medium (DMEM) with gentle stirring. Sterile filter at 0.22 μm and dispense 10 mL of the dissolved solution in 15 mL centrifuge tubes and store at 4 °C for short term storage (up to a week). For long term storage, keep at -20 °C (*see* **Note 1**).
2. 0.1% Collagenase-0.25% Dispase II mix: Warm the 10 mL tube of 2.5 mg/mL Dispase II to 37 °C prepared in **step 1**. Dissolve 10 mg collagenase powder in the warmed Dispase II. Sterile filter the mix and it is ready for use to isolate cells from the hair follicle dermal sheath.
3. 0.1% Trypsin solution: Add 0.4 g glucose, 0.4 g trypsin powder, 1.2 mL phenol red (0.5%) into 400 mL 1x Phosphate Buffered Saline (PBS). Adjust final pH to 7.45 just before filtration with up to 1 mL 1 N NaOH. Sterile filter at 0.22 μm and dispense into 50 mL centrifuge tubes. Store at –20 °C and thaw to 37 °C before use.
4. 0.02% Ethylene diamine tetra-acetic acid (EDTA) solution: To prepare 2 L, add 0.4 g NaEDTA powder into 2000 mL of 1x PBS. Sterile filter at 0.22 μm into 500 mL or 1 L bottles, store at room temperature.

2.2 Culture Medium

Sterilize all culture media after mixing all components using 0.22 μm filtration.

1. DS-MSC growth medium: mix 1% L-glutamine, 20% fetal calf serum (FCS) and 1x penicillin/streptomycin (P/S) in DMEM.
2. Adipogenic differentiation medium: mix 1% L-glutamine, 10% FBS, 0.5 mM isobutylmethylxanthine, 1 μM dexamethasone, 10 μM insulin, 200 μM Indomethacin and 1x penicillin/streptomycin (P/S) in DMEM.
3. Osteogenic differentiation medium: mix 10% FBS, 0.1 μM dexamethasone, 50 μM ascorbate-2-phosphate, 10 mM β-glycerolphosphate and 1x penicillin/streptomycin (P/S) in DMEM.

4. Chondrogenic differentiation medium: mix 1% FBS, 6.25 μg/mL insulin, 10 ng/mL TGF-β1, 50 nM ascorbate-2-phosphate and 1x penicillin/streptomycin (P/S) in DMEM.

2.3 Staining and Fixation

1. 2% crystal violet solution: Dilute 4 g crystal violet powder to 200 mL 95% methanol (*see* **Note 2**).
2. 0.7% Oil red O staining solution: Dissolve 0.7 g Oil red O in 100 mL of propylene glycol, slowly stir, heat to 100 °C for 10 min. Filter through Whatman #2 filter paper.
3. Von Kossa staining solution (5% Silver Nitrate solution): Dissolve 5 g silver nitrate in 100 mL of distilled water, mix well, store in clean brown bottle, protect from light.
4. Alcian blue solution: Dissolve 1 g Alcian blue in 97 mL distilled water, add 3 g acetic acid, mix well.
5. 10% buffered formalin.

2.4 Equipment

1. Horizontal laminar flow hood with an embedded dissecting microscope.
2. Class II Biosafety cabinet (BSC) for cell culture.
3. Chemical fume hood.
4. Phase contrast microscope with digital imaging capture system.
5. CO_2 incubator with controlling and monitoring system for CO_2, humidity, and temperature.
6. Cell culture centrifuge.
7. Glass hemocytometer (with Improved Neubauer slide).
8. Cell culture disposables: 30 g needles, 1 mL syringes, 100 mm cell culture dishes, multiwell plates, 15 mL and 50 mL centrifuge tubes, pipetman, pipettes, micropipette and pipette tips.

3 Methods

3.1 Isolation of Hair Follicle Sheath (HFS) Cells from Human Hair Follicle

1. Collect excess human scalp tissues with embedded hair follicles wrapped in sterile gauze and soaked with DMEM as transport medium in a sterile container (*see* **Note 3**).
2. Remove human scalp tissues from transport container and wash them in 100 mm cell culture dish with 20 mL 1x PBS twice to wash off blood and debris in Class II BSC for cell culture.
3. Gently aspirate away all the PBS using a pipette.
4. Add 10 mL of fresh 1x PBS and gently spread the reagent out to wet the entire surface of the dish.
5. Bring covered 100 mm dish to dissecting microscope in horizontal laminar flow hood.

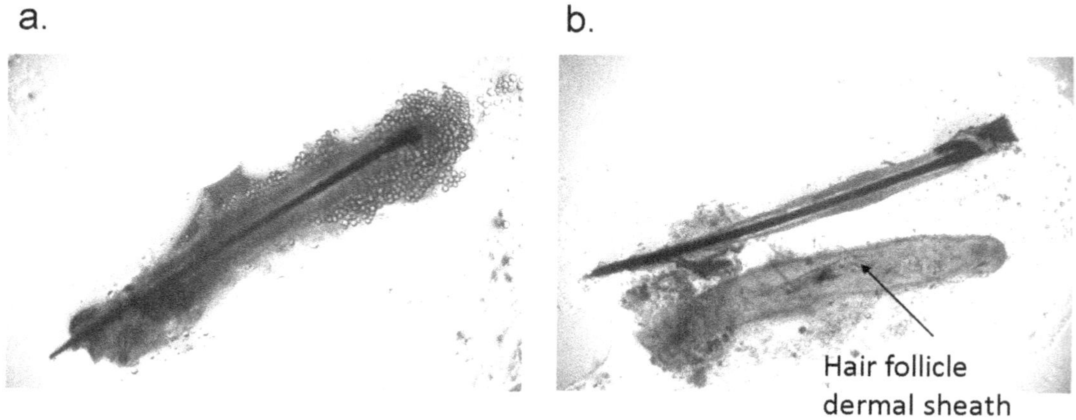

Fig. 1 (**a**) Single human hair follicle isolated from scalp tissue viewed under an inverted microscope. Reproduced from Ma et al. [2] with permission from Elsevier. (**b**) The hair follicle dermal sheath that was carefully separated by microdissection away from the main shaft of the hair follicle after treatment with 0.1% collagenase–0.25% Dispase II. (Reproduced from Ma et al. [2] with permission from Elsevier)

6. Remove cover of 100 mm dish. Gently micro-dissect out each strand of hair follicle from the scalp using a 30 g needle attached to a 1 mL syringe under a dissecting microscope, ensuring the dermal sheath is intact as shown in Fig. 1a.
7. After micro-dissecting out at least 10 strands, transfer all these strands of hair follicle with intact dermal sheath into another empty 100 mm cell culture dish.
8. Immediately transfer this covered 10 mm culture dish with hair follicles back into the BSC and aseptically add 10 mL of 0.1% Collagenase-0.25% Dispase II mix into the dish.
9. Transfer this dish into the CO_2 incubator at 37 °C and incubate for 30 min.
10. Remove dish from incubator and under the dissecting microscope in the laminar flow hood, gently separate the dermal sheath from the hair follicle shaft with the 30 g needle, ensuring that the sheath is in one intact piece as shown in Fig. 1b.
11. Aspirate out all the 0.1% Collagenase-0.25% Dispase II mix from the dish with a pipette.
12. Gently wash with 10 mL of 1x PBS for three times and gently aspirate out the PBS (*see* **Note 4**).
13. Remove and discard all hair follicle shafts using sterile forceps.
14. Add 1 mL of 0.02% EDTA followed by 1 mL of 0.1% trypsin solution (all pre-warmed to 37 °C) into the dish.
15. Mince the sheath tissues with scalpel blades for 5 min and then incubate these minced tissues in the same 0.1%trypsin/0.02% EDTA mix for further 30 min at 37 °C.

16. Add 1 mL of DS-MSC growth medium to quench the trypsinization process.
17. Collect cell suspension from the dish into a 50 mL centrifuge tube.
18. Repeat **steps 7** to **17** until all the hair follicles are dissected or until the centrifuge tube is full.
19. Centrifuge the cell suspension at 200 *g* for 5 min.
20. Aspirate out the supernatant and re-suspend with appropriate medium volume of DS-MSC growth medium (*see* **Note 5**).
21. Do cell count with hemocytometer to obtain cell density (number of cells per mL) and total number of cells based on the dilution volume above (*see* **Note 6**).
22. Seed 2×10^5 of the isolated cells in a 100 mm cell culture dish each (*see* **Note 7**).

3.2 Cell Culture of Isolated DS-MSCs

1. Change to fresh DS-MSC growth medium (10 mL for 100 mm plate) every 3 days and check the cultures daily (*see* **Note 8**).
2. When the DS-MSCs are about 90% confluent in 8–12 days, subculture the DS-MSCs using 5 mL of 0.02% EDTA and 5 mL 0.1% trypsin solution in the 100 mm cell culture dish.
3. Add 5 mL of DS-MSC growth medium to quench the trypsinization process.
4. Collect cell suspension from the dish into a 50 mL centrifuge tube.
5. Perform **steps 20** and **21** described in Subheading 3.1.
6. Do cell count with hemocytometer, obtain cell density (number of cells per mL) and total number of cells.
7. Seed 3.5×10^5 of the isolated cells in a 100 mm cell culture dish each for further subculture or analysis (*see* **Note 9**).

3.3 Colony Forming Efficiency (CFE) Assay of DS-MSCs

1. After either **step 21** in Subheading 3.1 or **step 6** in Subheading 3.2, do step dilution of cell suspension after cell counting.
2. Briefly, draw out 1 mL of cell suspension in fresh centrifuge tube and dilute to 10^5 cells per mL with DS-MSC growth medium.
3. Next draw out another 1 mL from the tube in above step into another fresh centrifuge and add 9 mL of DS-MSC growth medium to obtain 10^4 cells per mL.
4. Repeat **step 3** above to obtain final concentration of 10^3 cells per mL.
5. Seed 200 μL of medium from tube with 10^3 cells per mL into 100 mm culture dish (work in duplicate) with 10 mL of DS-MSC growth medium to start the CFE assay. This is considered Day 0.

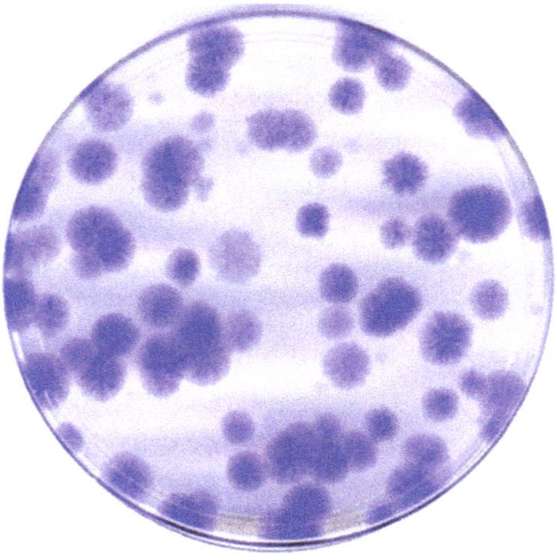

Fig. 2 A representative plate with 200 seeded DS-MSCs that were stained with 2% crystal violet solution after 12 days of culture

6. Change to fresh DS-MSC growth medium on Day 4 and Day 8.
7. On day 12, aspirate all cell culture medium from 100 mm plate and add 6 mL of 10% buffered formalin to cover the culture surface of the dish for 30 min at room temperature to fix the cultures.
8. Aspirate out all the buffered formalin, rinse the dish with 1X PBS.
9. Add 10 mL of 2% crystal violet solution for 15 min to stain the culture in the 100 mm culture dish.
10. Aspirate out the 2% crystal violet solution in hazardous waste bottle, rinse the cultures with distilled water and allow the dish to dry. The staining will have the appearance as shown in Fig. 2.
11. The number of colonies will be counted under a dissecting microscope and the CFE is determined as percentage relative to the 200 cells seeded.

3.4 Induction of DS-MSCs Towards Adipogenic Differentiation

1. After **step 6** in Subheading 3.2, seed isolated DS-MSCs at a cell density of $5 \times 10^4/cm^2$ into six-well culture plate and culture with 2 mL of DS-MSC growth medium in each well. Work in duplicates (*see* **Note 10**).
2. Allow the seeded DS-MSCs proliferate and be confluent. This will occur typically between 1 and 3 days.
3. When cultures are confluent, change to 2 mL adipogenic differentiation medium each of the wells meant for induction and 2 mL of DS-MSC growth medium each for wells which serve as negative control.

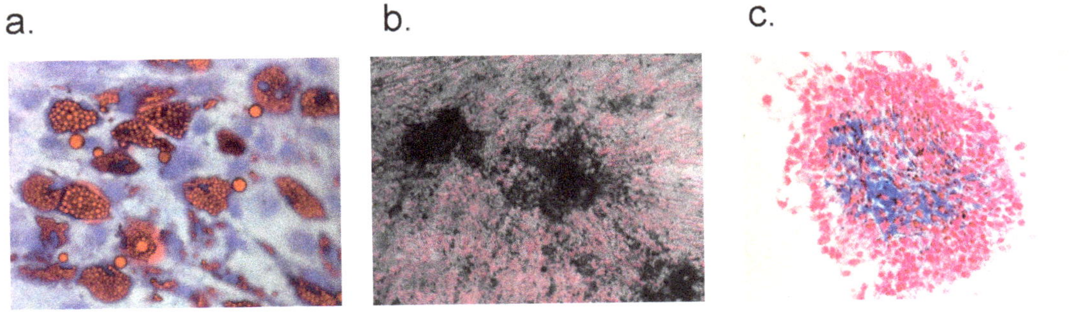

Fig. 3 In vitro differentiation of DS-MSCs. Reproduced from Ma et al. [2] with permission from Elsevier. (**a**) Positive Oil Red O staining, indicating the presence of lipid vacuoles within the adipocytes after adipogenic induction of DS-MSCs. (**b**) Positive von Kossa staining, which represented the formation of black calcium after osteogenic induction of DS-MSCs. (**c**) Positive Alcian blue staining, indicating cartilage extracellular matrix formation after chondrogenic induction of DS-MSCs

4. Continue to change media every 3 days and culture the cells for 2 weeks.
5. After 2 weeks, aspirate away culture media and wash cultures 2 times with 1x PBS.
6. Add 1.5 mL of 10% buffered formalin to cover the culture surface of each well with seeded DS-MSCs for 15 min at room temperature to fix the cultures.
7. Aspirate out all the buffered formalin and rinse the wells with distilled water.
8. Add 2 mL of 2% Oil red O staining solution to the wells for 30 min at room temperature (*see* **Note 11**).
9. Aspirate out the staining solution and wash the wells with 1x PBS twice.
10. Examine for generation of oil droplets in the cytoplasm of the differentiated cells under phase contrast microscope (Fig. 3a).

3.5 Induction of DS-MSCs Towards Osteogenic Differentiation

1. After **step 6** in Subheading 3.2, seed isolated DS-MSCs at a cell density of $5 \times 10^4/\text{cm}^2$ in six-well culture plate and culture with 2 mL of DS-MSC growth medium in each well. Work in duplicates.
2. Allow the seeded DS-MSCs proliferate and be confluent. This will occur typically between 1 and 3 days.
3. When cultures are confluent, change to 2 mL osteogenic differentiation medium each of the wells meant for induction and 2 mL of DS-MSC growth medium each for wells which serve as negative control.
4. Continue to change media every 3 days and culture the cells for 2 weeks.

5. After 2 weeks, aspirate away culture media and wash cultures two times with PBS.
6. Add 1.5 mL of 10% buffered formalin to cover the culture surface of each well with seeded DS-MSCs for 15 min at room temperature to fix the cultures.
7. Aspirate out all the buffered formalin and rinse the wells with distilled water.
8. Add 2 mL of Von Kossa staining solution for 30 min at room temperature (*see* **Note 11**).
9. Aspirate out the staining solution and wash the wells with 1x PBS twice.
10. Examine for mineral formation in the differentiated cells as indicted by the dark stains (Fig. 3b).

3.6 Induction of DS-MSCs Towards Chondrogenic Differentiation

1. After **step 6** in Subheading 3.2, seed isolated DS-MSCs at a cell density of $5 \times 10^4/\text{cm}^2$ in six-well culture plate and culture with 2 mL of DS-MSC growth medium in each well. Work in duplicates.
2. Allow the seeded DS-MSCs proliferate and be confluent. This will occur typically between 1 and 3 days.
3. When cultures are confluent, change to 2 mL chondrogenic differentiation medium each of the wells meant for induction and 2 mL of DS-MSC growth medium each for wells which serve as negative control.
4. Continue to change media every 3 days and culture the cells for 3–4 weeks.
5. After 2 weeks, aspirate away culture media and wash cultures two times with PBS.
6. Add 1.5 mL of 10% buffered formalin to cover the culture surface of each well with seeded DS-MSCs for 15 min at room temperature to fix the cultures.
7. Aspirate out all the buffered formalin and rinse the wells with distilled water.
8. Add 2 mL of Alician staining solution for 30 min at room temperature (*see* **Note 11**).
9. Aspirate out the staining solution and wash the wells with 1x PBS twice.
10. Examine for cartilage extracellular matrix formation as indicted by the blue stains (Fig. 3c).

4 Notes

1. The Dispase II solution works best at neutral pH; therefore make sure the Dispase II powder is dissolved in fresh DMEM to prevent elevated pH and make sure cap of the holding tubes are quickly closed after dispensing.
2. Methanol is hazardous and flammable. Dilute in chemical fume hood and wear appropriate personal protective equipment (PPE).
3. These tissues which would otherwise be discarded in plastic surgery hair transplantation procedures were collected with consent from patients and approval from the Institutional Review Board (IRB) of the hospital.
4. Beware not to aspirate out the hair follicles and the sheath pieces especially if vacuum pump is used.
5. Typically, 2–3 mL of growth medium is used to re-suspend the DS-MSCs isolated from ten strands of hair follicle.
6. For more accurate cell count, keep average cell count number of the four corner squares (1 mm^2 each) to between 20 and 90 cells per square. Otherwise, readjust to appropriate dilution volume and repeat the cell count.
7. Cell yield direct from dermal sheath in primary culture is typically low and that is why seeding density is lower at 2×10^5 cells per 100 mm cell culture dish. Subsequent cell seeding density is 3.5×10^5 cells per 100 mm cell culture dish during cell passaging.
8. The DS-MSCs will start to form colonies with sparse density of cells within each colony (Fig. 4a), observable after 4–5 days and each colony will be densely packed in 10–12 days' time (Fig. 4b).
9. Freshly isolated, cultivated human DS-MSCs—characterized with flow cytometry analysis for cell surface phenotype—were found to highly express for CD105, CD29, CD49b, and CD49d, while weakly expressed for CD90, CD44, CD34, and CD45 (Fig. 5).
10. In this case, working in duplicate means that 2 wells will be used to induce differentiation, and 2 wells will serve as negative control.
11. Do not let the cells dry for more than 30 s during the entire staining procedure.

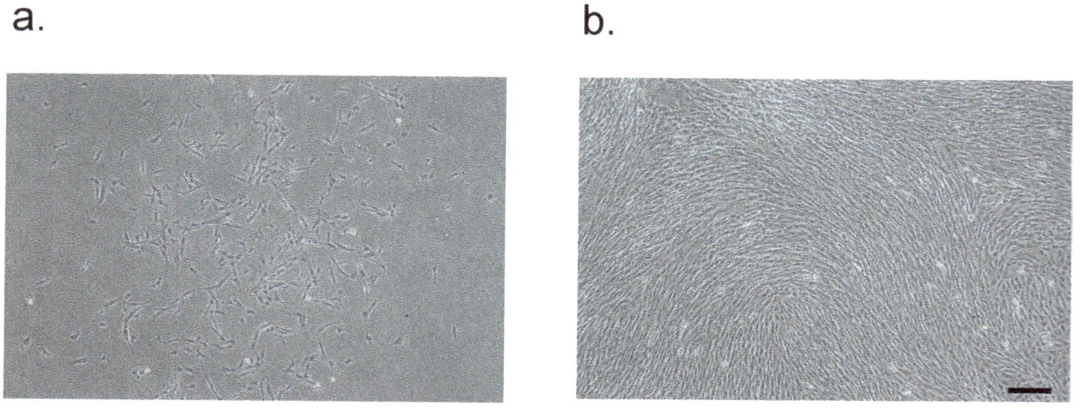

Fig. 4 Phase-contrast image of a colony of HF dermal sheath mesenchymal stromal cells (DS-MSCs) after isolation and plated for (**a**) 4–5 days and (**b**) 10–12 days. Scale bar 50 μm

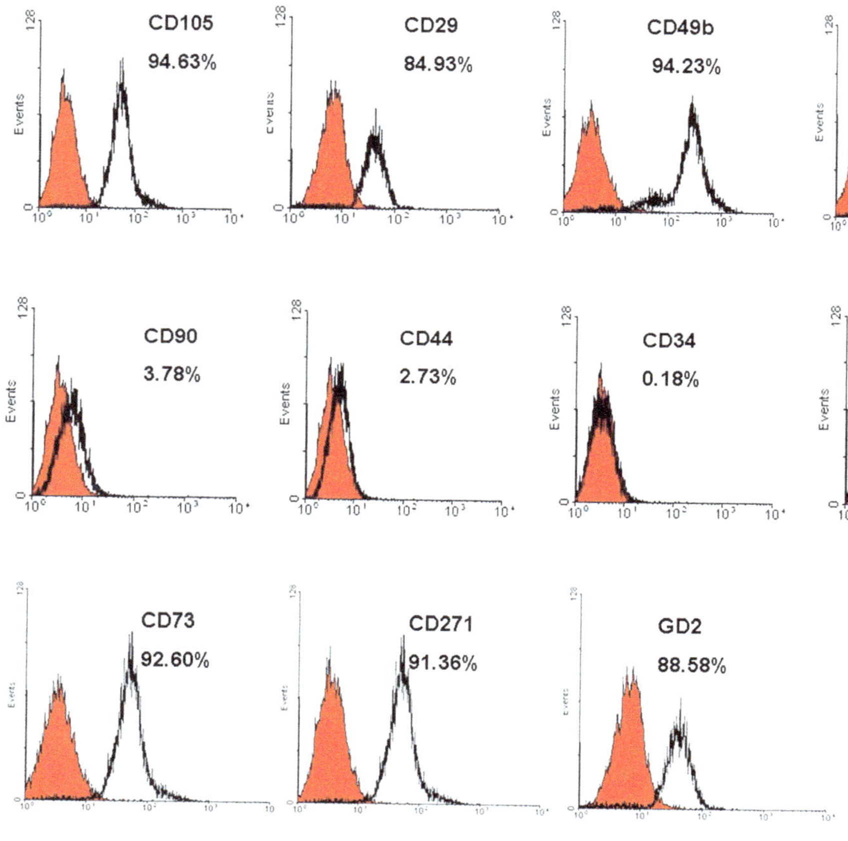

Fig. 5 Representative flow cytometry results of DS-MSCs stained with antibodies against CD105, CD29, CD49b, CD49d, CD90, CD44, CD34, and CD45. DS-MSCs (black peaks) strongly expressed CD105, CD29, CD49b, and CD49d but not CD90, CD44, CD34, and CD45. Isotype controls are shown in red peaks. (Reproduced from Ma et al. [2] with permission from Elsevier)

References

1. Richardson GD, Arnott EC, Whitehouse CJ et al (2005) Plasticity of rodent and human hair follicle dermal cells: implications for cell therapy and tissue engineering. J Investig Dermatol Symp Proc 10:180–183
2. Ma D, Kua JE, Lim WK et al (2015) In vitro characterization of human hair follicle dermal sheath mesenchymal stromal cells and their potential in enhancing diabetic wound healing. Cytotherapy 17:1036–1051
3. Richardson GD, Arnott EC, Whitehouse CJ et al (2005) Cultured cells from the adult human hair follicle dermis can be directed toward adipogenic and osteogenic differentiation. J Invest Dermatol 124:1090–1091
4. Li H, Masieri FF, Schneider M et al (2020) Autologous, non-invasively available mesenchymal stem cells from the outer root sheath of hair follicle are obtainable by migration from plucked hair follicles and expandable in scalable amounts. Cells 9:2069
5. Hernaez-Estrada B, Gonzalez-Pujana A, Cuevas A et al (2022) Human hair follicle-derived mesenchymal stromal cells from the lower dermal sheath as a competitive alternative for immunomodulation. Biomedicines 10:253

Chapter 6

Isolation and Culture of Human Dermal Fibroblasts

Fabiana Ziliani, Katarzyna Michalak-Micka, and Agnes S. Klar

Abstract

Dermal fibroblasts are the main cell type present in skin connective tissue (dermis). They are responsible for the synthesis of dermal extracellular matrix (ECM), providing mechanical stability for the dermal layer. In addition, fibroblasts interact with epidermal cells during hair development and in interfollicular skin cells through bidirectional interactions and secretion of growth factors and cytokines. The changes in the fibroblast population correlate with structural, compositional, quantitative, and qualitative transformations of the main ECM proteins in the dermis, as well as with the age-related disorders, and the state and proliferative activity of basal keratinocytes that are the main cells in the epidermis. Moreover, fibroblasts play an essential role during cutaneous wound healing and in bio-engineering of skin. Hence, culture of primary fibroblast is gaining in importance. In addition, fibroblast cultures established from skin biopsies provide a powerful tool for investigating normal skin physiology or specific disease states. In this chapter, detailed procedures for the establishing and maintaining primary cultures of adult human dermal fibroblasts are described.

Key words Skin, Fibroblasts, Dermis, ECM, Collagen, Organotypic culture, Skin equivalent, Regenerative medicine, Tissue engineering

1 Introduction

Human skin fibroblasts are mesenchymal/stromal cells derived from the embryonic mesoderm. They are located in the dermal layer of skin, where they produce extracellular matrix proteins to strengthen the dermal compartment and interact with epidermal cells. The extracellular matrix synthesized by fibroblasts is rich in type I and/or type III collagen [1, 2]. The mesenchymal–epithelial interactions regulate cell differentiation, pigmentation [3–7], and maintain tissue homeostasis of the epidermis [8]. Dermal fibroblasts are characterized by their spindle morphology and expression of stromal-specific markers such as fibronectin, vimentin, and TH-1 (CD90) [9]. However, fibroblasts from various organs, body sites, and spatial locations exhibit differences regarding marker expression and functional specialization [10–12].

Thomas Biedermann and Sophie Böttcher-Haberzeth (eds.), *Skin Tissue Engineering: Methods and Protocols*, Methods in Molecular Biology, vol. 2922, https://doi.org/10.1007/978-1-0716-4510-9_6,

Further, dermal fibroblasts are key players for orchestrating physiological tissue repair. During skin wound healing, dermal fibroblasts switch from a proliferative and migratory phenotype to a contractile, matrix-remodeling myofibroblast [13–18].

Myofibroblasts are identified on the basis of their ultrastructural morphology, with characteristic microfilament bundles in their cytoplasm distinguishing them from quiescent dermal fibroblasts. Myofibroblasts also demonstrate fibronexus junctions with the surrounding extracellular matrix (ECM) showing some of the morphological characteristics of smooth muscle (SM) cells [13]. Myofibroblasts regenerate connective tissue after an injury by collagen deposition and wound contraction. Afterwards, myofibroblasts disappear by apoptosis, thereby preventing excess contraction. However, under pathological conditions, excessive wound contraction may lead to fibrosis and scarring [19].

A promising development to improve wound healing and reduce scarring is the bio-engineering of skin substitutes, which has made significant progress over the last few decades [20, 21].

The establishment of appropriate culture conditions for fibroblasts pave the way for a successful application of those cells in bio-engineered skin substitutes. The addition of dermal fibroblasts improved consistently graft take, reduced susceptibility to infection, and graft contracture as compared to the previously used only epidermal sheets, such as autologous cultured epidermal autografts (CEA) [22].

Furthermore, the addition of the dermal component to the skin substitutes resulted in a more stable skin graft with improved functional and esthetic properties [20].

For several years, our laboratory has developed and extensively characterized dermo-epidermal skin substitutes (DESS) to treat full-thickness skin defects in clinical applications [4, 23–31].

Most recently, we reconstituted pigmented dermo-epidermal skin analogs from different pigmentation types and achieved very satisfactory in vivo results restoring the patients' native skin color [31–33].

To summarize, our three-dimensional DESS offer an exceptional model to study cell-cell interactions, as we demonstrated, e.g., the importance of different fibroblast populations in the context of reproducing the donor skin color in our pigmented skin substitutes [4, 32–34], as well as specific mechanisms involved in pathologies such as fibrosis [29].

This chapter describes methods and materials needed for all aspects of fibroblast culture, namely tissue preparation, isolation, cell culture, passaging, and cryopreservation.

2 Materials

2.1 DMEM

DMEM: Dulbecco's modified Eagle's medium (DMEM), store in the dark at 4 °C.

2.2 Sera

1. Fetal calf serum. Thaw sera at 4 °C or in cold water (*see* **Note 1**). Gently swirl to resuspend its components. Inactivate in a 56 °C water bath for 30 min and distribute in single use aliquots. Store at −20 °C or at −80 °C for long term storage.

2.3 Additives

1. 10,000 IU/mL penicillin/streptomycin.
2. 1 M HEPES (4-(2-hydroxyethyl)-1-piperazineethanesulfonic acid).
3. 250 μg/mL Amphotericin B.
4. 10 mg/mL Gentamycin.

2.4 Complete Media

1. Tissue transport medium (tDMEM): 10% v/v of fetal calf serum, 1.5 mL Amphotericin B, and 15 mL penicillin/streptomycin in 500 mL DMEM, store in the dark at 4 °C for up to 6 weeks.
2. Complete fibroblast culture medium (DMEM complete): 10% v/v of fetal calf serum, 5 mL penicillin/streptomycin, 5 mL HEPES in 500 mL DMEM, store in the dark at 4 °C for up to 4 weeks.
3. Cryopreservation medium: 10% v/v of dimethyl sulfoxide (DMSO) in fetal calf serum, keep on ice or store at 4 °C and use within the day.

   All frozen components can be thawed at 4 °C (*see* **Note 2**).

2.5 Other Solutions

1. Phosphate buffered saline (PBS) without Ca^{2+}/Mg^{2+}, store at room temperature.
2. Collagenase solution: 10 mg Collagenase + 1 mL DMEM/10% FCS + 4 mL DMEM, sterilize by filtration through a 0.22 μm low binding disposable filter.
3. Dispase solution: 2.5 mL dispase (50 U/mL), 2.5 mL PBS, 2.5 μL gentamycin.
4. Trypsin/EDTA: 0.05% w/v trypsin, 0.01% w/v EDTA in PBS, store at −20 °C to −80 °C.
5. Trypan blue solution.
6. 100% isopropyl alcohol.

2.6 Tissues and Cells

1. Human skin biopsy: surgically removed 3–6 cm^2 skin biopsy.

2.7 Labware

1. 0.22 μm low binding disposable filters.
2. 100 μm strainers.
3. Sterile containers.
4. 50 mL tubes.
5. 100 × 15 mm cell culture Petri dishes.
6. Dissecting curved forceps.
7. Size 3 scalpel and size 22 blade.
8. 25 or 75 cm^2 tissue culture flasks.
9. Sterile cryogenic vials.
10. Freezing containers.
11. Laboratory incubator shaker.
12. Cell culture microscope.
13. −80 °C freezer.
14. Liquid nitrogen tank for cell storage.
15. Hemocytometer.

3 Methods

Manipulations must be performed under a sterile laminar flow cabinet.

3.1 Tissue Sampling and Transport

1. Immediately following aseptic surgical removal (*see* **Note 3**), put the skin biopsy into a sterile container filled with cold (4 °C) tDMEM.
2. Samples must be kept at 4 °C (*see* **Note 4**).

3.2 Isolation

1. With sterile forceps, transfer the skin sample (*see* Subheading 3.1) into a 100 × 15 mm cell culture Petri dishes.
2. Use curved forceps to gently spread out the skin sample with the epidermis facing top in a Petri dish.
3. Use a scalpel with a size 22 blade to cut the sample into small pieces of approximately 4–6 mm diameter and transfer into a 50 mL tube.
4. Add 2.5 mL of Dispase solution.
5. Incubate overnight at 4 °C.
6. Use two curved forceps to carefully separate the dermis from the epidermis.
7. Incubate the dermis pieces with Collagenase solution for 60 min at 37 °C, 5% CO_2; shake regularly.

8. Add 15 mL of DMEM complete medium to the dissolved dermis.
9. Place a 100 μm strainer on a new 50 mL tube and pass the dermis suspension through the strainer (discard the strainer with left dermis pieces), yielding the fibroblast (NHF) suspension in the 50 mL falcon.
10. Centrifuge (200× *g*) the NHF suspension for 5 min at room temperature.
11. Resuspend cells in 10 mL of DMEM complete medium, stain with trypan blue solution and count NHF using a hemocytometer (*see* **Note 5**).

    Cell viability is expected to be greater than 80%.
12. Seed approx. 1–3 × 10^6 NHF into an uncoated 75 cm^2 tissue flask in DMEM complete medium. Total medium volume should not exceed 15 mL per 75 cm^2 (*see* **Note 6**).

3.3 Culture

1. Incubate NHF in a 5% CO_2 and 100% humidity atmosphere at 37 °C (*see* **Note 7**, Fig. 1).
2. Change the culture medium three times a week. Remove the medium from the culture flask. Replace it with warm (37 °C) DMEM complete medium.
3. Monitor NHF morphology and confluence under a microscope (*see* **Note 8**).
4. When NHF reach 70–90% confluence, subculture (*see* Subheading 3.4) or cryopreserve (*see* Subheading 3.5) them.

3.4 Subculture (Passage)

1. Remove the culture medium.
2. Rinse NHF in 75 cm^2 culture flasks with 5 mL PBS w/o Ca^{2+}/Mg^{2+}. Remove it.
3. Add 4 mL Trypsin/EDTA and 1 mL PBS into the 75 cm^2 culture flask.

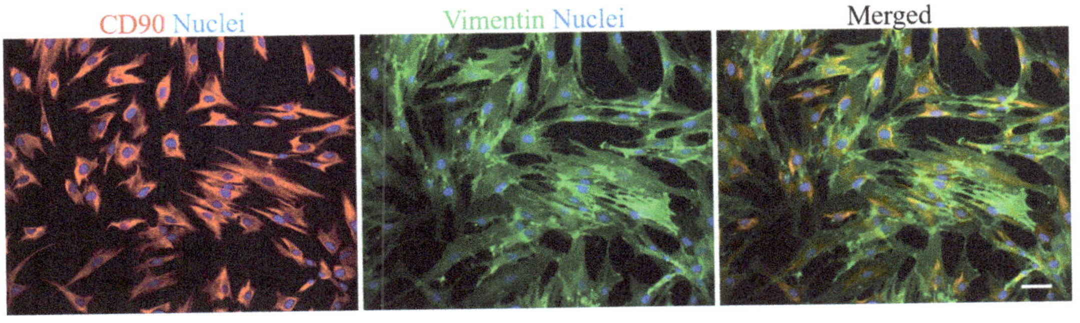

Fig. 1 Immunofluorescence analysis of first passage of NHF in vitro. NHF were stained with CD90 (red; left), vimentin (green; middle), and merged (right). Cell nuclei are stained blue. Scale bar is 50 μm

4. Incubate for 5 min at 37 °C until all NHF are completely detached from the flask (verify cell detachment under a microscope) (*see* **Note 9**).
5. Neutralize trypsin activity by adding 5 mL of DMEM complete medium, rinse rest from plate.
6. Vigorously pipette the NHF suspension up and down at least five times to ensure suspension homogeneity.
7. Transfer the NHF suspension into a 15 mL tube.
8. Centrifuge (200× *g*) the NHF suspension for 5 min at room temperature (*see* **Note 10**).
9. Remove the supernatant and resuspend NHF in DMEM complete medium.
10. Count cells using a hemocytometer. Use trypan blue staining to estimate cell viability. Cell viability is expected to be greater than 95%.
11. Seed NHF at 1–3 × 10^6 cells per 75 cm^2 or more into an uncoated culture flask. Seed directly into the culture medium Use 10 mL of total medium volume per 75 cm^2.

3.5 Cryopreservation

1. Fill a freezing container with 100% isopropyl alcohol and store it at 4 °C until cool.
2. Follow **steps 1** through **11** from Subheading 3.4.
3. Remove the supernatant and resuspend NHF at the desired cell concentration in cryopreservation medium. Put the tube on ice (*see* **Note 11**).
4. Aliquot in cryogenic vials on ice.
5. Put the cryogenic vials in the freezing container.
6. Store the container overnight at −80 °C. In these conditions, cell temperature should drop 1 °C/min.
7. Store cryogenic vials in liquid nitrogen for long term storage.

3.6 Thawing

1. Put the cryogenic vial in a 37 °C water bath. Do not let the NHF suspension thaw completely. A small ice pellet should remain.
2. Transfer the cell suspension immediately from the cryogenic vial into a 50 mL tube containing 5–10 mL of warm (37 °C) DMEM complete medium.
3. Rinse the cryogenic vial with 1–1.5 mL of the suspension.
4. Centrifuge (200× *g*) the NHF suspension for 5 min at room temperature.
5. Remove the supernatant and resuspend NHF in 10 mL of warm (37 °C) DMEM complete medium.

6. Use a hemocytometer to count the NHF.
7. Use trypan blue staining and a hemocytometer to estimate cell viability. Cell viability is expected to be greater than 80%.
8. Centrifuge (200× *g*) the NHF suspension for 5 min at room temperature.
9. Remove the supernatant and resuspend NHF at the desired concentration in DMEM complete medium.
10. Seed NHF at 1–3 × 10^6 cells per 75 cm^2 or more into an uncoated culture flask. Seed directly into the culture medium. Use 10 mL of total medium volume per 75 cm^2.

4 Notes

1. Sera and other frozen components can also be thawed more rapidly at room temperature or in a 37 °C water bath. However, we do not recommend this procedure.
2. Culture media are supplemented with antibiotics and antimycotics. However, they should be rinsed with 70% ethanol after prewarming to 37 °C, in water bath, to prevent contamination.
3. Adult skin is mostly contaminated with (mainly) nonpathogenic bacteria or molds. Therefore, these micro-organisms must be removed prior to skin biopsy to insure good sterility of the tissue.
4. To achieve maximum cell yield and viability, skin tissue should be processed directly after surgery. However, when necessary, tissue can be kept at 4 °C for up to 5–7 days without significant loss of viability.
5. The yield of cells per cm^2 skin is affected by the age and health of donor. Younger donors generate more stem cells than older ones.
6. Ensure proper distribution of cells by quickly moving the flask back and forth several times. Open and close the incubator door gently after cell plating.
7. There may be only a few adherent cells at 24 h. Wait for at least 2 days to allow cells to attach to the bottom of the flask.
8. It is helpful to examine routinely cells under microscope to determine their status and health. It is also recommended to record the details concerning cell cultures, including the subculture ratios and the passage number.
9. For optimal trypsinization efficiency, do not stack culture flasks on top each other in the incubator. Temperature is typically higher on the flask surface directly in contact with the incubator shelf.

10. Centrifugation following trypsin detachment of cells removes trypsin and cell debris. This removal helps to maintain viability and longevity of cultures.
11. DMSO is a toxic oxidative agent at temperatures above 10 °C. Therefore, work with this agent quickly and keep cells on ice.

Acknowledgments

This study was supported by research funding from the Swiss National Science Foundation (SNSF) grant no. 310030_207793 ("Role of LGR5 and LGR6 in the regulation of human epithelial stem cell niche in the interfollicular epidermis"), IZCOZ0_213354 ("Role of CD200-CD200 receptor interactions in human diabetic skin wounds"). The authors are particularly grateful to the Fondation Gaydoul for their generous financial support and interest in our work.

References

1. Tracy LE, Minasian RA, Caterson EJ (2016) Extracellular matrix and dermal fibroblast function in the healing wound. Adv Wound Care 5(3):119–136
2. Li W, Chi NW, Rathnayake RAC et al (2021) Distinctive roles of fibrillar collagen I and collagen III in mediating fibroblast-matrix interaction: a nanoscopic study. Biochem Biophys Res Commun 560:66–71
3. Biedermann T, Böttcher-Haberzeth S, Klar AS et al (2015) The influence of stromal cells on the pigmentation of tissue-engineered dermo-epidermal skin grafts. Tissue Eng Part A 21(5–6):960–969
4. Klar AS, Biedermann T, Michalak K et al (2017) Human adipose mesenchymal cells inhibit melanocyte differentiation and the pigmentation of human skin via increased expression of TGF-beta1. J Invest Dermatol 137(12): 2560–2569
5. Michalak-Micka K, Tenini C, Böttcher-Haberzeth S et al (2024) The expression pattern of cytokeratin 6a in epithelial cells of different origin in dermo-epidermal skin substitutes in vivo. Biotechnol J 19(1):e2300246
6. Klar AS, Michalak K, Böttcher-Haberzeth S et al (2018) The expression pattern of keratin 24 in tissue-engineered dermo-epidermal human skin substitutes in an in vivo model. Pediatr Surg Int 34(2):237–244
7. Michalak-Micka K, Klar AS, Böttcher-Haberzeth S et al (2019) Impact of human mesenchymal cells of different body site origins on the maturation of dermo-epidermal skin substitutes. Pediatr Surg Int 35(1):121–127
8. Yamaguchi Y, Hearing VJ, Itami S et al (2005) Mesenchymal-epithelial interactions in the skin: aiming for site-specific tissue regeneration. J Dermatol Sci 40(1):1–9
9. Alberts BAJJL (2002) Fibroblasts and their transformations: the connective-tissue cell family, 4th edn. Garland Science, New York
10. Driskell RR, Watt FM (2015) Understanding fibroblast heterogeneity in the skin. Trends Cell Biol 25(2):92–99
11. Lynch M, Watt F (2018) Fibroblast heterogeneity: implications for human disease. J Clin Invest 128(1):26–35
12. Philippeos C, Telerman SB, Oules B et al (2018) Spatial and single-cell transcriptional profiling identifies functionally distinct human dermal fibroblast subpopulations. J Invest Dermatol 138(4):811–825
13. Mascharak S, des Jardins-Park HE, Longaker MT (2020) Fibroblast heterogeneity in wound healing: hurdles to clinical translation. Trends Mol Med 26(12):1101–1106
14. Schwarz F, Jennewein M, Bubel M et al (2013) Soft tissue fibroblasts from well healing and chronic human wounds show different rates of myofibroblasts in vitro. Mol Biol Rep 40(2):1721–1733
15. Li B, Wang JHC (2011) Fibroblasts and myofibroblasts in wound healing: force generation and measurement. J Tissue Viability 20(4): 108–120

16. Novotny M, Vasilenko T, Varinska L et al (2011) ER-α agonist induces conversion of fibroblasts into myofibroblasts, while ER-® agonist increases ECM production and wound tensile strength of healing skin wounds in ovariectomised rats. Exp Dermatol 20(9): 703–708
17. Dvořánková B, Szabo P, Lacina L et al (2011) Human galectins induce conversion of dermal fibroblasts into Myofibroblasts and production of extracellular matrix: potential application in tissue engineering and wound repair. Cells Tissues Organs 194(6):469–480
18. Barisic-Dujmovic T, Boban I, Clark SH (2010) Fibroblasts/Myofibroblasts that participate in cutaneous wound healing are not derived from circulating progenitor cells. J Cell Physiol 222(3):703–712
19. Darby IA, Desmouliere A (2020) Scar formation: cellular mechanisms. In: Teot L, Mustoe T, Middelkoop E, Gauglitz G (eds) Textbook on scar management: state of the art management and emerging technologies. Springer, Cham, pp 19–26
20. Biedermann T, Boettcher-Haberzeth S, Reichmann E (2013) Tissue engineering of skin for wound coverage. Eur J Pediatr Surg 23(5): 375–382
21. Klar AS, Zimoch J, Biedermann T (2017) Skin tissue engineering: application of adipose-derived stem cells. Biomed Res Int 2017: 9747010
22. Wood FM, Kolybaba ML, Allen P (2006) The use of cultured epithelial autograft in the treatment of major burn injuries: a critical review of the literature. Burns 32(4):395–401
23. Rütsche D, Michalak-Micka K, Zielinska D et al (2022) The role of CD200-CD200 receptor in human blood and lymphatic endothelial cells in the regulation of skin tissue inflammation. Cells 11(6):1055
24. Michalak-Micka K, Rütsche D, Johner L et al (2022) Expression profile of CD157 reveals functional heterogeneity of capillaries in human dermal skin. Biomedicines 10(3):676
25. Ademi H, Michalak-Micka K, Moehrlen U et al (2023) Effects of an adipose mesenchymal stem cell-derived conditioned medium and TGF-®1 on human keratinocytes in vitro. Int J Mol Sci 24(19):14726
26. Iqbal MZ, Riaz M, Biedermann T et al (2024) Breathing new life into tissue engineering: exploring cutting-edge vascularization strategies for skin substitutes. Angiogenesis. https://doi.org/10.1007/s10456-024-09928-6
27. Braziulis E, Diezi M, Biedermann T et al (2012) Modified plastic compression of collagen hydrogels provides an ideal matrix for clinically applicable skin substitutes. Tissue Eng Part C Methods 18(6):464–474
28. Pontiggia L, Biedermann T, Meuli M et al (2009) Markers to evaluate the quality and self-renewing potential of engineered human skin substitutes in vitro and after transplantation. J Invest Dermatol 129(2):480–490
29. Klar AS, Guven S, Biedermann T et al (2014) Tissue-engineered dermo-epidermal skin grafts prevascularized with adipose-derived cells. Biomaterials 35(19):5065–5078
30. Bottcher-Haberzeth S, Klar AS, Biedermann T et al (2013) "Trooping the color": restoring the original donor skin color by addition of melanocytes to bioengineered skin analogs. Pediatr Surg Int 29(3):239–247
31. Bottcher-Haberzeth S, Biedermann T, Pontiggia L et al (2013) Human eccrine sweat gland cells turn into melanin-uptaking keratinocytes in dermo-epidermal skin substitutes. J Invest Dermatol 133(2):316–324
32. Michalak-Micka K, Büchler VL, Zapiórkowska-Blumer N et al (2022) Characterization of a melanocyte progenitor population in human interfollicular epidermis. Cell Rep 38(9):110419
33. Zielinska D, Fisch P, Moehrlen U et al (2023) Combining bioengineered human skin with bioprinted cartilage for ear reconstruction. Sci Adv 9(40):eadh1890
34. Biedermann T, Bottcher-Haberzeth S, Klar AS et al (2015) The influence of stromal cells on the pigmentation of tissue-engineered Dermo-epidermal skin grafts. Tissue Eng Part A 21(5–6):960–969

Chapter 7

Isolation and Culture of Human Dermal Microvascular Endothelial Cells

Jennifer Bourland, Dominique Mayrand, Nathalie Tremblay, Véronique J. Moulin, Julie Fradette, and François A. Auger

Abstract

Primary endothelial cells are needed for angiogenesis studies, and more particularly in the field of tissue engineering, to generate pre-vascularized tissues in vitro. Investigations often use human umbilical vein endothelial cells due to their extensive characterization, but also because they are easy to obtain and isolate. An alternative is the use of human dermal microvascular endothelial cells, more representative of adult skin angiogenic processes. This chapter presents a detailed methodology to isolate and culture microvascular endothelial cells from human skin biopsies based on enzymatic digestion and mechanical extraction.

Key words Skin, Tissue engineering, Endothelial cells, Microvascularization, Capillary, Cell isolation, Angiogenesis, Immunoselection

1 Introduction

Endothelial cells are widely used in vitro to investigate angiogenic processes, but also to prevascularize biomaterials or tissue-engineered constructs. Different human endothelial cells are available for these studies, originating either from large-diameter vessels or capillaries. Many experiments use human umbilical vein endothelial cells (HUVEC), which are easy to isolate and culture. However, these cells differ from native skin endothelial cells, in particular by the size of the vessels they can form in vitro and by their interactions with extracellular matrix components [1]. For example, human microvascular endothelial cells (HMVEC) are in direct contact with a basement membrane composed of proteins such as collagen type IV and laminin and thus are more prone to the production of these molecules compared to HUVEC. They are also more responsive to factors modulating angiogenesis such as angiopoietins [2]. HMVEC can be extracted from either adult or newborn skin using a method based on enzymatic digestion and

Thomas Biedermann and Sophie Böttcher-Haberzeth (eds.), *Skin Tissue Engineering: Methods and Protocols*, Methods in Molecular Biology, vol. 2922, https://doi.org/10.1007/978-1-0716-4510-9_7,

extrusion. The epidermis is first removed after enzymatic digestion. The capillaries are then isolated by mechanical extrusion from the dermis and seeded as explants. Colonies of endothelial cells can develop from endothelial cells originating from the tubules. The resulting population needs to be subsequently purified to remove residual fibroblasts, usually based on selection methods using CD31 expression. Indeed, CD31 (or platelet endothelial cell adhesion molecule-1, PECAM-1) is a molecule expressed at the surface of both blood and lymphatic microvascular endothelial cells [3], allowing their adhesion. CD31 also mediates angiogenic processes through the formation of filopodia during vessel sprouting [4, 5]. Two main methods of enrichment are available using immunomagnetic beads [6] or flow-assisted cell sorting (FACS) [3]. This chapter will detail a protocol using immunomagnetic positive selection to obtain endothelial cell populations from human post-natal dermal tissues.

2 Materials

Use ultrapure deionized and apyrogenic ultrapure water to prepare the solutions and to rinse the tools before sterilization. Store all reagents at 4 °C unless indicated otherwise. Properly dispose of all materials coming in contact with skin samples and primary cells as they are considered a biolevel two safety hazard. All work needs to be performed under aseptic conditions, using sterile solutions and materials.

2.1 Cell Extraction and Culture

1. Skin resection: human skin samples must be obtained with the consent of the donor or of his legal representative. The research project must have been previously approved by the institutional ethic committee.
2. Wash buffer: dissolve 80 g NaCl, 2 g KH_2PO_4, 14.4 g Na_2HPO_4 and 2 g KCl in 1 L water for 10× stock solution, adjust the pH between 7.4 and 7.6, dilute with water to 1× working solution, ensure complete dissolution, filter solution with a 0.22 μm membrane under a biological hood, sterile solution can be stored at 4 °C for up to a month, right before use, add 100 U/mL penicillin G potassium, 25 μg/mL gentamicin sulfate, 0.5 μg/mL fungizone®.
3. HEPES (4-(2-hydroxyethyl)-1-piperazineethanesulfonic acid) solution: 0.01 M HEPES, 1 mM $CaCl_2$, adjust the pH to 7.45, filter with a 0.22 μm membrane.
4. Hank's Balanced Salt Solution (HBSS): dissolve 0.98 g of HBSS powder in 100 ml H_2O, supplement with 35 mg of sodium bicarbonate and adjust the pH to 7.4, filter with a 0.22 μm membrane.

5. Dispase: prepare 1% weight/volume dispase neutral II solution in HBSS, add fetal bovine serum (FBS) to a final concentration of 2%, prepare 10 mL for 12 cm^2 of skin or less.
6. EGM™-2MV: use commercially available endothelial medium such as EGM™-2MV. Prepare it freshly by adding the Single-Quot kit supplements to the EBM-2 basal medium. The 5% FBS concentration adequately supports endothelial cell growth.
7. Sterile Petri dishes: at least five 100 mm Petri dishes and one 60 mm Petri dish for each 12 cm^2 of skin. Petri dishes should not be tissue culture treated to prevent cell adhesion.
8. Two sterile scalpels (size number 4) with blades (size #22).
9. Two sterile stainless steel curved tweezers.
10. A set of sterile stainless steel long tweezers.
11. Sterile curved scissors.
12. Parafilm®.
13. 50 mL conical sterile polypropylene tubes.
14. Gelatin-coated flasks: 0.1% gelatin solution and filter with a 0.22 μm membrane. Add 4 mL of the gelatin solution to a 25 cm^2 T-flask or 10 mL to a 75 cm^2 T-flask and be careful to spread the gelatin over the entire flask area. Incubate at 4 °C overnight. Be careful to place the flasks on a flat surface to ensure homogeneity of the coating. The next day, aspirate the gelatin solution, without touching the coated surface. Coated flasks can be stored for 1 month at 4 °C.
15. Dulbecco's Modified Eagle Medium (DMEM) supplemented with 10% FBS and antibiotics (100 U/mL penicillin G and 25 μg/mL gentamicin sulfate).

2.2 Endothelial Cell Selection

1. Dynabeads™ CD31 Endothelial Cell: the number of beads needed can vary but processing 12 cm^2 of skin usually requires 1–5 million beads.
2. DynaMag™-5 Magnet.
3. 5 mL round bottom polystyrene tubes compatible with the DynaMag™-5 Magnet.
4. HBSS with 2% FBS (*see* Subheading 2.1, **item 4**).
5. PBS supplemented with 0.1% (weight/volume) bovine serum albumin (BSA).
6. Ice.
7. Orbital shaker.
8. 15 mL conical sterile polypropylene tubes.

2.3 Cell Amplification and Banking

1. Gelatin-coated culture dishes (*see* Subheading 2.1, **item 14**).
2. EGM™-2MV medium (*see* Subheading 2.1, **item 6**).
3. Cryopreservation medium (FBS with 10% dimethyl sulfoxide (DMSO)): to prepare 10 mL of cryopreservation medium, mix 9 mL of cold FBS with 1 mL of DMSO. Be careful to use cold solutions while freezing cells.
4. Cryogenic storage tubes.
5. Freezing container.

3 Methods

Carry out all steps under a biological safety cabinet and under sterile conditions unless otherwise specified. Skin samples should be collected and transported in DMEM supplemented with 10% FBS and 100 U/mL of penicillin G and 25 μg/mL gentamicin sulfate. Keep at 4 °C until processing (maximum 48 h).

3.1 Day 1: Skin Sample Preparation

1. Remove the sample from the storage buffer and place in a sterile 100 mm Petri dish. A smaller Petri dish can be used for small samples.
2. Inspect the general aspect of the skin on both sides and note any particular characteristic (*see* **Note 1**).
3. Then, with the dermis facing upwards, remove any remaining adipose tissue with curved scissors. If damaged zones are present, they should be removed as well using a scalpel.
4. With clean sterile tweezers, retrieve the sample and place it in a new sterile Petri dish. The use of a new sterile dish or instruments contributes to reduce the risks of contamination. Place the epidermis on top and trace the shape of the skin to calculate its surface area (*see* Fig. 1).
5. Using long tweezers, transfer the sample to a 50 mL tube containing 40 mL of wash buffer. Depending on the sample size, add enough PBS to bring the total volume in the tube to 45 mL.
6. Add Parafilm® or a sterile gauze around the cap to prevent leakage.
7. Mix vigorously by shaking the tube for 30 s.
8. Discard the PBS by inversion and add a fresh PBS solution.
9. Repeat the **steps 6–8** with a new wash buffer nine times (*see* **Note 2**).
10. With a scalpel, cut the skin in pieces of approximately 0.3 cm × 1 cm.

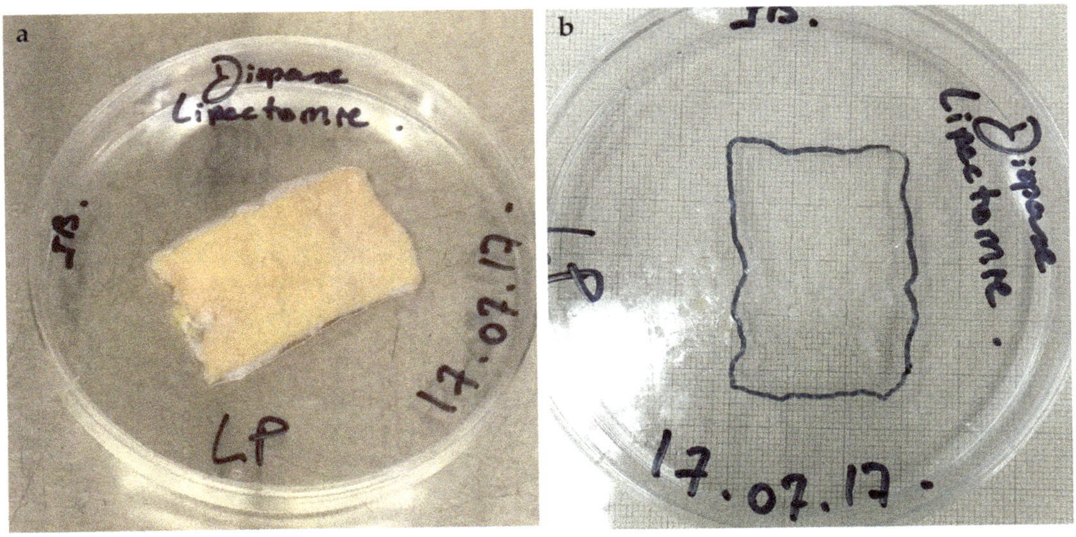

Fig. 1 Appearance of skin sample and surface area calculation. (**a**) Skin sample appearance once spread in a Petri dish and (**b**) example of how to trace the corresponding contours to measure the skin surface area. Knowing the tissue size allows to calculate cell yield per cm^2

11. Recommended: take one piece of the skin for embedding in optimal cutting temperature (OCT) compound for further characterization (*see* **Note 3**).
12. The next steps are optimized for 12 cm^2 of skin or less. If a larger sample is processed, duplicate the process in separate dishes and adapt all the following steps.
13. Transfer the pieces of skin in a new 60 mm Petri dish.
14. Add 10 mL of 1% dispase solution for 12 cm^2 of skin or less. Be careful to cover the whole sample (*see* **Note 4**).
15. Carefully seal the Petri dish with Parafilm® and place it inside a 100 mm Petri dish to avoid spillage and surface contamination with biological material.
16. Incubate overnight in the refrigerator (4–8 °C).

3.2 Day 2: Capillary Extrusion

1. After the overnight incubation, transfer one piece of skin into a new 60 mm Petri dish.
2. While maintaining the dermis in place with tweezers, use thin curved tweezers to delicately take hold of the epidermis and slowly peel it off the dermis. Place the epidermis in a sterile Petri dish containing some DMEM medium if you want to extract the keratinocytes (*see* **Note 5**), otherwise discard it.
3. Place the dermis in a 100 mm Petri dish containing 25 mL of DMEM with 10% FBS and antibiotics. Repeat with each skin piece.

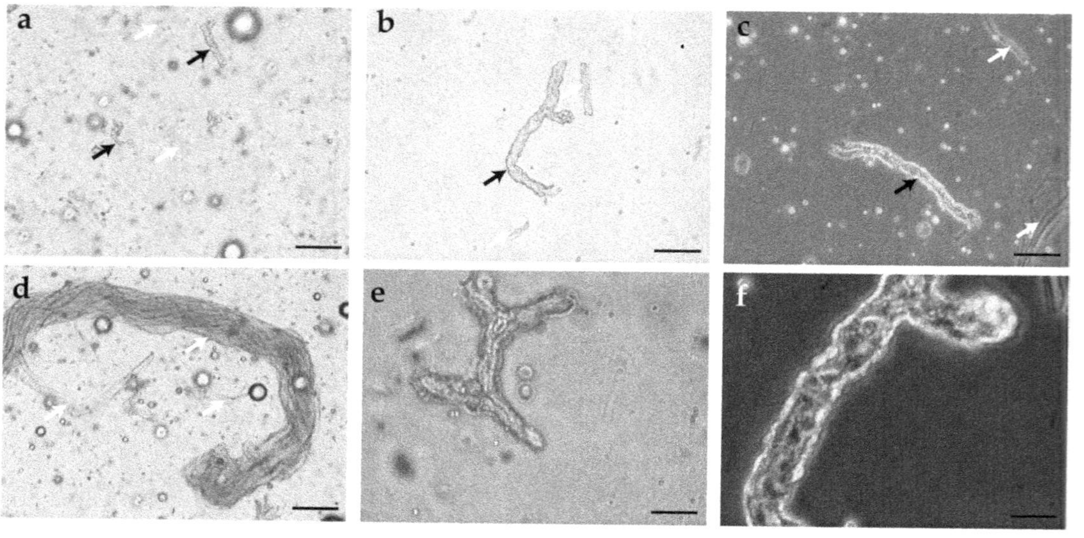

Fig. 2 Aspect of tubules obtained after extrusion from skin. Light-phase microscopy aspect of tubules during the extrusion phase. (**a**–**f**) Many cells and tissue fragments are visible. (**a**–**d**) Tubules are indicated by black arrows, whereas extracellular matrix fragments are indicated by white arrows. (**b**-**c**, **e**–**f**) Different views of extracted tubules featuring a linear or branching morphology. Scale bars: (**a**; **d**) 250 μm, (**b**, **c**) 100 μm, (e) 25 μm, (**f**) 10 μm

4. Transfer one piece of dermis in a 60 mm Petri dish containing 2 mL of EGM™-2MV and extrude the capillaries by applying regular pressure on the dermis using the curved part of the instrument and sliding towards the outside of the piece. The efficacy of the extrusion process can be verified using an optic microscope, as illustrated in Fig. 2 (Fig. 2).
5. After having confirmed the efficacy of your technique for extruding capillaries, proceed similarly for each piece of the dermis. (If you need to isolate the fibroblasts, keep the dermis for further processing after extrusion, *see* **Note 6**).
6. With a pipette, transfer the capillary and cell suspension obtained by extrusion in a 15 mL conical tube.
7. Centrifuge for 10 min at 300 g.
8. Resuspend the pellet, and if possible, proceed to count with a hemacytometer (*see* **Note 7**).
9. Seed 1×10^6 cells per 25 cm^2 flask coated with gelatin (usually seed in one 25 cm^2 flask for foreskin, and in 10–25 cm^2 for 12 cm^2 of adult skin).
10. Incubate at 37 °C and 5–8% CO_2.
11. The following day, rinse the previously seeded cells up to three times with 5 mL EGM™-2MV (for 25 cm^2). This allows to detach dermal pieces and fibroblasts which are weakly attached.

Maintain the cells in culture with 5 mL EGM™-2MV (for 25 cm^2). Change medium three times a week (every 48 h or 72 h).

12. In the subconfluent cultures, fibroblast contamination can be prevented by elimination of colonies of fibroblasts: if possible, identify fibroblast colonies located far enough from neighboring endothelial colonies using a microscope and then, using Pasteur pipets with an aspiration system, remove the supernatant and then aspirate carefully the fibroblast colony. Add 5 mL EGM-2MV and check for the presence of remaining fibroblasts with a microscope. Repeat if necessary (*see* **Note 8**).

3.3 Endothelial Cell Selection/Isolation

Endothelial cell purification can be performed either using immunomagnetic beads or by FACS using an antibody directly coupled to fluorescence and directed against cell surface markers such as CD31 (positive selection). Immunomagnetic purification will be detailed in **steps 3–17**.

1. Remove the culture medium and rinse with 1 mL of trypsin per 25 cm^2 flask. Add 3 mL of trypsin to the cells and incubate at 37 °C until complete cell detachment (3 mL for 25 cm^2 and 8 mL for 75 cm^2 flask).
2. After detachment, inhibit trypsin activity with the same volume of EGM-2MV. Transfer cells into a 15 mL conical sterile tube. Rinse the flask. Centrifuge cells at 300 g for 10 min and perform a cell count.
3. To calculate the bead concentration needed, consider the proportion of fibroblasts in the culture (Fig. 3). If endothelial cells are prominent and appear as numerous large colonies (*see* Fig. 3f), consider a ratio of one bead for one cell. If more fibroblasts are present (*see* Fig. 3), consider reducing the number of beads to a ratio between 1:2 and 1:4 beads per total cells (*see* **Note 9**).
4. Pipet a minimum of 25 μL of beads (equivalent to 1×10^7 beads) and place them in a 5 mL round bottom polystyrene tube containing 2 mL of PBS 0.1% BSA.
5. Mix carefully and place the tube in the magnetic support to trap the beads. Incubate 2 min.
6. While in the support, discard the supernatant by aspiration or pipetting. Be careful not to disturb the beads accumulated along the magnet.
7. Rinse twice with PBS 0.1% BSA.
8. Caution: The following steps need to be performed on ice with cold solutions (*see* **Note 10**).

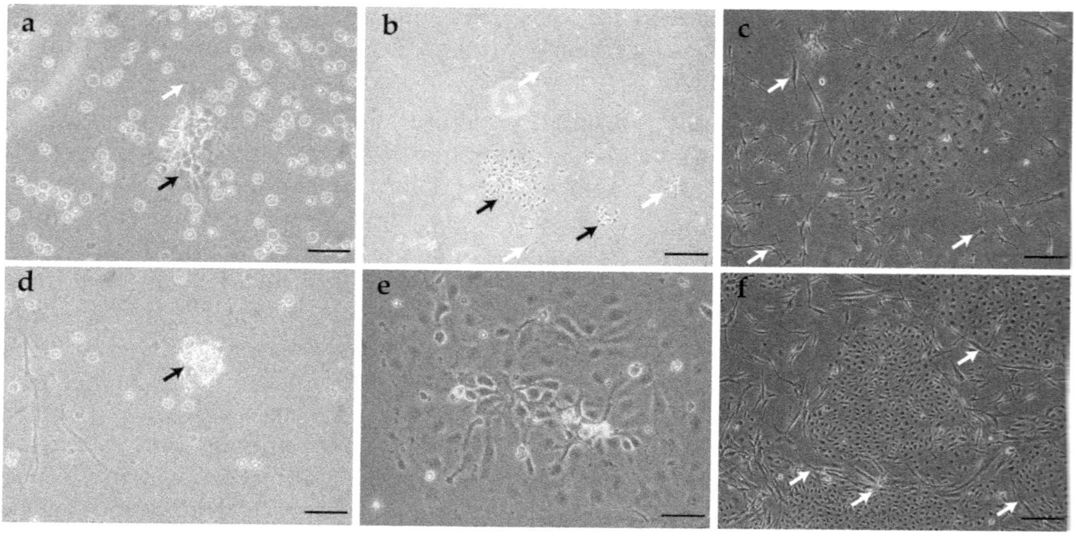

Fig. 3 Formation of endothelial cell colonies from extruded tubules. Migration and proliferation of endothelial cells from the previously extracted tubules after (**a**, **d**) 2 days, (**b**, **e**) 4 days and (**c**, **f**) 10 days. (**e**) Endothelial cells were extracted from a lipectomy skin sample. (**f**) Endothelial cells were extracted from skin obtained after breast surgery. Black arrows indicate endothelial cells whereas white arrows indicate fibroblasts. (**e**) Many fibroblasts are present, surrounding a small endothelial colony. (**f**) Endothelial cells represent the main population but the colonies are surrounded by fibroblasts. Scale bars: (**a**, **c**, **d**, **f**) 100 μm, (**b**) 250 μm, (**e**) 10 μm

9. Resuspend the cells in HBSS with 5% FBS at the minimum concentration of 2×10^6 cells/mL.
10. Add the amount of previously rinsed beads as determined at **step 3**.
11. Incubate on ice for 30 min under slight agitation.
12. After 30 min, dilute two-fold with PBS 0.1% BSA.
13. Place the tube in the magnetic support and incubate 2 min.
14. Aspirate the supernatant (or pipet it). Be careful not to disturb the beads accumulated along the magnet.
15. Remove the tube from the magnetic support.
16. Add 2 mL of PBS 0.1% BSA.
17. Repeat **steps 13–16** four times.
18. Count the cells and plate them at a minimum of 5000 cells/ cm^2 in EGM2-MV (be careful to use culture dishes coated with gelatin). Cell number can be as low as 20% of the previously estimated endothelial cell number (*see* **Note 11**).
19. Change culture medium three times per week. Be careful to passage the cells by trypsinization before they reach confluency to maintain them in a proliferative state.

20. Cells can be frozen at 1–4 × 10^6 cells per mL in freezing medium (we recommend 90% FBS and 10% DMSO, Subheading 2.3, **item 3**).
21. It is recommended to characterize the resulting HMVEC populations according to parameters important for their intended use (*see* **Note 12**).

4 Notes

1. Endothelial cell extraction and efficacy can greatly vary between skin sample types: newborn foreskin, adult mammary skin, or abdominal skin from lipectomy procedures, which is why it is necessary to inspect the sample and note its characteristics such as the presence of hairs or naevi, as well as the color of the dermis (a pink dermis indicates high vascularity). The method described in this chapter can be adapted to pathological skin: at the LOEX center, this protocol was applied to skin from hypertrophic scars as well as from scleroderma patients [7, 8]. Skin surface can be calculated by placing the skin in a Petri dish and drawing its periphery. Cell yield can be evaluated by calculating the number of endothelial cells obtained after immunomagnetic isolation per cm^2. The sample can also be weighted (after adipose tissue removal) to calculate cell yield. Cells can be isolated from a sample as small as 0.5 cm^2.
2. Precautions must be taken to avoid bacterial and fungal contamination since human skin samples may contain many microorganisms. Fresh antibiotics must be added in the solutions and media shortly before use. The sample needs to be thoroughly washed and tools need to be changed as often as necessary to ensure sterility.
3. If the sample is large enough, it might be useful to embed a piece in OCT for cryosectioning and immunolabeling, or fix a sample with formol to perform in toto immunolabelings followed by confocal imaging [9]. This will allow to assess the baseline microvascular density of the native tissue and compare it with results obtained for engineered tissues prevascularized in vitro with endothelial cells, if necessary.
4. An alternative to the dispase digestion step is the use of thermolysin. Instead of incubating the skin overnight in dispase (Subheading 3.1, **step 14**), it can be incubated with a thermolysin solution: dissolve thermolysin at the concentration of 500 μg/mL in 0.01 M HEPES supplemented with 1 mM $CaCl_2$ at pH 7.45 [10]. The epidermis can then be peeled in the same way as with dispase digestion and further processed similarly. The dermis tends to be stiffer after the thermolysin digestion.

5. Keratinocytes can be isolated from the peeled epidermis. At **step 17**, place the epidermis in a Petri dish containing a small volume of trypsin solution (0.05% with 0.01% EDTA). Transfer in a trypsinization unit with trypsin and incubate under agitation at 37 °C for 20 min. To review the complete protocol [11].
6. Instead of discarding the dermis after extrusion, it can be digested with collagenase H, allowing fibroblast isolation from the same sample [11].
7. Cells may be difficult to count as many of them will still be within intact tubules. If you need a precise cell count, you can add a step of trypsin digestion (5 min at 37 °C in a 0.05% trypsin—0.01% ethylenediaminetetraacetic acid (EDTA) solution), on the cell pellet to obtain a higher proportion of single cells in the suspension as this step helps detach the cells from the extracellular matrix and from each other.
8. Fibroblasts feature high proliferation rates and can overgrow and displace endothelial cell colonies. To prevent this, the culture needs to be monitored daily before purification. Endothelial cells migrating from the explants will form colonies. These colonies should not be surrounded by confluent fibroblasts or the latter will proliferate and finally displace endothelial cells. If fibroblasts are growing too fast, an elimination step must be carried out before they overgrow the endothelial cells. If applicable, fibroblasts can be carefully removed by aspiration (*see* Subheading 3.3, **step 1**).
9. For the magnetic selection, the ratio of beads per endothelial cells is a critical parameter. This involves estimating the proportion of contaminating fibroblasts. For example, if mid-size endothelial colonies are present but surrounded by many fibroblasts, you need to consider that you may have as many fibroblasts as endothelial cells. In this case, add one bead for two cells of the total cell count, as you estimate that only one of two cells is an endothelial cell. If the bead number is too low, some endothelial cells will be lost. On the contrary, if too many beads are present, they may impede proper cell growth after selection and more fibroblasts could be present. In your estimation, you need to consider that HMVEC are smaller than fibroblasts and grow at a higher density (for example, in a confluent 75 cm^2 flask you can obtain 2–3 million of fibroblasts whereas you will obtain 4–6 million of endothelial cells).
10. It is of utmost importance to keep cells in cold solutions on ice during the whole purification process. Indeed, fibroblasts tend to intake beads by phagocytosis, which will lead to a mixed population of endothelial cells and fibroblasts, reducing purification efficacy. If needed, the immunoselection process can be repeated after 1–2 passages to increase cell purity.

11. The viability of the isolated endothelial cells can be validated using Trypan blue during cell counting or using a viability marker compatible with flow cytometry analysis.
12. HMVEC can be characterized prior to use by seeding on gelatin-coated glass coverslips followed by immunolabeling for endothelial cell markers such as CD31, von Willebrand factor and VE-cadherin [12]. Cells obtained by the method described in this chapter have been used to produce vascularized tissue-engineered skin and 3D constructs presenting blood capillaries which can be modulated by growth factors, the surrounding microenvironment and even feature associated pericyte-like cells [13–24]. These endothelial cells can also be used for angiogenesis assays [25].

Acknowledgement

The authors acknowledge the precious help and expertise of Sebastien Larochelle, as well as the contribution of Adele Mauroux for the cell isolation. We acknowledge the support of the Centre de recherche du CHU de Québec-Université Laval, of the Fonds de recherche du Québec-Santé (FRQS) and of the Quebec Network for cell, tissue and gene therapies –ThéCell (a thematic network supported by the FRQS).

References

1. Chi J-T, Chang HY, Haraldsen G et al (2003) Endothelial cell diversity revealed by global expression profiling. Proc Natl Acad Sci 100(19):10623–10628
2. Thurston G, Daly C (2012) The complex role of angiopoietin-2 in the angiopoietin–tie signaling pathway. Cold Spring Harb Perspect Med 2(9):a006650
3. Kriehuber E, Breiteneder-Geleff S, Groeger M et al (2001) Isolation and characterization of dermal lymphatic and blood endothelial cells reveal stable and functionally specialized cell lineages. J Exp Med 194(6):797–808
4. Albelda SM (1991) Molecular and cellular properties of PECAM-1 (endoCAM/CD31): a novel vascular cell-cell adhesion molecule. J Cell Biol 114(5):1059–1068
5. Cao G, Fehrenbach ML, Williams JT (2009) Angiogenesis in platelet endothelial cell adhesion Molecule-1-null mice. Am J Pathol 175(2):903–915
6. Richard L, Velasco P, Detmar M (1998) A simple Immunomagnetic protocol for the selective isolation and long-term culture of human dermal microvascular endothelial cells. Exp Cell Res 240(1):1–6
7. Bellemare J, Roberge CJ, Bergeron D et al (2005) Epidermis promotes dermal fibrosis: role in the pathogenesis of hypertrophic scars. J Pathol 206(1):1–8
8. Corriveau M-PP, Boufaied I, Lessard J et al (2009) The fibrotic phenotype of systemic sclerosis fibroblasts varies with disease duration and severity of skin involvement: reconstitution of skin fibrosis development using a tissue engineering approach. J Pathol 217(4):534–542
9. Mayrand D, Fradette J (2018) High definition confocal imaging modalities for the characterization of tissue-engineered substitutes. Methods Mol Biol 1773:93–105
10. Germain L, Rouabhia M, Guignard R et al (1993) Improvement of human keratinocyte isolation and culture using thermolysin. Burns J Int Soc Burn Inj 19(2):99–104
11. Rochon MH, Gauthier MJ, Auger FA et al (2001) Simultaneous isolation of keratinocytes and fibroblasts from a human cutaneous biopsy

for the production of autologous reconstructed skin. Can J Chem Eng 79(4):663–667

12. Arif S, Richer M, Larochelle S et al (2024) Microvesicles derived from dermal myofibroblasts modify the integrity of the blood and lymphatic barriers using distinct endocytosis pathways. J Extracell Biol 3(5):e151
13. Moulin VJ, Mayrand D, Laforce-Lavoie A et al (2011) In vitro culture methods of skin cells for optimal skin reconstruction by tissue engineering. In: Eberli D (ed) Regenerative medicine and tissue engineering - cells and biomaterials, 1st edn. InTech, London
14. Aubin K, Vincent C, Proulx M et al (2015) Creating capillary networks within human engineered tissues: impact of adipocytes and their secretory products. Acta Biomater 11: 333–345
15. Mayrand D, Laforce-Lavoie A, Larochelle S et al (2012) Angiogenic properties of myofibroblasts isolated from normal human skin wounds. Angiogenesis 15(2):199–212
16. Gibot L, Galbraith T, Huot J et al (2010) A preexisting microvascular network benefits in vivo revascularization of a microvascularized tissue-engineered skin substitute. Tissue Eng Part A 16(10):3199–3206
17. Gibot L, Galbraith T, Huot J et al (2013) Development of a tridimensional microvascularized human skin substitute to study melanoma biology. Clin Exp Metastasis 30(1): 83–90
18. Berthod F, Symes J, Tremblay N et al (2012) Spontaneous fibroblast-derived pericyte recruitment in a human tissue-engineered angiogenesis model in vitro. J Cell Physiol 227(5):2130–2137
19. Tremblay P-LL, Hudon V, Berthod F et al (2005) Inosculation of tissue-engineered capillaries with the host's vasculature in a reconstructed skin transplanted on mice. Am J Transplant 5(5):1002–1010
20. Rochon M-HH, Fradette J, Fortin V et al (2010) Normal human epithelial cells regulate the size and morphology of tissue-engineered capillaries. Tissue Eng A 16(5):1457–1468
21. Tremblay P-LL, Berthod F, Germain L et al (2005) In vitro evaluation of the angiostatic potential of drugs using an endothelialized tissue-engineered connective tissue. J Pharmacol Exp Ther 315(2):510–516
22. Proulx M, Safoine M, Mayrand D et al (2016) Impact of TNF and IL-1β on capillary networks within engineered human adipose tissues. J Mater Chem B 4(20):3608–3619
23. Bourland J, Fradette J, Auger FA (2018) Tissue-engineered 3D melanoma model with blood and lymphatic capillaries for drug development. Sci Rep 8(1):13191
24. Proulx M, Mayrand D, Vincent C et al (2018) Short-term post-implantation dynamics of in vitro engineered human microvascularized adipose tissues. Biomed Mater 13(6):065013
25. Merjaneh M, Langlois A, Larochelle S et al (2017) Pro-angiogenic capacities of microvesicles produced by skin wound myofibroblasts. Angiogenesis 20(3):385–398

Chapter 8

Isolation of Stromal Vascular Fraction by Fractionation of Adipose Tissue

Joeri van Boxtel, Mustafa Uguten, Martin C. Harmsen, Hieronymus P. Stevens, and Joris A. van Dongen

Abstract

Adipose tissue-derived stromal cells (ASCs) are promising candidates for cellular therapy in the field of regenerative medicine. ASCs are multipotent mesenchymal stem cell-like and reside in the stromal vascular fraction (SVF) of adipose tissue with the capacity to secrete a plethora of pro-regenerative growth factors. Future applications of ASCs may be restricted through (trans)national governmental policies that do not allow for use of non-human-derived (non-autologous) enzymes to isolate ASC. Besides, enzymatic isolation procedures are also time-consuming. To overcome this issue, non-enzymatic isolation procedures to isolate ASCs or the SVF have been developed, such as the fractionation of adipose tissue procedure (FAT). This standardized procedure to isolate the stromal vascular fraction can be performed within 10–12 min. The short procedure time allows for intra-operative isolation of 1 mL of stromal vascular fraction derived from 10 mL of centrifuged adipose tissue. The stromal vascular fraction mostly contains blood vessels, extracellular matrix, and ASCs. However, based on the histological stainings, an interdonor variation exists which might result in different therapeutic effects. The existing interdonor variations can be addressed by histological stainings and flow cytometry. Furthermore, the re-usable open system has been replaced by a validated disposable semi-closed system with a one-hole fractionator.

Key words SVF, Adipose tissue, Stromal vascular fraction, Adipose tissue-derived stromal cells, ASCs, Fractionation of adipose tissue procedure (FAT)

1 Introduction

Adipose tissue-derived stromal cells (ASCs) are a promising therapeutic cell type for regenerative purposes because of their ability to differentiate in multiple cell lineages and their ability to secrete a plurality of pro-regenerative growth factors [1, 2]. ASCs are multipotent stem cell-like stromal cells, which are abundantly present in adipose tissue and are easily isolated. In adipose tissue, ASCs are attached around vessels as pericytes and supra-adventitial cells in the stromal vascular fraction (SVF) [3, 4]. The SVF of adipose tissue

Thomas Biedermann and Sophie Böttcher-Haberzeth (eds.), *Skin Tissue Engineering: Methods and Protocols*, Methods in Molecular Biology, vol. 2922, https://doi.org/10.1007/978-1-0716-4510-9_8,

contains all non-adipocyte cells (e.g., immune cells, fibroblasts, endothelial cells, smooth muscle cells, ASCs) [5].

The therapeutic potential of ASCs is thoroughly investigated clinically for bone and cartilage repair [6], dermal wound healing and fibrosis [7, 8] and myocardial infarction [9, 10], as well as in non-clinical research for tissue engineering purposes like skin tissue [11] or engineered blood vessels [12]. However, the clinical use of ASCs has become a major challenge because the "classical" enzyme-based isolation methods are legally restricted in many countries. Enzymatic isolation methods are time-consuming procedures which require non-autologous materials such as enzymes and animal derived products [13]. For those reasons, there is an inherent risk of contamination of the isolated ASCs or SVF cells. Moreover, to generate sufficient numbers of ASCs, culturing and expansion of ASCs is needed. The expansion of ASCs for clinical use requires specialized culture labs (Good Manufacturing Practice facilities (cGMP)) which renders the clinical application of ASCs a costly business.

Therefore, non-enzymatic intraoperative isolation procedures to yield a therapeutic cell fraction from adipose tissue are being developed to date [13]. Non-enzymatic, mostly mechanical, isolation procedures are faster and less expensive than enzymatic isolation procedures. Furthermore, non-enzymatic isolation procedures do not require non-autologous biological materials and can therefore be used intraoperatively. Non-enzymatic isolation procedures should not be confused with emulsification procedures that are used to increase the injectability of adipose tissue [13, 14]. In contrast to non-enzymatic isolation procedures, emulsification procedures are not able to disrupt adipocytes. Non-enzymatic isolation procedures often result in a SVF with most of the cell-cell and cell-matrix communications intact (the so-called tissue SVF or tSVF), while enzymatic isolation procedures result in a SVF with only single cells because enzymes disrupt all communication between cells and matrix (the so-called cellular SVF or cSVF) [13]. Clinically, the tSVF and cSVF might have a different therapeutic effect as single cells tend to migrate out of the injection area within the first 24 h after injection [15]. In tSVF, the ASCs are still attached around vessels and embedded in the extracellular matrix, which might result in higher retention rates.

The fractionation of adipose tissue procedure (FAT) yields the tSVF in a non-enzymatic manner [16]. This tSVF is an enrichment of blood vessels, extracellular matrix as well as ASCs by the disruption of adipocytes. The ASCs isolated from the tSVF are not affected in their function, phenotype, and colony formation capability. Moreover, the high amount of extracellular matrix present in the tSVF may serve as a natural scaffold to deliver and guide cells (e.g., ASCs) in their proliferation as well as differentiation. The extracellular matrix in tSVF contains a large number of vessels as well, which can augment vascularization and perfusion. These latter two are important for appropriate wound healing which is frequently impaired in patients suffering from systemic diseases such as diabetes. Therefore, the isolated tSVF by the FAT procedure might be suitable for skin tissue engineering in vivo to augment (diabetic) wound or ulcer healing. Several clinical studies have already shown the beneficial influence of adipose tissue or the stromal vascular fraction on dermal wound healing [8, 17–19].

The first developed FAT procedure was an open-procedure using standard syringes and a reusable three-hole fractionator to disrupt adipocytes. The initial FAT procedure appeared to benefit from several improvements. The three holes of the fractionator were off-centered causing clogging during shuffling. A newly designed disposable fractionator with a centered single hole was validated and compared to the original three-hole fractionator and showed comparable compositions of the obtained SVF [20, 21]. In the initial FAT procedure, the plunger from standard syringes had to be removed before centrifugation. To date, a double-syringe is used to remove oil and infiltration fluid without removing the plunger. A special centrifuge is developed to spin these double syringes with the plunger inside.

2 Materials

2.1 Liposuction of Adipose Tissue

1. Human subcutaneous liposuction derived adipose tissue.
2. 3 mL luer lock syringe.
3. 25 G needle.
4. 2% Lidocaine.
5. Scalpel 11.
6. Spiral multiple-hole infiltration cannula.
7. 50 or 20 mL luer lock syringes to infiltrate.
8. Modified Klein's solution (per 500 mL of saline, 30 mL of xylocaine 2% (epinephrine 1:200000) and 3 mL of natriumbicarbonate 8.4%.
9. 3-port harvesting cannula.
10. VacLok 30 mL syringe.
11. Arthrex Autologous Conditioned Plasma (ACP)® Double-Syringe System.

2.2 Fractionation of Adipose Tissue Procedure

1. Arthrex ACP® Double-Syringe System.
2. Centrifuge (swing out rotor) with the capability to go to $960 \times g$.

3. Luer to luer lock connector transparent transfer adapter with one hole of 1.4 mm inside.
4. Luer to luer lock connector transparent transfer adapter with one hole of 2.4 mm inside.
5. Regular luer lock 10 mL syringe.

2.3 Live/Dead Assay of tSVF

1. 0.001% carboxyfluorescein diacetate succinimidyl ester (CFDA-SE)/serum free Dulbecco's modified eagle's medium.
2. 0.001% propiumiodide (PI)/serum free DMEM.
3. 2% PFA.
4. 4′,6-diamidino-2-phenylindole (DAPI).

2.4 Histological Characterization of tSVF

1. 10% formalin.
2. 1.8% agarose solution.
3. Xylol.
4. 100%, 96%, 70% alcohol.
5. 0.1 M Tris/hydrochloric acid (HCl) buffer pH 9.0.
6. 10 mM Tris/1 mM ethylenediaminetetraacetic acid (EDTA) buffer pH 9.0.
7. 3% hydrogen peroxidase/phosphate buffered saline (PBS).
8. α-Smooth Muscle Actin (SMA, clone 1A4).
9. Perilipin A (Abcam #ab3526).
10. Von Willebrandfactor (vWF, Dako #A0082).
11. 1% Bovine serum albumin (BSA).
12. 1% Human serum.
13. 1% Swine serum.
14. Polyclonal Rabbit anti-Mouse.
15. Polyclonal Goat anti-Rabbit.
16. Polyclonal Swine anti-Rabbit.
17. 3,3'-Diaminobenzidine.
18. Hematoxylin.
19. Mounting solution.

2.5 Masson's Trichrome Staining of tSVF

1. 4% paraformaldehyde (PFA).
2. 1% triton X-100.
3. Bouin fixative:

   36 mL of picric acid (saturated), 12 mL of 37% formalin, 2 mL of acetic acid (glacial).
4. Weighert's iron hematoxylin:

(a) 25 mL of stock solution A.

(b) 25 mL of stock solution B.

(c) Stock solution A:

(i) 0.5 g of hematoxylin, 50 mL of 100% alcohol.

(d) Stock solution B:

(i) 0.6 g of iron chloride, 49.5 mL of demiwater, 0.5 mL of hydrochloric acid.

5. Biebriech scarlet-acid fuchsine:
   0.5 g of Biebriech scarlet, 0.5 g of acid fuchsine, 49.5 mL of demiwater, 0.5 mL of acetic acid (glacial).
6. Phosphomolybdic-phosphotungstic acid:
   2.5 g of phosphomolybdic acid, 2.5 g of phosphotungstic acid, 50 mL of demiwater.
7. Aniline Blue:
   1.3 g of aniline blue, 1 mL of acetic acid (glacial), 49 mL of demiwater.
8. 1% Acetic acid:
   0.5 mL of acetic acid (glacial), 49.5 ml of demiwater.

2.6 Analysis of Histology and Histochemical Stainings

1. Fiji image processing package with ImageJ.
2. NDP.view Image viewing software.
3. Microsoft Excel spreadsheet program.

2.7 Enzymatic Isolation of Adipose Derived Stromal (ASCs) Cells from tSVF

1. 0.1% bacterial collagenase A solution:
   50 mg bacterial collagenase A, 50 mL PBS/1% BSA.
2. 100 μm filters.
3. Lymphoprep.
4. Lysis buffer.
5. Bürker Türk counting chamber.
6. Low glucose culture medium: DMEM, 1.0 g/L glucose, 1% L-glutamin, 1% penicillin/streptomycin, 10% non-heating inactivated fetal bovine serum.

2.8 ASC Differentiation Assays

1. Flask.
2. Pipet.
3. 24-well plate.
4. Adipogenic medium:
   98 mL DMEM +10% Fetal Calf Serum (FCS), 1% p/s, 1% l-glutamin, 1 mL 1% insulin (1 nM), 100 μL dexamethasone (0,1 μM), 1 mL IBMX (0.5 mM).

5. Osteogenic medium:
   98 mL DMEM +10% FCS, 1% p/s, 1% l-glutamin, 100 μL dexamethasone 0,1 μM, 1 mL β-glycerophosphate 10 mM, 50 μL ascorbic acid 0.05 mM.
6. Smooth muscle cells medium:
   100 mL DMEM +10% FCS 1% p/s, 1% l-glutamin, 100 μL Transforming Growth Factor-β.
7. 3% hydrogen peroxidase/PBS.
8. 2% paraformaldehyde (PFA).
9. 60% isopropanol (in dH2O).
10. Phalloidin-AF488 (1:250).
11. 0.5% Alizarin Red.
12. DAPI (1:5000).
13. Hematoxylin.
14. Distilled water.
15. Mounting media.
16. Fluorescent microscope.

2.9 Colony Formation Assay

1. Pipet.
2. 6-well plate.
3. 3% hydrogen peroxidase/PBS.
4. 2% PFA.
5. 0.05% Crystal Violet.
6. Demi water.
7. Tap water.
8. High resolution scanner.
9. ImageJ (public domain).

3 Methods

3.1 Liposuction of Adipose Tissue

1. Infiltrate incision area with 1 mL of lidocaine 2% using the 3 mL luer lock syringe +25 G needle.
2. Make a small stab incision of 5 mm on the donor site (preferably legs or abdomen).
3. Infiltrate the donor site with a modified Klein's solution using the spiral multiple-hole infiltration cannula.
4. Harvest lipoaspirate with the 3-port harvesting cannula and a VacLok 30 mL syringe. Pull the plunger backwards to allow for negative pressure and lock to maintain the negative pressure.

5. Start harvesting by moving the harvesting cannula forward and backwards.
6. Replace the full VacLok 30 mL with a new one and start again from **step 4**.

3.2 Fractionation of Adipose Tissue Procedure

1. Transfer the lipoaspirate to the disposable Arthrex ACP Double-Syringe System from the VacLok 30 mL syringe (Fig. 1).
2. Centrifuge at 960× g for 4 min at room temperature (swing out rotor).
3. After centrifugation, the upper oily fraction is removed by pulling the inner syringe.

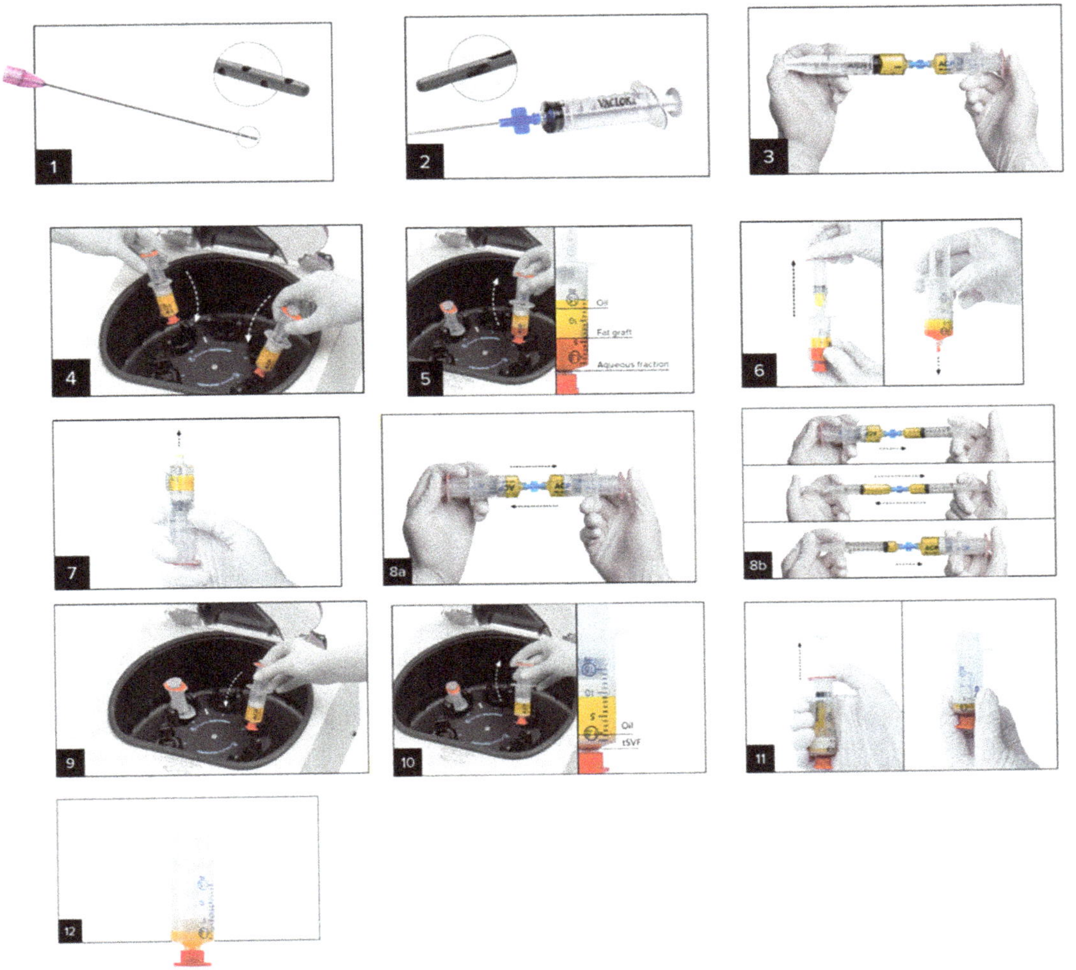

Fig. 1 A general overview of the fractionation of adipose tissue procedure (the FAT procedure, Arthrex GmbH, 2020 ©) to mechanically isolate tSVF

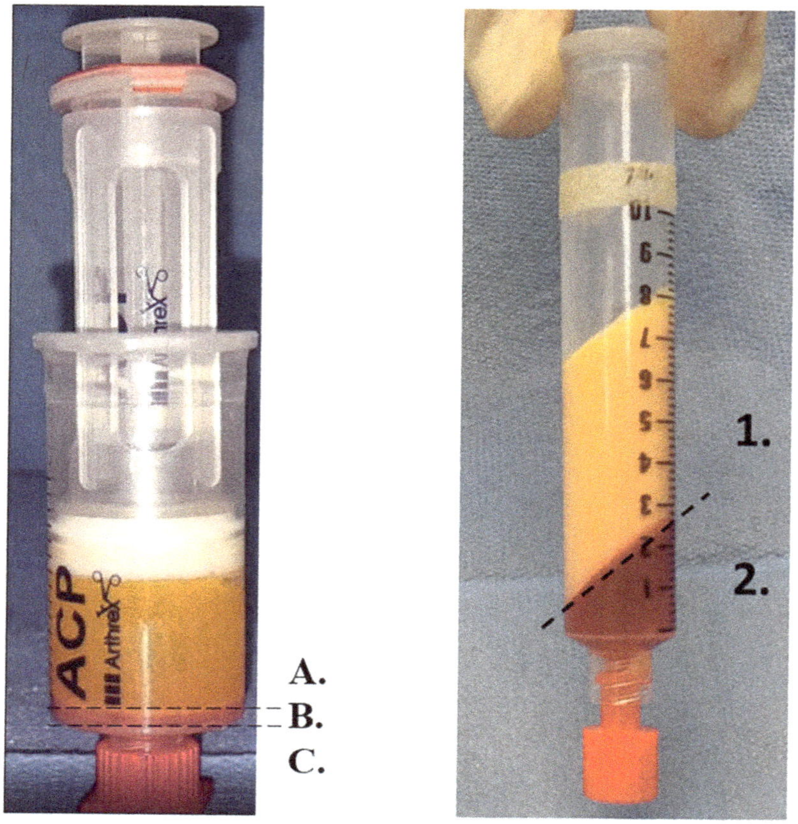

Fig. 2 tSVF in the ACP double Syringe versus emulsified adipose tissue. (**a**) Oily fraction, (**b**) tSVF, (**c**) Pellet layer versus (**1**) Adipose tissue and (**2**) Infiltration fluid

4. The lower aqueous fraction is discarded by opening the red closing cap. Remove oil (disrupted adipocytes) by pulling up the small syringe system in an upright position (*see* **Note 1**).
5. Connect the small syringes to a 2.4 mm transfer adapter and fill a regular 10 mL syringe with 10 mL of centrifuged adipose tissue.
6. Connect the 1.4 mm transfer adapter between an empty 10 mL syringe and a 10 mL syringe with centrifuged adipose tissue. Transfer the fat between the syringes at least 30×. Push the adipose tissue extensively forward and backwards (*see* **Note 2**).
7. Repeat **step 5** and transfer the fractionated fat from the regular 10 mL syringe to the ACP Double-Syringe System.
8. Centrifuge at 960× *g* for 4 min. at room temperature.
9. Transfer the oil into the smaller syringe by slowly pushing down on the outer syringe and simultaneously pulling up the plunger of the inner syringe.

10. Remove the cap to remove the small pellet layer at the bottom.
11. tSVF is left in the outer syringe (Fig. 2).

3.3 Live/Dead Assay of tSVF

1. tSVF is mixed with pre-heated (37 °C) 0.001% CFDA-SE and 0.001% (PI) in serum free DMEM and allow for 30 min. of incubation under normal culture conditions (37 °C).
2. Wash the tSVF with PBS three times.
3. Fix the tSVF with 2% PFA for 30 min.
4. Wash the tSVF with PBS three times.
5. Stain the nuclei DAPI in the dark for 30 min.
6. Wash the tSVF with PBS three times.

3.4 Histological Characterization of tSVF

1. Fix the isolated tSVF in 10% formalin overnight at 4 °C.
2. Embed the tSVF in a pre-heated (60 °C) 1.8% agarose solution (1:2).
3. Place the tSVF/agarose solution at 4 °C for 30 min.
4. Dehydrate samples with the following steps in sequence at room temperature:
   (a) 50% alcohol for 30 min.
   (b) 70% alcohol for 30 min.
   (c) 96% alcohol for 30 min.
   (d) Twice in 100% alcohol for 30 min.
   (e) Twice in Xylol for 30 min.
5. Embed the samples in paraffin.
6. Cut 4 mm sections and deparaffinize them in xylol for 15 min.
7. Refresh the xylol and place the samples for another 10 min in xylol.
8. Move the samples and place them in 100% alcohol for 10 min, then in 96% alcohol for 3 min, and finally in 70% alcohol for 3 min at room temperature.
9. Wash the samples in demi water for 3 min.
10. Incubate samples overnight with 0.1 M Tris/HCl buffer (pH 9.0) at 80 °C for α-SMA as well as for Perilipin A staining and with 10 mM Tris/1 mM EDTA buffer (pH 9.0) at 80 °C for vWF.
11. Cool down to room temperature for 30 min.
12. Wash samples with PBS three times.
13. Endogenous peroxidase activity is blocked with 3% hydrogen peroxidase in PBS at room temperature for 30 min.
14. Wash samples with PBS three times and incubate samples with primary antibodies at room temperature for 60 min.: for

α-SMA (1:200) + 1% BSA + 1% Human serum in PBS, for Perilipin A (1:200) + 1% BSA + 1% Human serum in PBS, for vWF (1:200) + 1%BSA + 1% Swine serum in PBS.

15. Wash samples with PBS three times.
16. Incubate samples with secondary antibodies for 30 min. at room temperature: polyclonal Rabbit anti-Mouse (1: 100) + 1% BSA + 1% Human serum in PBS for α-SMA, polyclonal Goat anti-Rabbit (1:100) + 1% BSA + 1% Human serum in PBS for Perilipin A and polyclonal Swine anti-Rabbit (1: 100) + 1% BSA + 1% Human serum in PBS for vWF.
17. Wash samples with PBS three times.
18. Incubate α-SMA sample with a third antibody at room temperature for 30 min with polyclonal Swine anti-Rabbit (1: 100) + 1% BSA + 1% Human serum in PBS.
19. Wash α-SMA samples with PBS three times.
20. Incubate all samples with DAB for 10 min at room temperature in the dark.
21. Wash all samples in demi water three times for 5 min.
22. Incubate all samples with hematoxylin for 1.5 min.
23. Rinse samples in water for 4 min.
24. Mount all the samples and place a coverslip (Subheading 3.6).

3.5 Masson's Trichrome Staining of tSVF

1. Follow **step 1** till 9. of Subheading 3.3.
2. Fix samples in 4% PFA for 60 min.
3. Permeabilize samples in 1% triton X-100 for 10 min.
4. Wash samples with PBS three times.
5. Fix samples in Bouins fixative at 51 °C for 5 min.
6. Wash samples in demiwater until color disappears.
7. Incubate samples with Weighert's iron hematoxylin for 20 min.
8. Wash samples in demiwater until color disappears.
9. Incubate samples with Beibrich Scarlet-acid fuchsine for 20 min.
10. Wash samples in demiwater for 2 min.
11. Incubate samples in phosphomolybdic-phosphotungstic acid for 12 min.
12. Wash samples in demiwater until color disappears.
13. Incubate samples in aniline blue for 7 min.
14. Wash samples in demiwater for 2 min.
15. Incubate samples in 1% acetic acid for 5 s.
16. Wash samples in demiwater until color disappears.

17. Dry samples for 20 min.
18. Mount samples in Permount.

3.6 Analysis of Histology and Histochemical Stainings

1. Divide all the different microscopic images of different patients in different folders.
2. Make an excel file for every different staining (e.g., VWF, αSMA).
3. Open a patient in NDP.view (drag the patient file in the viewer).
4. Scroll and zoom around for a representative selection of cells (i.e., a sample) and safe it as a JPEG (three times per patient to rule out a selection bias).
5. Drag the JPEG files to ImageJ.
6. Use "polygon selections" in ImageJ to cut out your sample.
7. Select the RGB stack for the αSMA, Masson's Trichome, Perilipin A and vWF staining (Fig. 3).
8. Select a threshold, which has to be about the same for all the samples of all the patients of the same staining. Safe the threshold per patient in the Excel.
9. Analyze and measure your sample.
10. Save the measurements in your excel and then calculate a mean percentage area.

3.7 Enzymatic Isolation of Adipose Derived Stromal (ASCs) Cells from tSVF

1. Wash the isolated tSVF with PBS three times.
2. Add 0.1% collagenase A solution 1:1 with tSVF.
3. Stir the collagenase/tSVF mixture in a water bath at 37 °C for 1.5 h.
4. Centrifuge the sample at 600× *g* at room temperature for 10 min.
5. Remove supernatant.
6. Collect cell pellet in PBS/1% BSA.
7. Filter the collagenase/tSVF mixture through filters.
8. Centrifuge the sample at 600× *g* at room temperature for 10 min.
9. Repeat **step 6**, 7, and **8**.
10. Remove supernatant.
11. Collect cell pellet in 30 mL of PBS/1% BSA.
12. Put 15 mL of lymphoprep in a 50 mL tube.
13. Gently add the 30 mL of PBS/1% BSA with the sample.
14. Centrifuge the sample at 1000× *g* at 4 °C for 20 min and put the brake on 0.

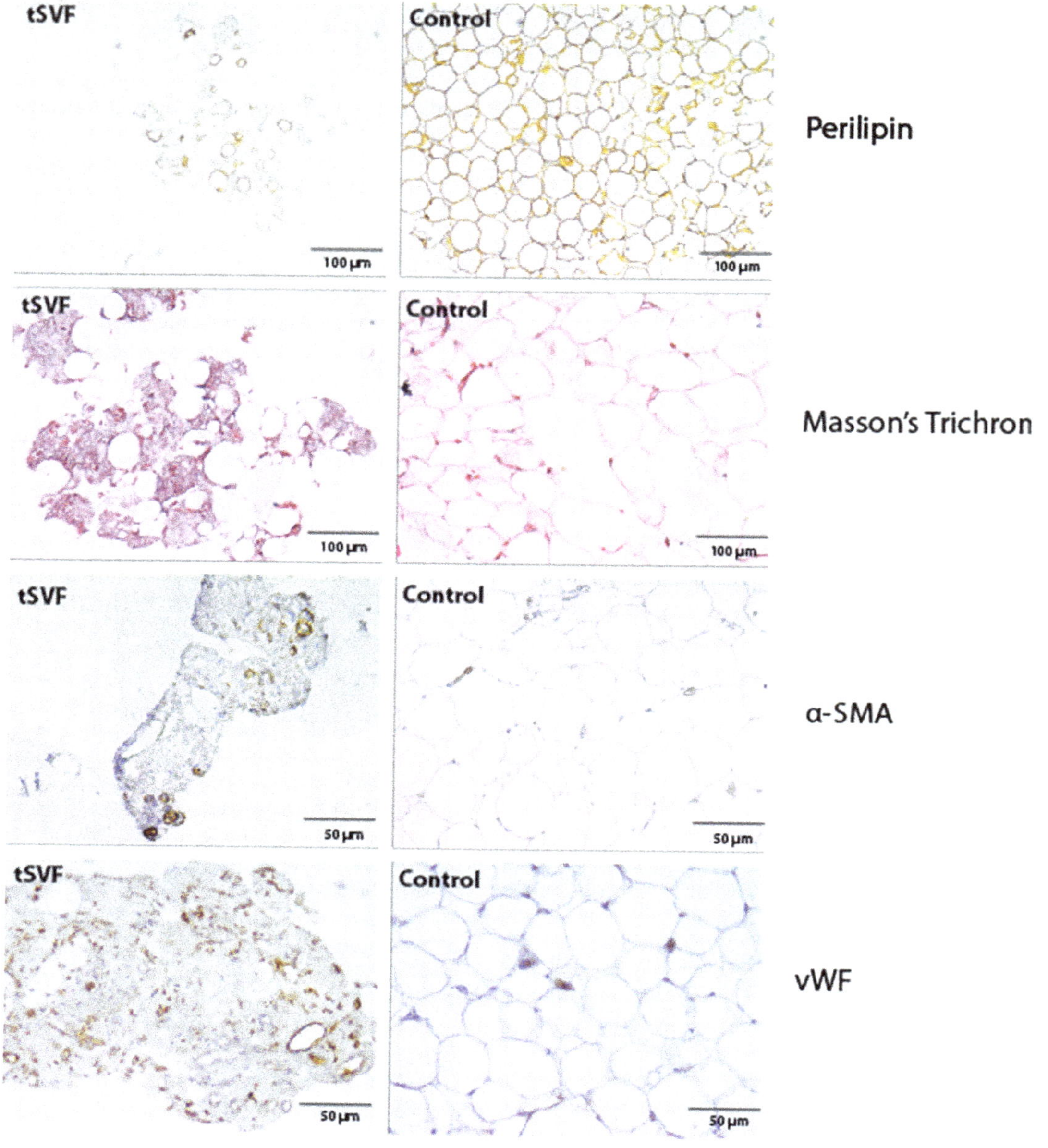

Fig. 3 Immunohistochemistry staining examples of how tSVF and unprocessed adipose tissue (control) should look like when a Perilipin A, Masson's Trichrome, α-Smooth Muscle Actin (α-SMA), and von Willebrandfactor (vWF) stainings are performed

15. Remove the upper layer.
16. Take cells from the interphase carefully.
17. Add PBS/1% BSA to the cells.
18. Centrifuge the sample at 800× g at 8 °C for 10 min.
19. Remove supernatant.

20. Resuspend cell pellet in lysis buffer and place the sample on ice for 5 min.
21. Centrifuge the sample at 800× *g* at 8 °C for 10 min.
22. Remove supernatant.
23. Repeat **step 19** and **23**. till all erythrocytes are disrupted and the red color has disappeared.
24. Resuspend cells in culture medium.
25. Dilute 40 μL cell suspension with 160 μL tryptan blue (filter before use).
26. Count the living cells with a Bürker Türk. Count 2 × 25 small squares and take the average.
27. Amount of cell/mL = outcome x 5 × 10,000.
28. Culture cells in culture medium.

3.8 ASC Differentiation Assays

1. Make 13 mL cell suspension of 40.000 cells/mL.
2. Pipet 0.5 mL in each well (± 2 cm2) of a 24-well plate (10.000 cells/cm^2).
3. Wait until they are confluent.
4. Add specific differentiation medium for adipocyte differentiation, osteoblast differentiation, and smooth muscle cells differentiation.
5. Culture for 2 weeks, refresh differentiation medium twice a week.
6. Fix cells with 2% PFA for 15 min.
7. Wash with PBS.

3.9 Staining for Adipocytes

1. Wash with tap water after fixation 3 × 2–3 min.
2. Rinse with 60% isopropanol (in dH_2O).
3. Rinse with 60% isopropanol.
4. Lightly stain nuclei with hematoxylin.
5. Rinse with distilled water.
6. Mount with aqueous mounting media.

3.10 Staining for Osteoblasts

1. Soak cells in 0.5% Alizarin Red 10 min.
2. Wash cells with PBS.
3. Lightly stain nuclei with hematoxylin.
4. Rinse with distilled water.
5. Mount with aqueous mounting media.

3.11 Staining for Smooth Muscle Cells

1. Wash cells with PBS after fixation.
2. Incubate with Phalloidin-AF488 (1:250) in DAPI (1:5000) for 30 min.
3. Wash with PBS 2x.
4. Keep PBS on the samples.

3.12 Colony Formation Assay

1. Make 8 mL cell suspension of 500 cells/mL.
2. Pipet 2 mL in the lower 3 wells of a 6-well plate.
3. Take 800 μL of the left over cell suspension and pipet it into 8 mL fresh medium (50 cells/mL).
4. Pipet 2 mL in the other 3 wells.
5. Leave the cells for 2 weeks, refresh medium twice a week.
6. Wash the cells with PBS.
7. Fixate for 15 min with 2% PFA.
8. Wash with PBS.
9. Stain for 30 min with 0.05% Crystal Violet in demi water (filter at 0.45 μm before use).
10. Wash with tap water.
11. Let the plate dry inverted for a couple of min.
12. The plate can then be photographed or scanned for colony counts (add LMP low melting point agarose to visualize colonies).
13. Count the number of colonies in each well consisting of more than 50 cells (should be 0.1–5%) using ImageJ.

4 Notes

1. The fractionation of adipose tissue procedure only works when the harvested adipose tissue is separated from all the infiltration fluid and oil by centrifugation at a high speed (960× *g*). When small amounts of infiltration fluid are left behind, none of the adipocytes will be disrupted and therefore no oil will appear after the final centrifugation step. In case fibrotic adipose tissue is processed by the fractionation of adipose tissue procedure, the fractionator can clog (Fig. 1).
2. In case some infiltration fluid is left behind, the harvested adipose tissue will turn white when the harvested adipose tissue is pushed forward and backwards through the Luer-to-Luer connector. After the second round of centrifugation, no oil will appear and therefore the stromal vascular fraction cannot be isolated. This is called emulsified adipose tissue; the liquid content is mixed with the adipose tissue content and allows for a better injectable fraction (Fig. 2).

References

1. Zuk PA, Zhu M, Mizuno H et al (2001) Multilineage cells from human adipose tissue: implications for cell-based therapies. Tissue Eng 7(2):211–228
2. Roux S, Bodivit G, Bartis W et al (2015) In vitro characterization of patches of human mesenchymal stromal cells. Tissue Eng Part A 21(3–4):417–425
3. Lin G, Garcia M, Ning H et al (2008) Defining stem and progenitor cells within adipose tissue. Stem Cells Dev 17(6):1053–1063
4. Corselli M, Chen CW, Sun B et al (2012) The tunica adventitia of human arteries and veins as a source of mesenchymal stem cells. Stem Cells Dev 21(8):1299–1308
5. Bourin P, Bunnell BA, Casteilla L et al (2013) Stromal cells from the adipose tissue-derived stromal vascular fraction and culture expanded adipose tissue-derived stromal/stem cells: a joint statement of the International Federation for Adipose Therapeutics and Science (IFATS) and the International Society for Cellular Therapy (ISCT). Cytotherapy 15(6):641–648
6. Grayson WL, Bunnell BA, Martin E et al (2015) Stromal cells and stem cells in clinical bone regeneration. Nat Rev Endocrinol 11(3): 140–150
7. Spiekman M, van Dongen JA, Willemsen JC et al (2017) The power of fat and its adipose-derived stromal cells: emerging concepts for fibrotic scar treatment. J Tissue Eng Regen Med 11(11) 3220–3235. https://doi.org/10.1002/term.2213
8. Bura A, Planat-Benard V, Bourin P et al (2014) Phase I trial: the use of autologous cultured adipose-derived stroma/stem cells to treat patients with non-revascularizable critical limb ischemia. Cytotherapy 16(2):245–257
9. Jiang Y, Chang P, Pei Y et al (2014) Intramyocardial injection of hypoxia-preconditioned adipose-derived stromal cells treats acute myocardial infarction: an in vivo study in swine. Cell Tissue Res 358(2):417–432
10. Chen L, Qin F, Ge M et al (2014) Application of adipose-derived stem cells in heart disease. J Cardiovasc Transl Res 7(7):651–663
11. Klar AS, Zimoch J, Biedermann T (2017) Skin tissue engineering: application of adipose-derived stem cells. Biomed Res Int 2017: 9747010
12. Parvizi M, Bolhuis-Versteeg LA, Poot AA et al (2016) Efficient generation of smooth muscle cells from adipose-derived stromal cells by 3D mechanical stimulation can substitute the use of growth factors in vascular tissue engineering. Biotechnol J 11(7):932–944
13. van Dongen JA, Tuin AJ, Spiekman M et al (2018) Comparison of intraoperative procedures for isolation of clinical grade stromal vascular fraction for regenerative purposes: a systematic review. J Tissue Eng Regen Med 12(1):e261–e274
14. Tonnard P, Verpaele A, Peeters G et al (2013) Nanofat grafting: basic research and clinical applications. Plast Reconstr Surg 132(4): 1017–1026
15. Parvizi M, Harmsen MC (2015) Therapeutic Prospect of adipose-derived stromal cells for the treatment of abdominal aortic aneurysm. Stem Cells Dev 24(13):1493–1505
16. van Dongen JA, Stevens HP, Parvizi M et al (2016) The fractionation of adipose tissue procedure to obtain stromal vascular fractions for regenerative purposes. Wound Repair Regen 24(6):994–1003
17. Stasch T, Hoehne J, Huynh T et al (2015) Debridement and autologous Lipotransfer for chronic ulceration of the diabetic foot and lower limb improves wound healing. Plast Reconstr Surg 136(6):1357–1366
18. Raposio E, Bertozzi N, Bonomini S et al (2016) Adipose-derived stem cells added to platelet-rich plasma for chronic skin ulcer therapy. Wounds Compend Clin Res Pract 28(4): 126–131
19. Conde-Green A, Marano AA, Lee ES et al (2016) Fat grafting and adipose-derived regenerative cells in burn wound healing and scarring: a systematic review of the literature. Plast Reconstr Surg 137(1):302–312
20. Van Dongen JA, Gostelie OFE, Vonk LA et al (2020) Fractionation of adipose tissue procedure with a disposable one-hole fractionator. Aesthet Surg J 40(4):NP194–NP201
21. van Boxtel J, Vonk LA, Stevens HP et al (2022) Mechanically derived tissue stromal vascular fraction acts anti-inflammatory on TNF alpha-stimulated chondrocytes in vitro. Bioengineering (Basel) 9(8):345

Part II

Tissue Engineering of Skin and Engineered Skin Models

Chapter 9

Engineering a Pigmented Skin Equivalent That Is Responsive to External Stimuli

Paola De Los Santos Gomez, Kirsty Goncalves, Victoria Maltman, Lucy Smith, and Stefan Przyborski

Abstract

Recently, dermatological research has more widely adopted the use of human skin equivalents (HSEs) to better recreate skin structure and function in vitro, which can act as a preclinical tool. However, many popular HSEs contain only two main cell types: keratinocytes and fibroblasts to model epidermal and dermal compartments respectively. This lack of supporting cell types can be associated with a wide range of limitations, most notably an inaccurate representation of skin's innate response to exogenous stressors such as UV radiation. We have adapted our novel full-thickness skin platform to incorporate human melanocytes that produce melanin and transfer melanin to neighboring keratinocytes, where it is located apically to the nucleus, forming typical supranuclear protective caps. The use of Alvetex® scaffold and the ability of dermal fibroblasts to secrete their own endogenous extracellular matrix (ECM) proteins provide a robust dermal foundation for the construction of a pigmented epidermis and are essential for supranuclear cap formation, a physiological phenomenon essential to melanin's protective function. This model system is responsive both to up and downregulation of melanogenesis, therefore providing an in vitro platform for a wide array of applications, ranging from industrial active testing for cosmetic formulations to academic insights into the cellular response to environmental stressors.

Key words Pigmentation, Skin tone, Ultraviolet radiation (UVR), Skin tissue engineering, Melanin, Skin equivalent, Melanocytes

1 Introduction

Human skin equivalents (HSEs) are a bioengineered tissue construct that offer an in vitro alternative to human or animal skin in vivo. They closely recapitulate many aspects of skin structure and function and offer a wide array of potential academic and industrial applications: disease modelling [1], drug discovery [2, 3], ageing research [4, 5], screening of actives [6, 7] and the impact of

---

Paola De Los Santos Gomez and Kirsty Goncalves have equally contributed to this chapter.

Thomas Biedermann and Sophie Böttcher-Haberzeth (eds.), *Skin Tissue Engineering: Methods and Protocols*, Methods in Molecular Biology, vol. 2922, https://doi.org/10.1007/978-1-0716-4510-9_9,

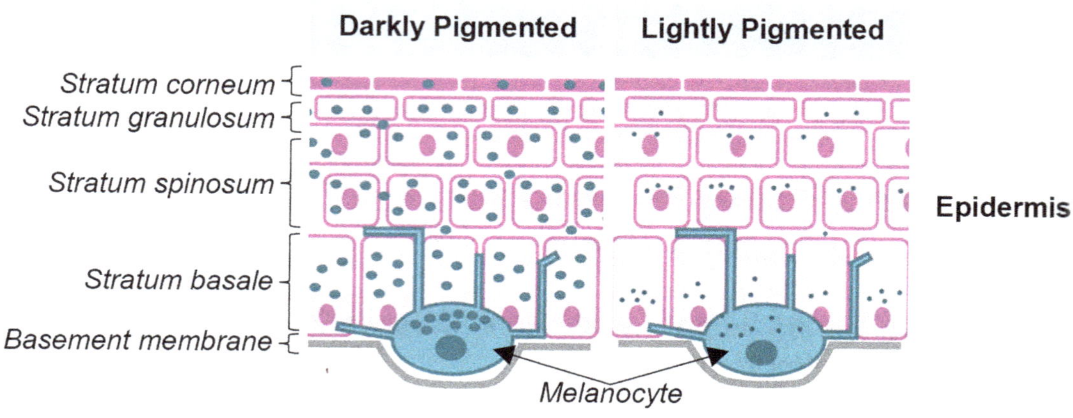

Fig. 1 Differences in melanin distribution between light and darkly pigmented skin. Melanocytes reside in the basal layer of the epidermis and exist as a unit with multiple keratinocytes. Melanocytes produce melanin, packaged in specialized organelles known as melanosomes that are transferred to neighboring keratinocytes. Melanokerasomes are much larger in keratinocytes of darker skin and are more resistant to breakdown and persist into the uppermost layers of the epidermis. In contrast in lightly pigmented skin, melanokerasomes are smaller, cluster above the nucleus and rarely persist to the *stratum granulosum* and *stratum corneum*

environmental stressors [4, 8] on skin health. Although HSEs provide a more physiologically relevant in vitro approach to dermatological research, the majority only contain two distinct cell types: keratinocytes and fibroblasts [9, 10]. While these represent the two main compartments of skin tissue (epidermis and dermis respectively), in vivo, human skin is a much more complex organ, comprising of a range of supporting cell types and appendages [11]. In particular, this limits their application to investigate the impact of environmental stressors, most notably ultraviolet radiation (UVR) on skin health, as they lack innate protection and their response is not representative of native tissue.

Melanocytes are specialized cells that produce eumelanin and pheomelanin pigments [12], are located in the *stratum basale* of the epidermis [13, 14] and are anchored at the basement membrane (Fig. 1). Melanin pigments are generated by melanocytes and packaged in specialized organelles termed *melanosomes* that are transferred to neighboring keratinocytes though the melanocyte dendrites [11, 15]. Once within the keratinocytes, melanin is organized in groups of pigment cores or as isolated pigment cores surrounded by a single outer membrane [16], and then is located at the perinuclear area as supranuclear caps (SPNCs) that protect DNA from UVR-induced damage [11, 17]. Recently, the melanin-containing organelles in keratinocytes have been termed melanokerasomes [55]. The size, number, and organization of such deposits, along with their distribution throughout the epidermal layers and resistance to degradation, are factors that differ among skin tones [18, 19] (Fig. 1). Skin pigmentation is not determined by differences in the number of melanocytes in the skin but from differences in the types of melanin produced, melanogenic activity, along with size, number, and packaging of melanokerasomes [20]. For example, darkly pigmented skin can contain three- to sixfold higher levels of melanin, more eumelanin and a higher number of large melanokerasomes compared with lightly pigmented skin. Melanokerasomes in dark skin are also more resistant to degradation and therefore remain intact throughout the epidermis [20].

The main function of melanin deposits is photoprotection, achieved through the formation of SPNCs within keratinocytes, which physically block the passage of UVR to the nucleus, where it can damage cellular DNA and impact skin health [21–23]. SPNCs absorb UVR, preventing negative consequences such as formation of pyrimidine dimers and DNA damage, which can ultimately lead to sunburn and skin cancers [24–26]. For this reason, there is a strong correlation between melanin content, skin tone, and photodamage susceptibility [27].

Advancing technology has allowed for increasing complexity of standard HSE platforms, most notably the incorporation of additional, supporting cell types into the basic keratinocyte/fibroblast/extracellular matrix (ECM) culture foundation [28, 29]. Incorporation of cell types such as sensory neurons and immune components has led to an expanding range of applications for HSEs, particularly within the space of skin inflammation [29–31]. However, most notably, the emergence of melanocyte-containing pigmented HSEs has provided tools capable of investigation into the impact of external stimuli on skin health [28, 31–34], including UVR, pollution, mechanical trauma, cosmetic screening, and academic insights.

1.1 Accurately Recapitulating Pigmented Epidermal Microanatomy In Vitro

Although the emergence of both non-pigmented HSEs and melanocyte-containing pigmented HSEs is becoming increasingly frequent, the re-creation of the correct microanatomy, melanin distribution, and SPNC formation is still a rarity despite being essential for melanin's primary function of photoprotection. The most ubiquitous pigmented HSEs model the epidermal compartment alone (epidermal-only HSE, EO-HSE) and are available from both commercial sources: SkinEthic™ RHPE (Episkin, L'oreal, France), and developed academically [35–40]. While EO-HSE offer a reductionist approach providing information pertaining to epidermal biology, skin in vivo is a complex tissue impacted by dynamic interactions between cell types.

The absence of a dermal compartment within pigmented EO-HSEs can have a significant impact on experimental outcomes as a consequence of the absence of cellular interactions between the epidermal and dermal compartments. At a structural level, the importance of melanocyte-fibroblast crosstalk can be demonstrated by a lack of melanin organization and most notably, absence of SPNC formation or localization of melanin to the perinuclear area of keratinocytes (Fig. 2). Multiple methods of interaction between melanocytes and fibroblasts have been reported, including the secretion of soluble factors from fibroblasts that impact melanocyte biology in a paracrine fashion [41] such as neurogulin-1 [42], *Dikkopf-1* [42], and keratinocyte growth factor (KGF) [43]. Other contributions of dermal fibroblasts to skin tone and pigment production have also been evidenced including physical interactions between melanocytes and fibroblast secreted ECM [44, 45] and basement membrane formation [45]. The true role of fibroblasts in establishment of skin pigmentation is likely to be multifactorial, combining multiple mechanisms of interaction, and supporting the notion that increasing a HSEs complexity through incorporation of additional cell types, adds to its physiological relevance.

While full-thickness HSEs (FT-HSEs) aim to recapitulate both epidermal and dermal compartments, many utilize the exogenous addition of ECM constituents which are often animal derived [8, 46]. Use of animal derived products often contributes to lack of reproducibility, batch to batch variation and raises questions as to the biocompatibility of interspecies cultures. Furthermore, due to legislation associated with the reduction in use of animals for cosmetic testing, use of animal derived components often limits the application of such tissue constructs for industrial pursuits, particularly as a pre-clinical tool.

An adaptation to the Alvetex® based FT-HSE described within Chap. 12 has allowed for the successful incorporation of melanocytes into the epidermal compartment [47]. The use of a porous, polystyrene scaffold such as Alvetex® is associated with many advantages such as being inert, therefore eliminating biochemical interactions between the substrate and cell types from which the tissue is constructed. The most notable property of this system is that the scaffold provides support and is populated by human dermal fibroblasts that neosynthesize endogenous ECM, eliminating the need for animal-derived ECM constituents. This feature improves the reproducibility of the system and promotes the formation of a de novo basement membrane, an event that is essential for correct melanocyte localization and ultimately functionality.

Incorporation of melanocytes into the epidermal compartment of the Alvetex® based FT-HSE was achieved through simultaneous seeding of keratinocytes and melanocytes within the same solution, onto a mature dermal foundation [28]. It is evident that the two epidermal cell types self-organize and orientate with melanocytes situated in the *stratum basale* and in contact with the basement membrane (Fig. 3). Melanin produced by melanocytes is then transferred to neighboring keratinocytes, where it localizes in the

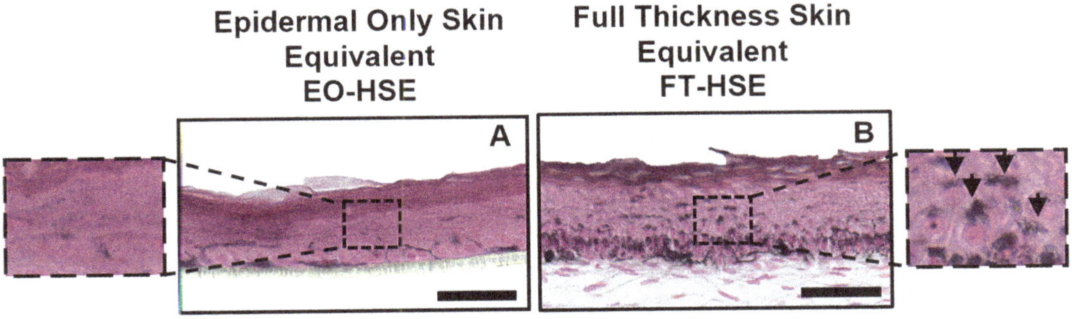

Fig. 2 Prescence of a solid dermal foundation is essential for physiological melanin localization. Fontana–Masson staining highlights melanin distribution and demonstrates diffuse, punctate staining within keratinocytes in epidermal only skin equivalent (EO-HSE) (**a**). Whereas in the full-thickness skin equivalent (FT-HSE) (**b**), in the presence of dermal fibroblasts and endogenous extracellular matrix, melanin staining appears in the perinuclear region in typical cap structures. Supranuclear caps (SPNCs) are visible in keratinocytes within the FT-HSE and are highlighted by arrows, formation of which is essential for melanin's primary role in photoprotection. Scale bars: 50 μm

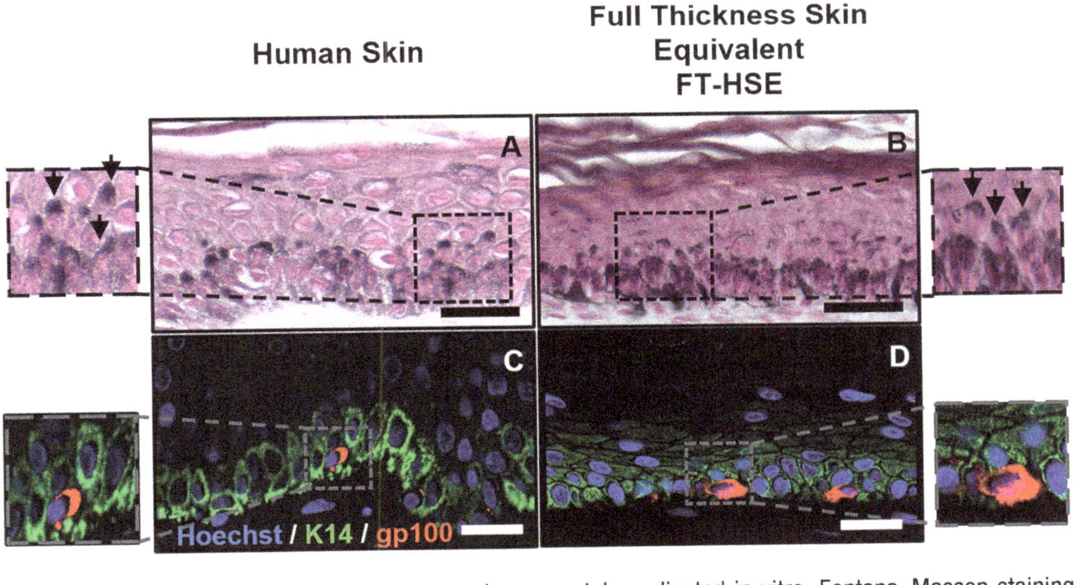

Fig. 3 Pigmented epidermal microanatomy can be accurately replicated in vitro. Fontana–Masson staining demonstrates the perinuclear localization of melanin within keratinocytes forming typical supranuclear caps (SPNCs) (arrows), present in both native human skin (**a**) and bioengineered full-thickness constructs (**b**). Immunoreactivity for the melanocyte-specific biomarker gp100 (red) demonstrates the correct position of melanocytes along the basement membrane and situated in the *stratum basale* in both native (**c**) and bioengineered tissues (**d**). Basal keratinocytes are highlighted in green through staining for keratin 14 and nuclei are highlighted in blue. Scale bars: 25 μm

perinuclear area as typical SPNCs, akin to that of native human skin (Fig. 3). This specific distribution of melanin granules within keratinocytes of the epidermis is essential for functionality and is a feature lacking in other pigmented HSEs described in the literature [32, 33, 48] or commercially available.

1.2 Tissue Responsiveness to Exogenous Stimuli: Melanogenesis Modulation

The relationship between cell/tissue structure and function is the foundation of a basic biological principle, applicable to the field of bioengineering. The successful recapitulation of pigmented epidermal architecture is essential for the bioengineered, pigmented FT-HSE to fulfill its function as an in vitro model system, capable of responding to external stimuli in a physiologically representative manner.

The presence of a strong dermal foundation comprising of dermal fibroblasts and endogenous ECM built upon Alvetex® scaffold is an essential step in the construction of the pigmented FT-HSE, which ultimately has a wide variety of applications. Not only does the melanin distribution reflect that of native human skin, but the tissue construct itself is dynamic and responsive to stimuli known to influence melanogenesis and melanin dynamics in vivo, an example of which is the ability of kojic acid, an active often incorporated into cosmetic products, to downregulate melanin production in the tissue construct. Kojic acid's mechanism of action is to inhibit tyrosinase, the rate limiting enzymatic step in melanin production, reducing the action of melanocytes to produce melanin pigments, therefore reducing tissue pigmentation [49, 50]. This is demonstrated in vitro through reduced pigmentation at a gross level and reduced melanin index (Fig. 4).

Conversely, external stimuli, most notably exposure to UVR, are known to increase melanin production, resulting in "tanning" [24, 25, 51]. In vitro, this is demonstrated as increased melanin content within the tissue and a gross increase in tissue pigmentation (Fig. 5). One of the causative factors that increases melanogenesis post-UVR exposure is the release of pro-inflammatory mediators, secretion of which is also induced by mechanical insult. This gives rise to a phenomenon known as post-inflammatory hyperpigmentation (PIHP), which is an unwanted consequence of wound healing in individuals with darker tones in vivo [52–54]. This has also been demonstrated in vitro following a scratch wound to the surface of the pigmented tissue and in mechanical exfoliation through tape stripping of the epidermis which gradually over a period of 7 and 4 days, respectively, resulted in hyperpigmentation at the lesion site accompanied by an increase in melanin granules (Fig. 5).

These applications not only demonstrate the dynamic nature of the bioengineered tissue construct to respond to known stimuli that impact melanogenesis in a physiological manner, but also demonstrate the extent to which a pigmented HSE can be utilized as a platform technology. The concept of re-creating the anatomy

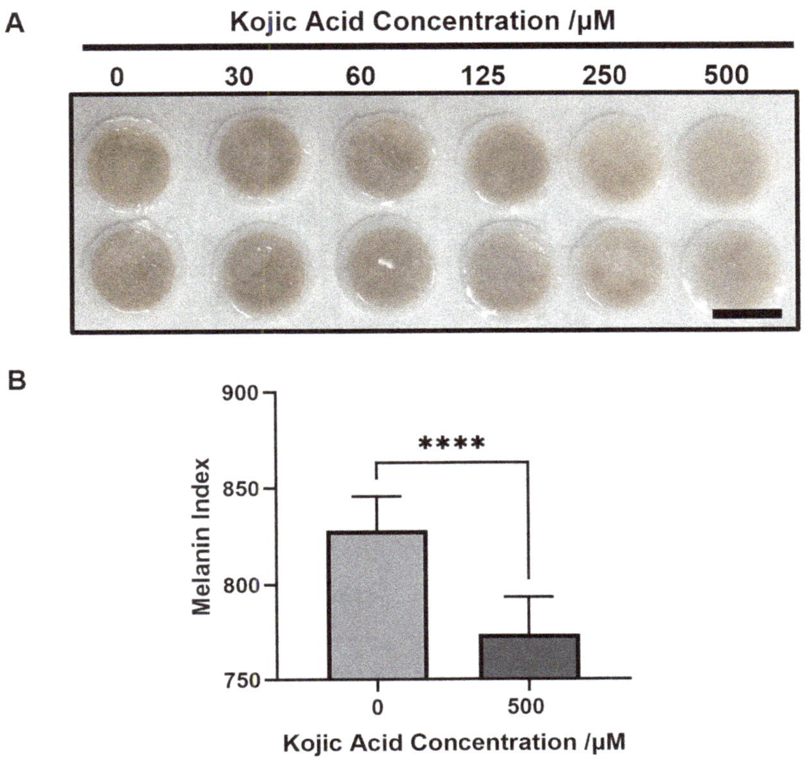

Fig. 4 Exogenous application of kojic acid reduces melanin content in vitro. Gross appearance (**a**) of pigmented FT-HSEs treated with kojic acid, a tyrosinase inhibitor that reduces melanin production, for 10 days. Tissues appear less pigmented in appearance, with increasing kojic acid concentration up to 500 μM, as demonstrated by a significantly reduced melanin index (**b**) with maximal kojic acid treatment. **** $= p < 0.0001$, $n = 9 \pm$ SEM

of a native tissue at a cellular and molecular level in vitro is an essential step to create a tissue whose function closely resembles that of its in vivo counterpart. Ultimately the development and advancement of bioengineering has led to the emergence of pigmented HSEs, which are rapidly becoming important tools in dermatology research. This is particularly important as focus shifts toward diversity and inclusivity in both academic and industrial research.

2 Materials

2.1 Bioengineering Pigmented Human Skin In Vitro

1. Human neonatal dermal fibroblasts (HDFn).
2. Human neonatal keratinocytes (HEKn).
3. Darkly pigmented human neonatal epidermal melanocytes (HEMn-Dp) (*see* **Note 1**).
4. T175 flasks.

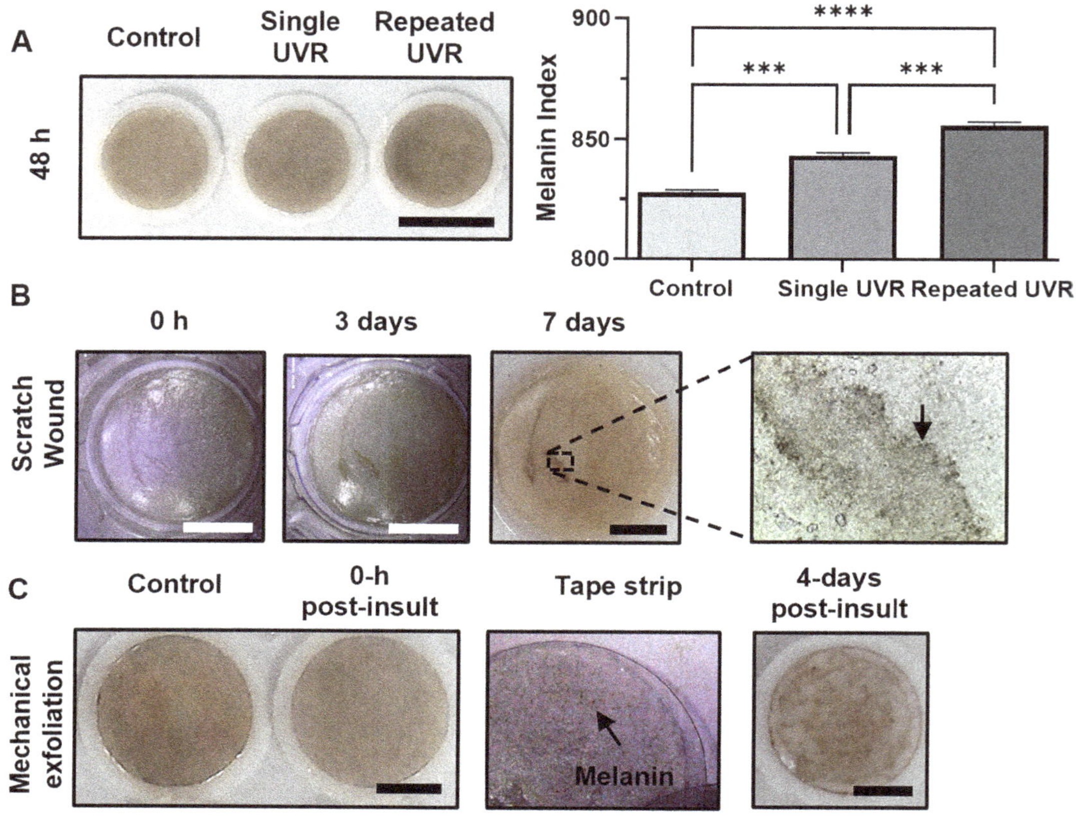

Fig. 5 External stimuli induce melanogenesis in pigmented skin equivalent. Gross appearance of pigmented FT-HSEs either non-irradiated (control), single (single UVR) or daily irradiated 5 times with 3.3 J cm^{-2} (96.5% UVA/3.5% UVB) (Repeated UVR) 48 h post-exposure (**a**). An increase in observable gross pigmentation accompanied by a significant increase in melanin index confirms a *tanning* response. A scratch wound to the surface of the pigmented tissue developed deeper pigmentation over a 7 day time period (**b**). At 7 days post-wounding, a deep pigmentation is identifiable in the wound area with melanin granules observable at higher magnification (arrow). Tape stripping of the epidermis reduces pigmentation through melanin removal from the *stratum corneum* (**c**). At 4 days post-exfoliation, a post-inflammatory hyperpigmentation effect is visible in some areas of the epidermis. *** $= p < 0.001$, **** $= p < 0.0001$, $n = 9 \pm$ SEM. Scale bars: 1 cm (**a**), 5 mm (**b**, **c**)

5. 12-well Alvetex® insert.
6. 6-well plate.
7. Alvetex® well insert holder and deep Petri dish.
8. Dulbecco's Modified Eagle Medium (DMEM).
9. Fetal Bovine Serum (FBS).
10. L-glutamine 200 mM.
11. EpiLife® Medium.
12. 100× human keratinocyte growth supplement (HKGS).
13. Medium 254®.

14. 100× human melanocyte growth supplement (HMGS).
15. Gentamicin.
16. Amphotericin B.
17. TGFβ1 recombinant human protein.
18. Ascorbic acid.
19. Keratinocyte growth factor (KGF).
20. Calcium chloride.
21. Trypsin-EDTA.
22. Trypsin neutralizer.
23. Sterile Dulbecco's phosphate-buffered saline (DPBS).
24. Phosphate-buffered saline (PBS).
25. Trypan blue solution.
26. Sterile forceps.

2.2 Analyzing Skin Topography and Gross Appearance of the Skin Equivalents

1. Andonstar AD210 10.1 Inch LCD Digital Microscope.

2.3 Analyzing Pigmentation Parameters of the Skin Equivalents

1. SkinColorCatch.
2. Parafilm wrapping film.

2.4 Processing the Pigmented Skin Equivalents for Paraffin Wax Embedding

1. Paraformaldehyde (PFA).
2. 30, 50, 70, 80, 90, 95, and 100% ethanols.
3. Histoclear.
4. Histoclear II.
5. Paraffin wax.
6. Tissue-processing cassettes.
7. Plastic molds.

2.5 Processing the Pigmented Skin Equivalents for Fontana–Masson Staining

1. Fontana–Masson stain kit.
2. 70, 95, and 100% ethanols.
3. Histoclear.
4. 27% ammonium hydroxide.
5. Glass Coplin jar and lid.
6. Omnimount mounting medium.

2.6 Processing the Pigmented Skin Equivalents for Epidermal-Whole Mount Staining

1. 4 mm Kai sterile dermal biopsy punch.
2. Ammonium thiocyanate.
3. Methanol.
4. Acetone.
5. Triton X-100.
6. Anti-TRP1 (clone TA99).
7. Donkey Anti-Mouse IgG Alexa Fluor® 488.
8. Charged Superfrost Plus microscope slides.
9. Vectashield antifade mounting medium.

2.7 Processing the Pigmented Skin Equivalents for Melanin Quantification Assay

1. 1.5 mL Polypropylene microcentrifuge tubes.
2. Sodium Hydroxide 1 N.
3. Cell culture grade Dimethyl Sulfoxide (DMSO).
4. Melanin from Sepia officinalis.
5. 15 mL Polypropylene centrifuge tubes.
6. 50% (w/v) hydrogen peroxide solution.

3 Methods

The development of a pigmented full-thickness skin equivalent involves the generation of a mature dermal equivalent by fibroblasts producing endogenous extracellular matrix proteins within the Alvetex® scaffold, followed by the generation of a well-differentiated pigmented epidermis (Fig. 6). A sterile, category 2 laminar flow hood at room temperature is required to perform this protocol. Follow your respective guidelines and COSHH (Control of Substances Hazardous to Health) forms when disposing of waste materials.

3.1 Bioengineering Pigmented Human Skin In Vitro

3.1.1 Revival of Fibroblasts

1. Prepare complete DMEM (500 mL DMEM supplemented with 50 mL FBS and 5 mL L-glutamine 200 mM). Revive and culture HDFn seeded at 2.5×10^3 cm^{-2} according to the supplier's instructions in 25 mL in complete DMEM in a T175 flask.
2. Incubate HDFn at 37 °C and 5% CO_2 in a humidified environment and change the medium the day after revival to remove the cryopreservative and every 2–3 days subsequently until 80% confluency is reached.

3.1.2 Preparation of Alvetex® Scaffold

The Alvetex® scaffold must be treated with ethanol to render it hydrophilic before use. Pretreat 12-well Alvetex® scaffolds in a sterile 6-well plate with:

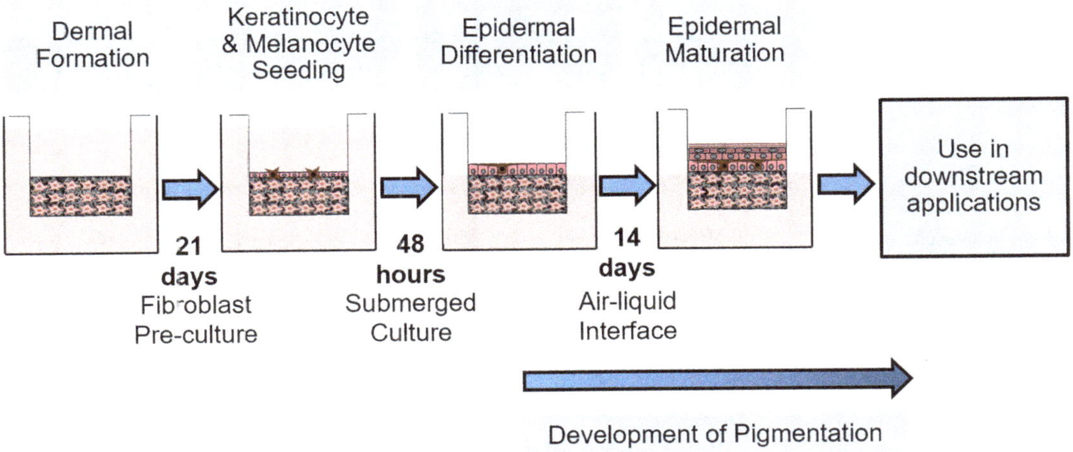

Fig. 6 Methodology used in the generation of the pigmented full-thickness skin equivalent. Initially, a dermal component is generated by seeding dermal neonatal fibroblasts into Alvetex® scaffold, which were allowed to populate the scaffold and neosynthesize endogenous extracellular matrix. Melanocytes and keratinocytes are then seeded in a single cell suspension on to the surface of the dermal compartment and cultured in submerged conditions for 48 h to promote epidermal cell proliferation. Epidermal differentiation is then encouraged by raising cultures to air-liquid interface for a minimum of 14 days to promote stratification and induction of pigmentation. After 14 days, a mature epidermis is present and the tissue can be used in downstream applications

(a) 70% Ethanol.
(b) 1 × sterile DPBS.
(c) Complete DMEM
(d) Leave scaffold submerged in complete DMEM until required.

3.1.3 Seeding Fibroblasts onto Alvetex® Scaffold

1. Prepare trypsin-EDTA by diluting to 1:10 in versene.
2. Discard media from the HDFn flask and wash with 5 mL trypsin-EDTA.
3. Trypsinize fibroblasts using 5 mL trypsin-EDTA dilution at 37 °C until cells have detached.
4. Neutralize the trypsin-EDTA with trypsin neutralizer, as per the manufacturer's instructions; or in 0.5% FBS in DPBS.
5. Centrifuge HDFn at 200 g for 5 min.
6. Perform cell count using a trypan blue exclusion assay.
7. Adjust cell density according to number of dermal equivalents to be seeded.
8. Seed 0.5 × 10^6 HDFn in 100 μL complete DMEM onto 12-well Alvetex® scaffold inserts and incubate at 37 °C in a 5% CO_2 humidified incubator for 2 h.

9. Add 10 mL complete DMEM supplemented with 5 ng mL^{-1} TGFβ1, 100 μg mL^{-1} ascorbic acid, 10 μg mL^{-1} gentamicin, and 0.25 μg mL^{-1} amphotericin B per dermal equivalent and incubate at 37 °C in a 5% CO_2 humidified incubator.
10. Maintain dermal equivalents for 21 days with 2 × medium changes a week.

3.1.4 Revival of Melanocytes

1. HEMn-Dp are to be revived 10–12 days before seeding onto dermal equivalents (*see* **Note 1**). Prepare a bottle of Medium 254® (500 mL Medium 254® supplemented with 5 mL HMGS).
2. Revive and culture HEMn-Dp seeded at 5 × 10^3 cm^{-2} according to the supplier's instructions in 25 mL in complete Medium 254® in a T175 flask.
3. Incubate HEMn-Dp at 37 °C and 5% CO_2 in a humidified environment and change the medium the day after revival to remove the cryopreservative and every 2–3 days subsequently until 80% confluency is reached.

3.1.5 Revival of Keratinocytes

1. HEKn are to be revived 6–8 days before seeding onto dermal equivalents. Prepare a bottle of EpiLife® Medium (500 mL EpiLife® Medium supplemented with 5 mL HKGS).
2. Revive and culture HEKn seeded at 2.5 × 10^3 cm^{-2} according to the supplier's instructions in 25 mL in complete EpiLife® Medium in a T175 flask.
3. Incubate HEKn at 37 °C and 5% CO_2 in a humidified environment and change the medium the day after revival to remove the cryopreservative and every 2–3 days subsequently until 80% confluency is reached.

3.1.6 Seeding of Keratinocytes and Melanocytes onto the Dermal Equivalent

1. Prepare trypsin-EDTA by diluting to 1:10 in versene.
2. Discard medium from the HEMn-Dp flask and wash with 5 mL sterile trypsin-EDTA.
3. Trypsinize HEMn-Dp using 5 mL trypsin-EDTA dilution at room temperature until cells have detached.
4. Neutralize the trypsin-EDTA with trypsin neutralizer, as per the manufacturer's instructions.
5. Centrifuge HEMn-Dp at 200 g for 5 min.
6. Perform cell count using a trypan blue exclusion assay.
7. Perform trypsinization and cell count steps with keratinocytes. HEKn require 37 °C for trypsinization.
8. Each dermal equivalent is seeded with 0.13 × 10^6 HEMn-Dp and 1.3 × 10^6 HEMn-Dp in 200 μL complete EpiLife® Medium. Adjust cell density accordingly.

9. Seed 200 μL prepared HEKn and HEMn-Dp suspension onto mature dermal equivalents and incubate at 37 °C in a 5% CO_2 humidified incubator for 2 h.
10. Add 10 mL complete EpiLife® Medium supplemented with 10 ng mL^{-1} KGF, 140 μM $CaCl_2$, 100 μg mL^{-1} ascorbic acid, 10 μg mL^{-1} gentamicin, and 0.25 μg mL^{-1} amphotericin B per skin equivalent and incubate at 37 °C in a 5% CO_2 humidified incubator for 48 h.

3.1.7 Raising Full-Thickness Equivalents to the Air-Liquid Interface

1. After 48 h, raise the full-thickness skin equivalents to the air-liquid interface by carefully aspirating medium from inside the insert and transferring to an Alvetex® well insert holder set on medium height in a deep Petri dish (up to 3 × skin equivalents per deep Petri dish).
2. Culture the full-thickness skin equivalents in 35 mL complete EpiLife® Medium supplemented with 10 ng mL^{-1} KGF, 1.64 mM $CaCl_2$, 100 μg mL^{-1} ascorbic acid, 10 μg mL^{-1} gentamicin, and 0.25 μg mL^{-1} amphotericin B per deep dish.
3. After 7 days at the air-liquid interface, the epidermis is fully formed; however, pigmentation is noticeable after 10–14 days and reaches a dark tone at day 21. Pigmented skin equivalents can be maintained for an extended period for long-term experiments. Maintain skin equivalents with 2 × medium changes a week.

3.2 Processing and Analyzing the Pigmented Skin Equivalents

Human skin equivalents must be characterized and validated prior to their use. The images obtained for this book chapter were generated using the following protocols. Alternative protocols and reagents can also be used.

3.2.1 Analyzing Skin Topography and Gross Appearance of the Skin Equivalents

1. Turn on the Andonstar LCD Digital Microscope.
2. Set top and side lights position, and adjust light controller. Keep these settings consistent with all the skin equivalents.
3. Place a white paper sheet on the surface of the base of the digital microscope to use as background.
4. Place the deep dish with the skin equivalents and position a skin model below the camera.
5. Press the M (mode) button and use the camera mode.
6. Adjust the focus controller.
7. Press the camera button to capture the picture.

3.2.2 Analyzing Pigmentation Parameters of the Skin Equivalents

The SkinColorCatch is a clinical instrument and colorimeter that allows the measuring of individual typological angle (ITA) and melanin index (MI). Such pigmentation parameters allow the classification of pigmentation of the generated pigmented skin equivalents in an objective manner.

1. Remove the skin equivalent from the plastic insert holders using a pair of flat-ended sterile forceps and maintain epidermal surface dry.
2. Place the skin equivalent with the dermal compartment facing down on a hydrophobic surface such as parafilm to avoid dryness. For consistent values, maintain the same background for all the measurements in all the skin equivalents.
3. Place the SkinColorCatch standing onto the surface of the skin equivalent for a few seconds until the reading is shown.
4. Repeat the measurement of the pigmentation parameters two times in different surface areas to get average values throughout the surface.
5. Move the skin equivalents to PBS and continue processing.

3.2.3 Processing the Pigmented Skin Equivalents for Paraffin Wax Embedding

1. After removing the skin equivalent from the plastic insert, wash the skin equivalent three times in 5 mL PBS.
2. Fix the skin equivalents in 4% PFA for 2 h at room temperature or at 4 °C overnight.
3. Wash the skin equivalent three times in 5 mL PBS.
4. Sequentially dehydrate the skin equivalent in 30%, 50%, 70%, 80%, 90%, and 95% ethanol (v/v) for 10 min each. Then, fully dehydrate in 100% ethanol for 30 min.
5. Transfer the skin equivalents to tissue-processing cassettes and place in Histoclear II for 30 min.
6. Incubate the skin equivalents in 1:1 ratio of Histoclear II: molten wax paraffin in a convection oven at 65 °C for 30 min.
7. Place the skin equivalents in 100% molten wax for at least 1 h at 65 °C, prior to cutting the models in half across the diameter and embedding in molten wax, with the flat edge facing down within the tissue-processing plastic molds. Allow the embedded samples to set at room temperature overnight prior to sectioning.

3.2.4 Processing the Pigmented Skin Equivalents for Fontana–Masson Staining

1. 5 μm thick sections were deparaffinized in Histoclear and gradually rehydrated through 100%, 95%, and 70% ethanols and distilled water.
2. Fontana–Masson staining kit includes gold chloride solution (0.2%), silver nitrate solution (10%), sodium thiosulfate solution (5%), and nuclear fast solution. (*see* **Note 2**).
3. Make up ammoniacal silver solution in a glass bottle by mixing 27 mL distilled water with 9 mL silver nitrate solution (10%) and adding ammonium hydroxide in a dropwise manner, swirling gently after each drop. The ammoniacal silver solution is ready to use once the formed sediment dissolves and the color turns transparent (*see* **Note 3**).

4. Transfer the fresh ammoniacal silver solution to a Coplin jar with lid and warm in a 58–60 °C water bath. Wait for the temperature to equilibrate.
5. Incubate sample sections in warmed ammoniacal silver solution for 30 min or until the tissue section becomes yellowish/ brown (*see* **Note 4**).
6. Rinse slides with two changes of distilled water.
7. Incubate slides in Gold Chloride Solution (0.2%) at room temperature for 30 s and rinse again in several changes of distilled water.
8. Incubate sections in sodium thiosulfate solution (5%) at room temperature for 1–2 min and rinse for 2 min in running tap water followed by 2 changes of distilled water.
9. Incubate slides in nuclear fast red solution for 5 min and rinse for 2 min in running tap water followed by 2 changes of distilled water.
10. Dehydrate sections very quickly in 3 changes of 100% ethanol.
11. Clear sections in Histoclear and mount on coverslips with Omnimount mounting medium.
12. Allow the slides to dry at room temperature prior to imaging with a brightfield microscope.

3.2.5 Processing the Pigmented Skin Equivalents for Epidermal-Whole Mount Staining

To visualize the distribution of melanocytes across the epidermis, the epidermal layer of a section of the skin equivalent is separated from the underlying dermis and stained with a melanocyte marker (Fig. 7). Processing for epidermal whole-mount staining includes a different fixation method and thus, the preparation of the sample must be done during the harvest of the skin equivalents from cell culture.

1. Make a punch biopsy per skin equivalent using a 4 mm Kai sterile dermal biopsy punch.
2. Place punch biopsies *stratum corneum* down in 200 μL 3.8% ammonium thiocyanate in PBS and incubate at room temperature for 20 min.
3. Peel off the epidermis from the dermis using forceps and wash twice in PBS.
4. Fix the epidermal samples in a 1:1 solution of methanol: acetone for 20 min at −20 °C and wash twice in PBS. Samples can be stored at 4 °C for a few days before staining if desired.
5. Block and permeabilize epidermal samples in blocking buffer consisting of 20% NCS and 0.4% Triton X-100 in PBS for 1 h.
6. Incubate epidermises in anti-TRP1 primary antibody diluted 1: 100 in blocking buffer at room temperature for 2 h.

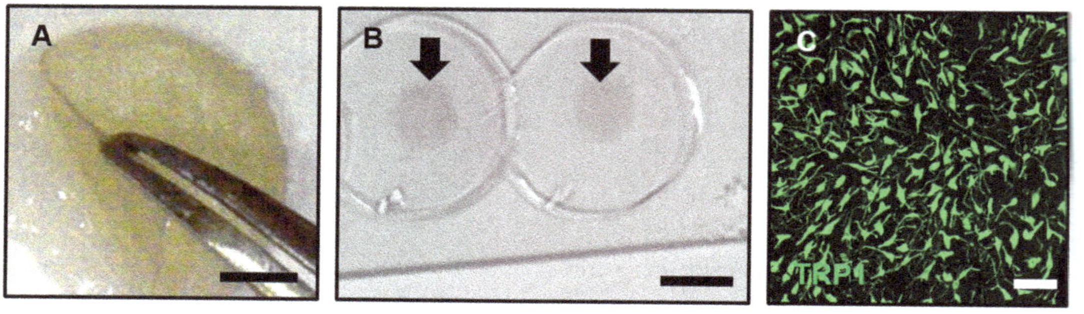

Fig. 7 Methodology used in the epidermal whole-mount staining protocol. A 4-mm punch biopsy of a 12-well pigmented skin equivalent is used for epidermal detaching. Chemical digestion with 3.8% ammonium thiocyanate in PBS enables epidermal detaching using forceps (**a**). Pigmented epidermises are stained and mounted on microscope slides with coverslips (**b**). TRP1-stained melanocytes are identified across the epidermis by confocal microscopy. Scale bars: 1 mm (**a**), 5 mm (**b**), 100 μm (**c**)

7. Wash epidermal samples three times in PBS for 10 min each.
8. Incubate epidermises with donkey anti-mouse Alexa Fluor 488 secondary antibody diluted 1:1000 in blocking buffer for 1 h at room temperature.
9. Wash epidermal samples three times in PBS.
10. Mount epidermal sheets on microscope slides with coverslips using Vectashield antifade mounting medium with the *stratum basale* facing up to allow imaging of this layer.
11. Image fluorescent melanocytes of the *stratum basale* using a confocal microscope.

3.2.6 Processing the Pigmented Skin Equivalents for Melanin Quantification Assay

1. After removing the skin equivalent from the plastic insert, make a punch biopsy using a 5 mm Kai sterile dermal biopsy punch. Keep the punch biopsy size consistent through all the skin equivalents.
2. Place each punch biopsy in an Eppendorf with 0.5 mL lysis buffer made of 10% (v/v) DMSO in sodium hydroxide 1 N. Samples in lysis buffer can be stored at −20 °C until the assay is performed.
3. Create a melanin standard curve by making a range of concentrations of Melanin from Sepia officinalis reconstituted in 0.5 mL lysis buffer in Eppendorf tubes and serially dilute from 200 μg mL^{-1} to 6.25 μg mL^{-1}.
4. Incubate samples and standards at 80 °C in water bath for 1 h and vortex every 20 min for melanin solubilization.
5. Centrifuge samples at 3000 g for 5 min for melanin solution clarification.
6. Collect 300 μL supernatant from each sample and standard in new 15 mL Falcon tubes for melanin quantification.

7. For melanin oxidation, add 450 μL 50% (w/v) hydrogen peroxide solution to the 300 μL supernatant to give a final concentration of 30% (v/v). Incubate samples for 4 h at room temperature, protected from light.
8. Centrifuge samples at 3000 g for 5 min.
9. Plate out 100 μL of the dilutions in triplicate into wells of a 96-well plate.
10. Measure fluorescence with a microplate reader using 470 nm and 550 nm as excitation and emission wavelengths, respectively.

4 Notes

1. Human pigmented melanocytes are available in different pigmentation, which can be used to generate various skin tones using the same cell seeding number in the development of the FT-HSE.
2. All reagents should be handled with care and disposed of properly. Wear gloves, lab coat, and protective eyewear. Please review the Safety Datasheet of Fontana–Masson stain kit for information. Gold chloride causes skin irritation, serious eye irritation, and may cause respiratory irritation. Silver nitrate solution may be corrosive to metals, may intensify fire (oxidizer), causes severe skin burns and eye damage, is very toxic to aquatic life, and may cause long-term adverse effects in the aquatic environment. Nuclear fast red solution causes serious eye damage.
3. Make up silver ammoniacal solution in a chemical fume hood.
4. After incubation of the slides in ammoniacal silver solution, inactivate the ammoniacal silver solution by the addition of hydrochloric acid.

References

1. Hill DS, Robinson NDP, Caley MP et al (2015) A novel fully humanized 3D skin equivalent to model early melanoma invasion. Mol Cancer Ther 14:2665–2673. https://doi.org/10.1158/1535-7163.MCT-15-0394
2. Wufuer M, Lee G, Hur W et al (2016) Skin-on-a-chip model simulating inflammation, edema and drug-based treatment. Sci Rep 6:37471. https://doi.org/10.1038/srep37471
3. Vörsmann H, Groeber F, Walles H et al (2013) Development of a human three-dimensional organotypic skin-melanoma spheroid model for in vitro drug testing. Cell Death Dis 4: e719–e719. https://doi.org/10.1038/cddis.2013.249
4. Lecas S, Boursier E, Fitoussi R et al (2016) In vitro model adapted to the study of skin ageing induced by air pollution. Toxicol Lett 259:60–68. https://doi.org/10.1016/j.toxlet.2016.07.026
5. Diekmann J, Alili L, Scholz O et al (2016) A three-dimensional skin equivalent reflecting some aspects of in vivo aged skin. Exp Dermatol 25:56–61. https://doi.org/10.1111/exd.12866

6. Bjerke DL, Li R, Price JM et al (2021) The vitamin A ester retinyl propionate has a unique metabolic profile and higher retinoid-related bioactivity over retinol and retinyl palmitate in human skin models. Exp Dermatol 30:226–236. https://doi.org/10.1111/exd.14219
7. Schlotmann K, Kaeten M, Black AF et al (2001) Cosmetic efficacy claims in vitro using a three-dimensional human skin model. Int J Cosmet Sci 23:309–318. https://doi.org/10.1046/j.1467-2494.2001.00098.x
8. Duval C, Schmidt R, Regnier M et al (2003) The use of reconstructed human skin to evaluate UV-induced modifications and sunscreen efficacy. Exp Dermatology, Suppl 12:64–70. https://doi.org/10.1034/j.1600-0625.12.s2.10.x
9. Reijnders CMA, van Lier A, Roffel S et al (2015) Development of a full-thickness human skin equivalent in vitro model derived from TERT-immortalized keratinocytes and fibroblasts. Tissue Eng Part A 21:2448–2459. https://doi.org/10.1089/ten.TEA.2015.0139
10. Karl F, Wußmann M, Kreß L et al (2019) Patient-derived in vitro skin models for investigation of small fiber pathology. Ann Clin Transl Neurol 6:1797–1806. https://doi.org/10.1002/acn3.50871
11. Montagna W, Carlisle K (1991) The architecture of black and white facial skin. J Am Acad Dermatol 24:929–937. https://doi.org/10.1016/0190-9622(91)70148-U
12. Del Bino S, Ito S, Sok J et al (2015) Chemical analysis of constitutive pigmentation of human epidermis reveals constant eumelanin to pheomelanin ratio. Pigment Cell Melanoma Res 28:707–717. https://doi.org/10.1111/pcmr.12410
13. Lin JY, Fisher DE (2007) Melanocyte biology and skin pigmentation. Nature 445:843–850. https://doi.org/10.1038/nature05660
14. Fitzpatrick TB, Breathnach AS (1963) The epidermal melanin unit system. Dermatol Wochenschr 147:481–489
15. Greatens A, Hakozaki T, Koshoffer A et al (2005) Effective inhibition of melanosome transfer to keratinocytes by lectins and niacinamide is reversible. Exp Dermatol 14:498–508. https://doi.org/10.1111/j.0906-6705.2005.00309.x
16. Hurbain I, Romao M, Sextius P et al (2018) Melanosome distribution in keratinocytes in different skin types: melanosome clusters are not degradative organelles. J Invest Dermatol 138:647–656. https://doi.org/10.1016/j.jid.2017.09.039
17. Choi W, Miyamura Y, Wolber R et al (2010) Regulation of human skin pigmentation in situ by repetitive UV exposure: molecular characterization of responses to UVA and/or UVB. J Invest Dermatol 130:1685–1696. https://doi.org/10.1038/jid.2010.5
18. Xiao K, Yates JM, Zardawi F et al (2017) Characterising the variations in ethnic skin colours: a new calibrated data base for human skin. Skin Res Technol 23:21–29. https://doi.org/10.1111/srt.12295
19. Castanedo-Cazares JP, Hernández-Blanco D, García-Cortés JD et al (2018) Analysis of cutaneous pigmentation in a sample. Gac Med Mex 154:68–73. https://doi.org/10.24875/GMM.17003050
20. Brenner M, Hearing VJ (2008) The protective role of melanin against UV damage in human skin. Photochem Photobiol 84:539–549. https://doi.org/10.1111/j.1751-1097.2007.00226.x
21. Tobin DJ (2006) Biochemistry of human skin—our brain on the outside. Chem Soc Rev 35:52–67. https://doi.org/10.1039/b505793k
22. Bohr VA, Smith CA, Okumoto DS et al (1985) DNA repair in an active gene: removal of pyrimidine dimers from the DHFR gene of CHO cells is much more efficient than in the genome overall. Cell 40:359–369. https://doi.org/10.1016/0092-8674(85)90150-3
23. Yoon JH, Lee CS, O'Connor TR et al (2000) The DNA damage spectrum produced by simulated sunlight. J Mol Biol 299:681–693. https://doi.org/10.1006/jmbi.2000.3771
24. Orazio J, Jarrett S, Amaro-Ortiz A et al (2013) UV radiation and the skin. Int. J. Mol, Sci, p 14
25. Chen H, Weng QY, Fisher DE (2014) UV signaling pathways within the skin. J Invest Dermatol 134:2080–2085. https://doi.org/10.1038/jid.2014.161
26. Douki T, Perdiz D, Gróf P et al (1999) Oxidation of guanine in cellular DNA by solar UV radiation: biological role. Photochem Photobiol 70:184–190
27. Visscher MO (2017) Skin color and pigmentation in ethnic skin. Facial Plast Surg Clin North Am 25:119–125. https://doi.org/10.1016/j.fsc.2016.08.011
28. Goncalves K, De Los Santos Gomez P, Costello L et al (2022) Investigation into the effect of skin tone modulators and exogenous stress on skin pigmentation utilizing a novel bioengineered skin equivalent. Bioeng Transl Med 8:e10415. https://doi.org/10.1002/btm2.10415

29. Freer M, Darling N, Goncalves K et al (2023) Development of a mammalian neurosensory full-thickness skin equivalent and its application to screen sensitizing stimuli. Bioeng Transl Med 8:e10484. https://doi.org/10.1002/btm2.10484
30. Ouwehand K, Spiekstra SW, Waaijman T et al (2011) Technical advance: langerhans cells derived from a human cell line in a full-thickness skin equivalent undergo allergen-induced maturation and migration. J Leukoc Biol 90:1027–1033. https://doi.org/10.1189/jlb.0610374
31. Gibbs S, Murli S, De Boer G et al (2000) Melanosome capping of keratinocytes in pigmented reconstructed epidermis—effect of ultraviolet radiation and 3-isobutyl-1-methylxanthine on melanogenesis. Pigment Cell Res 13:458–466. https://doi.org/10.1034/j.1600-0749.2000.130608.x
32. Duval C, Chagnoleau C, Pouradier F et al (2012) Human skin model containing melanocytes: essential role of keratinocyte growth factor for constitutive pigmentation-functional response to alpha-melanocyte stimulating hormone and forskolin. Tissue Eng Part C Methods 18:947–957. https://doi.org/10.1089/ten.TEC.2011.0676
33. Duval C, Cohen C, Chagnoleau C et al (2014) Key regulatory role of dermal fibroblasts in pigmentation as demonstrated using a reconstructed skin model: impact of photo-aging. PLoS One 9:e114182
34. De Los Santos Gomez P, Costello L, Goncalves K et al (2024) Comparison of photodamage in non-pigmented and pigmented human skin equivalents exposed to repeated ultraviolet radiation to investigate the role of melanocytes in skin photoprotection. Front Med 11. https://doi.org/10.3389/fmed.2024.1355799
35. Gendreau I, Angers L, Jean J et al (2013) Pigmented skin models: understand the mechanisms of melanocytes. Regen Med Tissue Eng. https://doi.org/10.5772/55565
36. Cario-André M, Pain C, Gauthier Y et al (2006) In vivo and in vitro evidence of dermal fibroblasts influence on human epidermal pigmentation. Pigment Cell Res 19:434–442. https://doi.org/10.1111/j.1600-0749.2006.00326.x
37. Cario-André M, Pain C, Taïeb A et al (2000) Studies on epidermis reconstructed with and without melanocytes: melanocytes prevent sunburn cell formation but not appearance of DNA damaged cells in fair-skinned caucasians. J Invest Dermatol 115:193–199. https://doi.org/10.1046/j.1523-1747.2000.00007.x
38. Cario-André M, Briganti S, Picardo M et al (2002) Epidermal reconstructs: a new tool to study topical and systemic photoprotective molecules. J Photochem Photobiol B Biol 68:79–87. https://doi.org/10.1016/S1011-1344(02)00328-7
39. Bessou S, Surlève-Bazeille J-E, Sorbier E et al (1995) Ex vivo reconstruction of the epidermis with melanocytes and the influence of UVB. Pigment Cell Res 8:241–249. https://doi.org/10.1111/j.1600-0749.1995.tb00670.x
40. Bessou-Touya S, Pain C, Taïeb A et al (1998) Chimeric human epidermal reconstructs to study the role of melanocytes and keratinocytes in pigmentation and photoprotection. J Invest Dermatol 111:1103–1108. https://doi.org/10.1046/j.1523-1747.1998.00405.x
41. Imokawa G, Yada Y, Morisaki N et al (1998) Biological characterization of human fibroblast-derived mitogenic factors for human melanocytes. Biochem J 330 (Pt 3):1235–1239. https://doi.org/10.1042/bj3301235
42. Choi W, Wolber R, Gerwat W et al (2010) The fibroblast-derived paracrine factor neuregulin-1 has a novel role in regulating the constitutive color and melanocyte function in human skin. J Cell Sci 123:3102 LP–3103111. https://doi.org/10.1242/jcs.064774
43. Hirobe T, Hasegawa K, Furuya R et al (2013) Effects of fibroblast-derived factors on the proliferation and differentiation of human melanocytes in culture. J Dermatol Sci 71:45–57. https://doi.org/10.1016/j.jdermsci.2013.03.012
44. Hedley S, Gawkrodger DJ, Weetman AP et al (1996) Investigation of the influence of extracellular matrix proteins on normal human melanocyte morphology and melanogenic activity. Br J Dermatol 135:888–897. https://doi.org/10.1046/j.1365-2133.1996.d01-1091.x
45. Hedley SJ, Layton C, Heaton M et al (2002) Fibroblasts play a regulatory role in the control of pigmentation in reconstructed human skin from skin types I and II. Pigment Cell Res 15:49–56. https://doi.org/10.1034/j.1600-0749.2002.00067.x
46. Ng WL, Qi JTZ, Yeong WY et al (2018) Proof-of-concept: 3D bioprinting of pigmented human skin constructs. Biofabrication 10:25005. https://doi.org/10.1088/1758-5090/aa9e1e
47. Roger M, Fullard N, Costello L et al (2019) Bioengineering the microanatomy of human skin. J Anat 234:438–455. https://doi.org/10.1111/joa.12942

48. Zoio P, Ventura S, Leite M et al (2021) Pigmented full-thickness human skin model based on a fibroblast-derived matrix for long-term studies. Tissue Eng Part C Methods 27:433–443. https://doi.org/10.1089/ten.TEC.2021.0069
49. Davis EC, Callender VD (2010) Postinflammatory hyperpigmentation: a review of the epidemiology, clinical features, and treatment options in skin of color. J Clin Aesthet Dermatol 3:20–31
50. Gold MH, Biron J (2011) Efficacy of a novel hydroquinone-free skin-brightening cream in patients with melasma. J Cosmet Dermatol 10:189–196. https://doi.org/10.1111/j.1473-2165.2011.00573.x
51. Marionnet C, Tricaud C, Bernerd F (2014) Exposure to non-extreme solar UV daylight: spectral characterization, effects on skin and photoprotection. Int J Mol Sci 16:68–90. https://doi.org/10.3390/ijms16010068
52. Tian BWCA (2015) Novel low fluence combination laser treatment of solar lentigines in type III Asian skin. J Cutan Aesthet Surg 8:230–232. https://doi.org/10.4103/0974-2077.172199
53. James AG, Pople JE, Parish WE et al (2006) Histological evaluation of hyperpigmentation on female Filipino axillary skin. Int J Cosmet Sci 28:247–253. https://doi.org/10.1111/j.1467-2494.2006.00328.x
54. de Angelis F, Kolesnikova L, Renato F et al (2011) Fractional nonablative 1540-nm laser treatment of striae distensae in fitzpatrick skin types II to IV: clinical and histological results. Aesthetic Surg J 31:411–419. https://doi.org/10.1177/1090820X11402493
55. Moreiras H, Seabra MC & Barral DC (2021) Melanin Transfer in the Epidermis: The Pursuit of Skin Pigmentation Control Mechanisms. Int J Mol 22(9): 4466. https://doi.org/10.3390/ijms22094466

Chapter 10

Engineering a Multilayered Skin Substitute with Keratinocytes, Fibroblasts, Adipose Derived Stem Cells and Adipocytes

Maike Keck, Alfred Gugerell, and Johanna Kober

Abstract

A variety of skin substitutes that restore epidermal and dermal structures are currently available on the market. While the main focus in research and clinical application lies in dermal and epidermal substitutes, the development of a subcutaneous replacement, the hypodermis, is often neglected. This chapter describes the use of fibrin sealant as a hydrogel scaffold to generate a three dimensional skin substitute. For the hypodermal layer, adipose-derived stem cells (ASCs) and mature adipocytes are seeded within a fibrin hydrogel. On top, another fibrin clot with incorporated fibroblasts is placed for the construction of the dermal layer. Keratinocytes are added on top of the two-layered construct to form the epidermal layer. The three layered construct is cultivated for up to 3 weeks with keratinocytes being exposed to air according to the air-liquid interface cultivation model.

Key words Three layered skin substitute, Keratinocytes, Fibroblasts, Adipose derived stem cells, Adipocytes, Regenerative medicine, Tissue engineering

1 Introduction

A variety of skin substitutes that restore epidermal and dermal structures are currently available on the market. While the main focus in research and clinical application lies in dermal and epidermal substitutes, the development of a subcutaneous replacement, the hypodermis, is often neglected. Nevertheless, this part of the skin is of major importance as it functions as an insulator, energy storage, and endocrine organ and defines the shape of the body.

This chapter describes the use of fibrin sealant as a hydrogel scaffold to generate a three dimensional skin substitute. The physical properties of fibrin hydrogel mimic the soft character of fatty tissue and also its mechanical properties resemble those of adipose

Thomas Biedermann and Sophie Böttcher-Haberzeth (eds.), *Skin Tissue Engineering: Methods and Protocols*, Methods in Molecular Biology, vol. 2922, https://doi.org/10.1007/978-1-0716-4510-9_10,

tissue to a satisfying extent. Therefore fibrin hydrogel is suggested to be a suitable candidate for adipose tissue-equivalent formation [1, 2].

For the hypodermal layer, adipose-derived stem cells (ASCs) and mature adipocytes are seeded within a fibrin hydrogel. On top, another fibrin clot with incorporated fibroblasts is placed for the construction of the dermal layer. Keratinocytes are added on top of the two-layered construct to form the epidermal layer. The three layered construct is cultivated for up to 3 weeks with keratinocytes being exposed to air according to the air-liquid interface cultivation model [3].

2 Materials

2.1 Biological Material

1. Skin and subcutaneous fat tissue is usually gained from healthy patients undergoing abdominoplasty or other body contouring surgery (*see* Fig. 1) (*see* **Notes 1** and **2**).

2.2 Culture Media

1. ASC proliferation medium: 10% FCS, 100 units/mL penicillin, 100 μg/mL streptomycin in Dulbecco's Modified Eagle Medium (DMEM) (*see* **Note 3**).
2. ASCs differentiation medium: Preadipocyte Differentiation Medium (Table 1).
3. Adipocyte cultivation medium: Adipocyte Nutrition Medium (Table 1).
4. Fibroblast culture medium: 10% fetal calf serum, 1% glutamine, 100 units/mL penicillin, 100 μg/mL streptomycin in DMEM.
5. Keratinocyte medium: DermaLife Keratinocyte medium.
6. Three-layered skin construct medium: 10% fetal calf serum in DMEM.

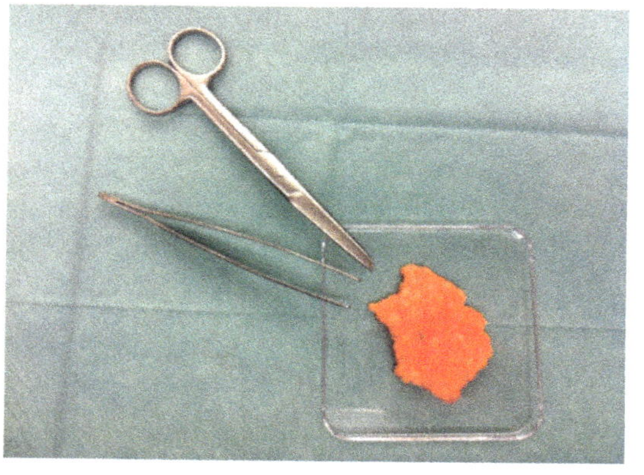

Fig. 1 Starting material: adipose tissue from abdominoplasties

Table 1
Composition of Promocell media

Final supplement concentrations (after addition to the medium)	Preadipocyte differentiation medium	Adipocyte nutrition medium
Fetal Calf Serum	–	0.03 mL/mL
d-Biotin	8 μg/mL	8 μg/mL
Insulin (recombinant human)	0.5 μg/mL	0.5 μg/mL
Dexamethasone	400 ng/mL	400 ng/mL
IBMX	44 μg/mL	–
L-Thyroxine	9 ng/mL	–
Ciglitazone	3 μg/mL	–

2.3 Reagents

1. PBS (phosphate buffered saline).
2. 2% Penicillin/Streptomycin.
3. Collagenase solution: 2 mg/mL Collagenase Type IV in Hanks' buffered salt solution.
4. AdipoRed staining.
5. 125 mg/mL fibrinogen.
6. 10 IU/mL thrombin.
7. Trypsin.

2.4 Equipment

1. 70 μm cell strainer.
2. Cell culture inserts.

3 Methods

3.1 Isolation of Human Adipose-Derived Stem Cells

1. Wash the adipose tissue in PBS (phosphate buffered saline) + 2% Penicillin/Streptomycin (*see* **Note 4**).
2. Mince the adipose tissue with scissors and/or scalpel (*see* Fig. 1). Remove large amounts of connective tissue and vessels.
3. Add the shredded fat tissue into the tubes to collagenase solution (25 mL collagenase + about 10 mL fatty tissue).
4. Incubate in a shaking bath at 37 ° C for 1 h.
5. Filter the dissolved fatty tissue including collagenase through gauze into a new 50 mL tube.
6. Centrifuge for 10 min at 1200 rpm at 4 ° C.

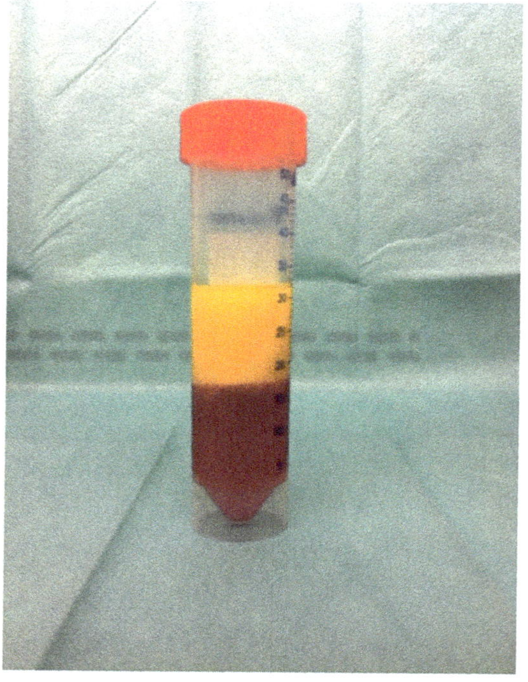

Fig. 2 Different phases after centrifugation: oil (destroyed adipocytes), adipocytes, collagenase, SVF (from top to bottom)

7. The resulting cell pellet (stromal vascular fraction, SVF) consists of ASCs, macrophages, erythrocytes, endothelial cells and other cell types (*see* Fig. 2).
8. Carefully aspirate the upper layer including adipocytes (protocol see below).
9. Resuspend pellet with 2 mL Red Blood Cell Lysing Buffer and incubate for 8 min on ice.
10. Add cold medium and filter the cells through a 70 μm cell strainer.
11. Centrifuge for 7 min at 1500 rpm at 4 °C.
12. Resuspend pellet with ASC proliferation medium (*see* **Note 5**). ASCs are cultured as a monolayer at 37 °C in ASC proliferation medium in a humidified atmosphere with 5% CO_2. ASC passages up to passage 2 or 3 are recommended for use in skin engineering approaches (*see* Fig. 3).

3.2 Isolation of Human Adipocytes

1. Collect freshly isolated adipocytes during the isolation process as described above from the usually discarded adipose phase after the first centrifugation step.
2. Carefully aspirate the upper layer including adipocytes. Possibly start with a 25 mL pipette, as for the upper fat layer the smaller Pasteur pipettes may be too narrow. Culture cells in adipocyte

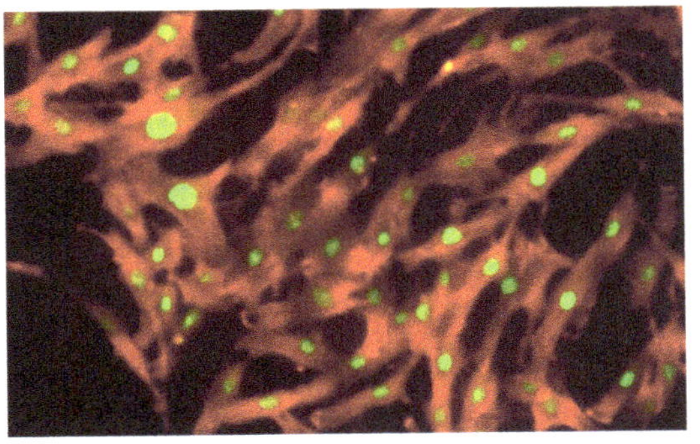

Fig. 3 Isolated ASC, TRITC-phalloidin (cytosceleton)/SYTOXgreen (nuclei) staining

Fig. 4 Isolated adipocytes, TRITC-phalloidin (cytosceleton)/SYTOXgreen (nuclei) staining

cultivation medium. The cultivation of the cells takes place in incubators at 37 °C, 5% CO_2 and 95% humidity. Isolated adipocytes are depicted in Fig. 4 (*see* **Notes 6** and 7).

3.3 Adipocyte Differentiation

1. Incubate ASCs for 3 days in differentiation medium.
2. Incubate cells further in adipocyte cultivation medium. Medium has to be changed every third day.
3. Evaluation of differentiation can histologically be performed by AdipoRed staining (*see* Fig. 5) or by RNA quantification.

3.4 Isolation of Human Dermal Fibroblasts

1. Cut human donor skin into small pieces of about 1 × 1 mm.
2. Put skin fragments in 6-well plates or dishes and incubate for 14 days in fibroblast culture medium.
3. Skin fragments need to stick to the bottom of the well for fibroblasts to grow out onto the bottom of the plate.
4. Culture upon confluency and split for further cultivation in culture flasks (*see* Fig. 6). The cultivation of the cells takes place in incubators at 37 °C, 5% CO_2 and 95% humidity.

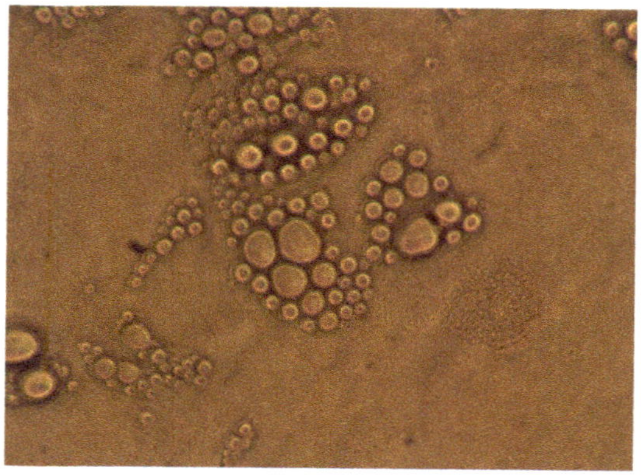

Fig. 5 ASC differentiated into adipogenic lineage (day 24). Intracellular lipid droplets are visible

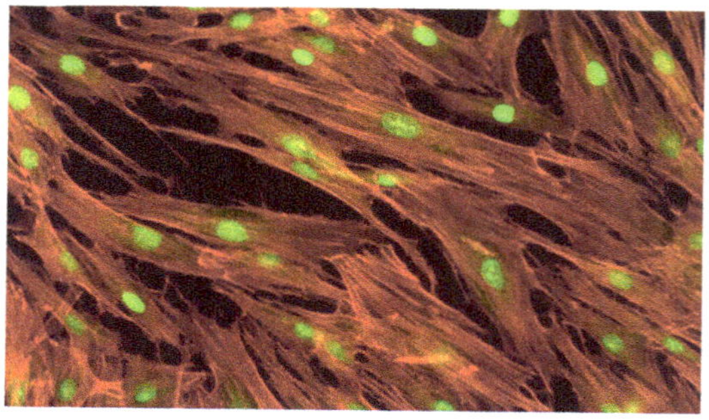

Fig. 6 Isolated fibroblasts, TRITC-phalloidin (cytosceleton)/SYTOXgreen (nuclei) staining

3.5 Isolation of Human Keratinocytes

1. Cut human skin into pieces of about 2 × 2 cm. Remove subcutis and most of the dermis.
2. Trypsinize for 16 h at 4 °C.
3. Filter the resulting suspension through a gauze and centrifuge for 5 min at 1200 rpm.
4. Discard supernatant and resuspend in keratinocyte medium. Culture cells in incubators at 37 °C, 5% CO_2 and 95% humidity.

3.6 Construction of a Three-Layered Skin Substitute with a Fibrin Hydrogel Matrix

For fibrin clot formation, equal volumes of fibrinogen and thrombin components have to be mixed in cell culture inserts (*see* **Note 8**).

1. For the hypodermal layer: mix 3×10^5 ASCs with 20 μL of mature adipocytes (about 1.4×10^6 cells/mL) in 100 μL fibrin clots in cell culture inserts (*see* **Note 9**).
2. Incubate clots at 37 °C for 20 min for full polymerization.

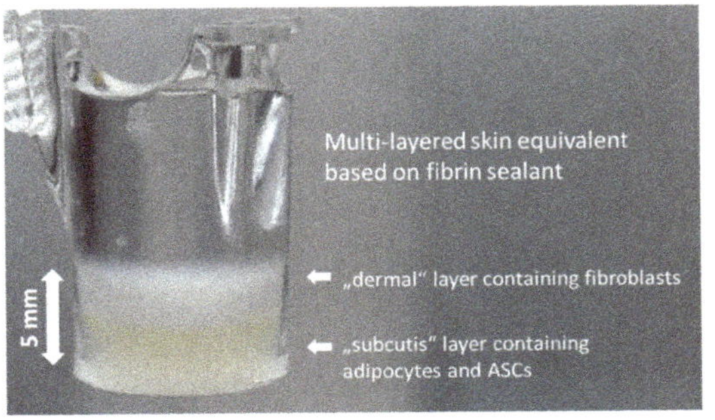

Fig. 7 Two-layered construct

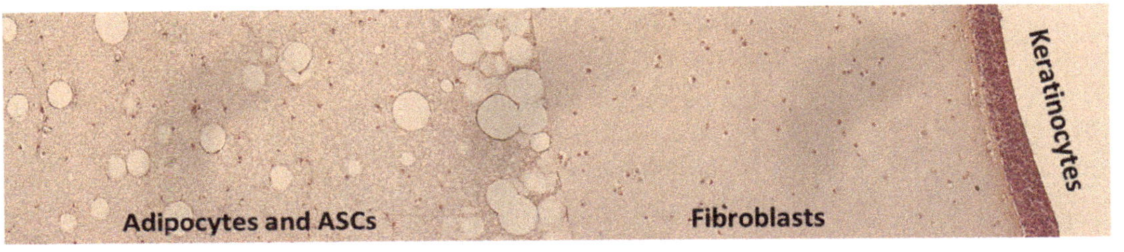

Fig. 8 Three-layered construct

3. For the dermal layer: polymerize another fibrin clot on top of the hypodermal layer with 3×10^5 fibroblasts (*see* Fig. 7).
4. Incubate clots at 37 °C for 15 min for full polymerization.
5. For the epidermal layer: add $0.6–1 \times 10^6$ keratinocytes per clot are added on top of the dermal layer.
6. The three-layered construct can be cultivated for up to 3 weeks (at 37 ° C, 5% CO_2 and 95% humidity) in three-layered skin construct medium where keratinocytes are exposed to air according to the air-liquid interface cultivation model (*see* Fig. 8).

4 Notes

1. There are differences in the viability of cells depending on the donation protocol, especially concerning liposuction methods. As fat tissue is a huge endocrine organ, patients with metabolic disorders like diabetes should be excluded due to eventually altered cell behavior [4].
2. Subjects have to give a written informed consent which has to be approved by the local ethics committee.
3. Required media and solutions are always freshly prepared before use.

4. All cell culture work is done with sterile materials and under sterile workbenches.
5. Isolated cells should be counted in a Bürker-Türk counting chamber with Trypan Blue stain to prove viability. By cultivation and passaging, the SVF selection and expansion of ASCs will be achieved, remaining blood cells or endothelial cells will be removed or overgrown in large parts. It is recommended to prove the purity of cells with FACS (e.g., CD90+, CD105+, CD29+, CD44+, CD45-) and also methodically by differentiating in desired lineages (i.e., adipogenic lineage) [5].
6. Adipocytes are sensitive and break easily resulting in an oil film. Adipocytes have to be washed two times with PBS to remove collagenase. Afterwards they can be used for the experiment right away.
7. A long-term cultivation of mature adipocytes is not recommended due to cell break and also possible de-differentiation [6].
8. Fibrinogen and thrombin need to be reconstituted according to the manufacturer´s protocol. As fibrinogen/thrombin is used primarily as hemostat or sealer, components should be diluted.
9. In a fibrin clot with high fibrinogen concentration, cells may be stressed due to a high stiffness and density of the hydrogel [7]. High thrombin concentrations may be toxic for cells [8].

References

1. Peterbauer-Scherb A, Danzer M, Gabriel C et al (2012) In vitro adipogenesis of adipose-derived stem cells in 3D fibrin matrix of low component concentration. J Tissue Eng Regen Med 6(6): 434–442
2. Siegel KR, Clevenger TN, Clegg DO et al (2018) Adipose stem cells incorporated in fibrin clot modulate expression of growth factors. Arthroscopy 34(2):581–591. https://doi.org/10.1016/j.arthro.2017.08.250
3. Kober J, Gugerell A, Schmid M et al (2015) Generation of a fibrin based three-layered skin substitute. Biomed Res Int. https://doi.org/10.1155/2015/170427
4. Keck M, Kober J, Riedl O et al (2014) Power assisted liposuction to obtain adipose-derived stem cells: impact on viability and differentiation to adipocytes in comparison to manual aspiration. J Plast Reconstr Aesthet Surg 67(1):E1–E8
5. Mildmay-White A, Khan W (2017) Cell surface markers on adipose-derived stem cells: a systematic review. Curr Stem Cell Res Ther 12(6): 484–492
6. Huber B, Kluger PJ (2015) Decelerating mature adipocyte dedifferentiation by media composition. Tissue Eng Part C Methods 21(12): 1237–1245
7. Nürnberger S, Wolbank S, Peterbauer-Scherb A et al (2010) Properties and potential alternative applications of fibrin glue. In: Janek von Byern IG (ed) Biological adhesive systems: from nature to technical and medical application. Springer, Vienna
8. Gugerell A, Schossleitner K, Wolbank S et al (2012) High thrombin concentrations in fibrin sealants induce apoptosis in human keratinocytes. J Biomed Mater Res Part A 100a(5): 1239–1247

Chapter 11

Isolation and Culture of Epidermolysis Bullosa Cells and Organotypic Skin Models

Yinghong He and Cristina Has

Abstract

Isolation and culture of keratinocytes from patients with different types of epidermolysis bullosa are sometimes challenging, because of the inherent adhesion defects of these cells. We routinely employ a well-established protocol for in vitro culture of these cells from small skin samples remaining left after diagnostic procedures are performed. Keratinocytes and fibroblast can be used for downstream expression and functional studies or for construction of in vitro organotypic cocultures. These cells maintain main common characteristics of spreading, adhesion, migration, and survival, which depend on the underlying molecular defect.

Key words Epidermolysis bullosa, Cell culture, Organotypic coculture, Adhesion, Blister, Keratinocyte, Fibroblast

Abbreviations

DMEM	Dulbecco's Modified Eagle Medium
EB	Epidermolysis bullosa
Fb	Fibroblasts
FCS	Fetal calf serum
PBS	Phosphate buffered saline

1 Introduction

Isolation and culture of human skin cells from the same skin biopsy are the basic requisites for the research in the field of skin biology [1]. Human monogenetic diseases provide perfect paradigms for the investigation of the function of the single molecules in the skin cells and skin.

Epidermolysis bullosa (EB) is a heterogeneous group of genetic disorders characterized by mechanically induced skin blistering. EB

Thomas Biedermann and Sophie Böttcher-Haberzeth (eds.), *Skin Tissue Engineering: Methods and Protocols*, Methods in Molecular Biology, vol. 2922, https://doi.org/10.1007/978-1-0716-4510-9_11,

is caused by mutations in genes encoding structural proteins with roles in mechanical stability, cell-cell or cell-matrix adhesion, such as, keratin5 and 14, type VII collagen, type XVII collagen, laminin 332, integrin α6β4, integrin α3, or kindlin-1 [2–6]. All proteins which are altered in EB are expressed by keratinocytes, with exception of collagen VII, which is expressed by both keratinocytes and fibroblast. Consequently, keratinocytes isolated from patients with EB display deficiencies in their adhesive function, the main property of epidermal keratinocytes. For example, laminin 332- and kindlin-1-negative cells hardly adhere on cell culture dishes and are frequently lost during isolation from skin samples [7, 8]. To enable their culture, seeding on pre-coated dishes with extracellular matrix proteins (such as laminin 332 or Matrigel) is helpful. Intriguingly, keratinocytes isolated from patients with *COL7A1* or *COL17A1* mutations adhere quite well, but demonstrate other more subtle anomalies. Adhesion defects result in additional anomalies of cell survival and migration, which may influence cell culture. Because of these limitations in the culture of primary EB keratinocytes and their rapid senescence [9], we routinely immortalize primary cells by viral transduction of a plasmid containing the E6E7 HPV genes [10, 11]. Primary fibroblasts isolated from most EB patients behave like fibroblasts isolated from non-EB individuals, except of those from patients with dystrophic EB and *COL7A1* mutations and of those isolated from scars, which will display the corresponding features.

This chapter details the methods for extraction and culture of fibroblasts and keratinocytes from human skin biopsies, especially from EB patients, as used in our laboratory. They are followed by a detailed description of the generation of reconstructed skin by the self-assembly approach [12]. Other laboratories employ different methods or modifications of these protocols. The protocols included in this chapter are designed to enable rapid isolation of the cells from small biopsies in a daily routine. Skin samples, usually 4 mm punches, are obtained for molecular diagnostics by immunofluorescence mapping. After written informed consent from the patient, the material remaining after the diagnostic procedure is used for cell isolation for scientific purposes.

2 Materials and Methods

Prepare and store the culture media and the general materials in proper conditions and keep in sterile. Be sure to use analytical grade reagents for all the materials. Keep the disinfection rules and perform all procedures under the sterile hood. Diligently follow all waste disposal regulations when disposing waste materials and the autoclave regulations.

2.1 Culture Media and Reagents

Sterile DMEM (Dulbecco's Modified Eagle Medium), 20 mM Hepes supplemented with 1% antibiotic/antimycotic mix.

2.1.1 Tissue Transport Medium

1. DMEM, 20 mM Hepes.
2. Antibiotic/antimycotic mix (100×).

Thaw in cold water. Inactivate complement proteins in hot water (56 °C) for 30 min. Distribute in 5 mL aliquots and store at −20 °C until use.

2.1.2 Fb Medium (Human Fibroblast Culture Medium)

DMEM, 10% FCS, 1% antibiotic/antimycotic mix, 2% L-Glutamin

1. DMEM.
2. Antibiotic/antimycotic mix (100×).
3. Fetal Calf Serum (FCS).
   Thaw in cold water. Inactivate complement proteins in hot water (56 °C) for 30 min. Distribute in 50 mL aliquots and store at −20 °C until use.
4. 100 mL 200 mM L-Glutamin.
   Thaw in cold water. Inactivate complement proteins in hot water (56 °C) for 30 min. Distribute in 10 mL aliquots and store at −20 °C until use.
5. Ascorbic acid. Dissolve 50 mg/mL ascorbic acid in sterile water to make a 1000× stock solution. Sterilize by filtration through a 0.22 μm low-binding disposable filter (*see* **Note 1**).

2.1.3 KGM Medium (Human Keratinocyte Culture Medium)

Keratinocyte SFM with supplements

1. Keratinocyte serum free medium: 500 mL SFM.
2. Supplements: 2.5 μg human recombinant EGF, 25 mg bovine pituitary extract.

2.1.4 OTC—Medium (Organotypic Coculture Medium)

Air-Liquid medium, Supplements

1. Air-Liquid medium: 3:1 DMEM/Ham's F12, 5% fetal bovine serum, 100 U penicillin, 100 μg/mL Streptomycin.
2. Supplements: 5 μg/mL insulin, 1×10^{-4} M adenine, 1×10^{-10} M choleratoxin, 0.4 μg/mL hydrocortisone, 1 ng/mL EGF.

2.1.5 Freezing Medium

Cell growth medium, 20% FCS, 10% DMSO

1. Fetal calf Serum.
2. Dimethyl sulfoxide (DMSO).

2.2 General Materials

1. Phosphate buffered saline (PBS).
2. 100 mL Trypsin/EDTA solution (0.05/0.02% w/v).

Thaw in cold water. Inactivate complement proteins in hot water (56 °C) for 30 min. Distribute in 10 mL aliquots and store at −20 °C until use.

3. Falcon tubes, Petri dish, and parafilm.
4. Dissecting curved forceps.
5. Scalpel and blade.
6. Tissue culture flask, 75 cm^2.
7. Sterile cryogenic vials.
8. Deep well plates.
9. Filter inserts, pore size 1 μm.
10. Glass rings (1, 2 cm diameter, 1 cm high).
11. Sterile glass beaker with magnetic stirrer.
12. Hank's balanced salt solution (HBSS).
13. Collagen type I (acid-extracted from rat tail, concentration 6 mg /mL).

3 Methods

3.1 Monolayer Culture of the Fibroblasts from the Skin Biopsy from the EB Patient

3.1.1 Extraction and Culture of Human Fibroblasts

1. Source of cells. A small piece of skin biopsy from EB patient (fresh blister or after rubbing the skin) removed by surgery following procedures approved by the institution's committee for the protection of human subjects can be stored in the tissue transport medium at 4 °C for a maximum of 3 days.
2. Remove excess fat and blood, rinse tissue many times with PBS supplemented with 8% antibiotic/antimycotic mix.
3. Cut skin into pieces of 2 × 2 mm size.
4. Transfer pieces into empty T25 cell culture flask and place dermis-side-down.
5. Incubate for 30–60 min at 37 °C in a cell culture incubator to allow tissue to adhere to the bottom of the flask.
6. After tissue is well attached, add 5 mL in Fb medium, incubate at 37 °C, 5% CO_2.
7. First medium change after 5 days, usually fibroblasts grow out of the tissue within 7 –14 days, sometimes it can take up to 30 days (Fig. 1).

3.1.2 Subculture of Human Fibroblasts (Passage)

1. Discard the medium, wash cells with PBS.
2. Add 1–2 mL trypsin solution (2.5% w/v, diluted 1:10 with PBS) per T25 cell culture flask, 3 mL per T75 cell culture flask.
3. Cells need about 5 min to detach, watch continuously to avoid overlong incubation.

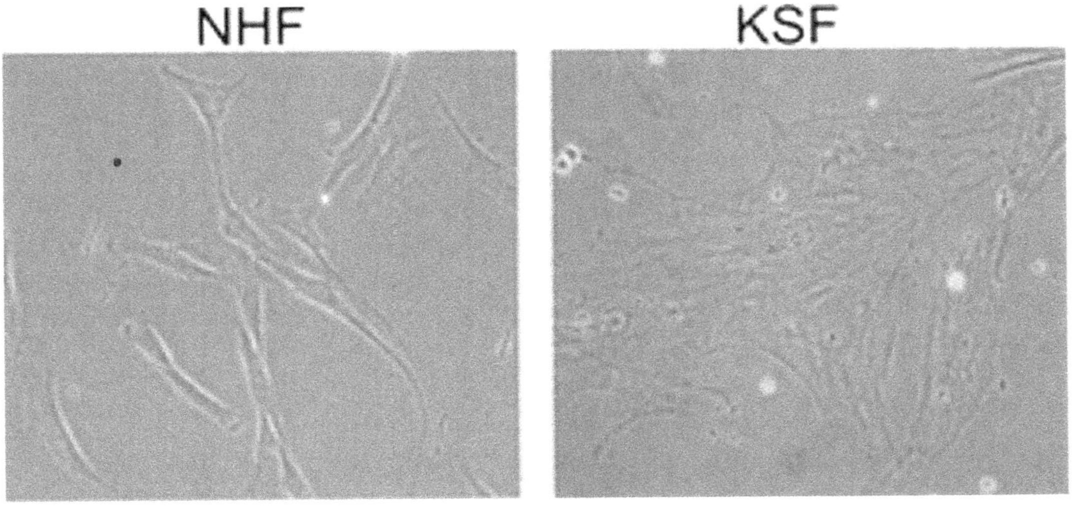

Fig. 1 Morphology of primary normal human (NHF, left) and Kindler EB (KSF, right) fibroblasts. Note that Kindler EB fibroblasts are larger and display morphology suggestive for myofibroblasts

4. Stop reaction by adding 2–3 times the volume of PBS with 10% FCS.
5. Transfer cell suspension into a Falcon tube, centrifuge 5 min at 1000× *g*, room temperature (RT).
6. Resuspend the pellet in Fb medium.
7. Plate approximately 0.5×10^6 cells per T25 cell culture flask in Fb medium.

3.1.3 Cryopreservation of Human Fibroblasts

1. Resuspend 1×10^6 cells in 1 mL freezing medium:
2. Freeze at −80 °C in an isopropanol-filled freezing container.
3. Transfer to liquid nitrogen within 48 h after freezing.

3.1.4 Thawing of Human Fibroblasts

1. Thaw quickly the 1 mL cryogenic vial containing the milliliter cell-liquid from liquid nitrogen storage in a 37 °C water bath until the milliliter is liquid.
2. Resuspend in 15 mL cell culture medium and wash cells, centrifuge the cells in 1100 rpm for 5 min.
3. Collect the cell pellet in the appropriate volume of culture medium and plate on a cell culture flask.

3.2 Monolayer Culture of the Keratinocytes from Skin Biopsies from EB Patients

3.2.1 Extraction and Culture of Human Keratinocytes

1. Source of cells. A small piece of skin biopsy from EB patient removed by surgery following procedures approved by the institution's committee for the protection of human subjects can be stored in the tissue transport medium at 4 °C for a maximum of 3 days.
2. Remove excess fat and blood, rinse tissue many times with PBS supplemented with 8% antibiotic/antimycotic mix.

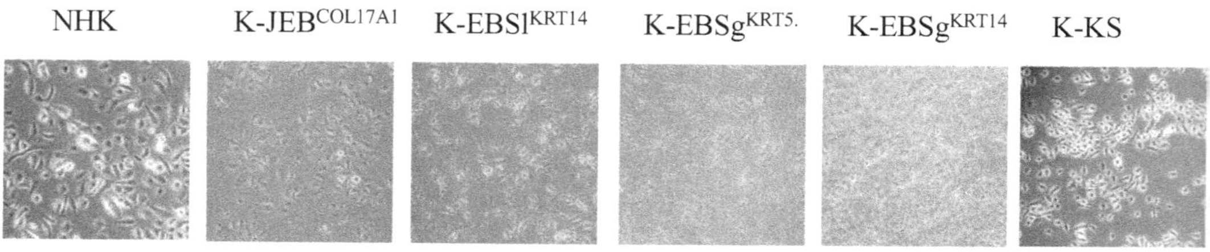

Fig. 2 Morphology of primary normal human (NHK) and EB keratinocytes: junctional EB with *COL17A1* mutations, K-JEB^{COL17A1}, EB simplex localized with *KRT14* mutations K-EBSI^{KRT14}, EB simplex generalized with *KRT5* mutations K-EBSg^{KRT5}, EB simplex generalized with *KRT14* mutations K-EBSg^{KRT14}, Kindler EB keratinocytes K-KS

3. Transfer skin sample to sterile Petri dish and remove dermis as well as possible.
4. Cut skin sample into 5 × 5 mm pieces and transfer them to a 10 cm Petri dish containing 30 mL of Trypsin/EDTA solution (0.05/0.02% w/v).
5. Incubate at 37 °C, 5% CO_2 for 60 min (*see* **Note 2**).
6. When epidermis loosens, transfer skin pieces to a Petri dish containing 30 mL PBS with 10% FCS.
7. Remove epidermis and tear into small fragments with forceps.
8. Carefully scratch dermis with forceps to remove residual keratinocytes.
9. Remove the dermal tissue.
10. Transfer cell suspension into a 50 mL Falcon tube and centrifuge 5 min at 1000× *g*, RT.
11. Resuspend pellet in KGM medium (*see* **Note 3**) and plate into T25 cell culture flask (approximately 1 cm^2 skin/T25) (*see* **Note 4**).
12. Incubate at 37 °C, 5% CO_2.
13. Observe cells daily and handle cells gently, especially keratinocytes deficient for proteins of the dermal-epidermal junction (Fig. 2).
14. Change medium three times a week.

3.2.2 Subculture of Human Keratinocytes (Passage)

1. Discard the medium.
2. Add 1–2 mL trypsin solution (2.5% w/v, diluted 1:10 with PBS) per T25 cell culture flask, 3 mL per T75 cell culture flask.
3. Cells need about 5 min to detach, watch continuously to avoid overlong incubation.
4. Stop reaction by adding 2–3 times the volume of PBS with 10% FCS.

5. Transfer cell suspension into a Falcon tube, centrifuge 5 min at 1000× *g*, RT.
6. Resuspend the pellet in KGM medium and count the cells.
7. Plate approximately 0.5×10^6 cells per T25 in KGM medium.

3.2.3 Cryopreservation of Human Keratinocytes

As described in Subheading 3.1.3.

3.2.4 Thawing of Human Keratinocytes

As described in Subheading 3.1.4.

3.3 Human Dermis Reconstruction by the Self-Assembly Approach

All further manipulations are performed under a sterile laminar flow hood cabinet.

3.3.1 Formation of Fibroblast-Collagen I Gel for the Dermal Sheets

1. Put the filter inserts in each well into the 6-well deep well culture plate (*see* **Note 5**).
2. Mix collagen I solution with HBSS, then neutralizes the mixture by adding 5 M NaOH.
3. Confluent fibroblasts between their second and sixth passages will be suspended in the FCS in a concentration with 3×10^6/mL gel.
4. Add FCS-fibroblast suspension to collagen solution; mix thoroughly in the sterile glass beaker with magnetic stirrer.
5. Pipet 2.5 mL of the mixture into one filter insert (*see* **Note 6**), incubates at 37 °C, 5% CO_2.
6. Gels solidify within 1 h at 37 °C, set glass ring into insert, press down gently without rupturing the gel.
7. After 1 h, remove the fluid and immerse gel in Fb medium containing 50 μg/mL ascorbic acid (Fig. 3a).

3.3.2 Human Skin Reconstruction (Keratinocyte Addition)

1. One day after the preparation of the gels, remove medium inside the glass ring.
2. Resuspend the trypsinized EB keratinocytes as described in KGM medium in 1×10^6/mL.
3. Remove the culture medium from reconstructed dermis. Seed 1×10^6 keratinocytes (1 mL) into the glass ring. Incubate at 37 °C, 5% CO_2.
4. One day later keratinocyte seeding, the glass ring can be gently removed with forceps (*see* **Note 7**).

3.3.3 Maturation of the Skin Model: Air/Liquid Interface Culture

1. One day to one week after the seeding of keratinocytes, remove culture medium above the insert.
2. Lift reconstructed skin and transfer it on the air–liquid stand.

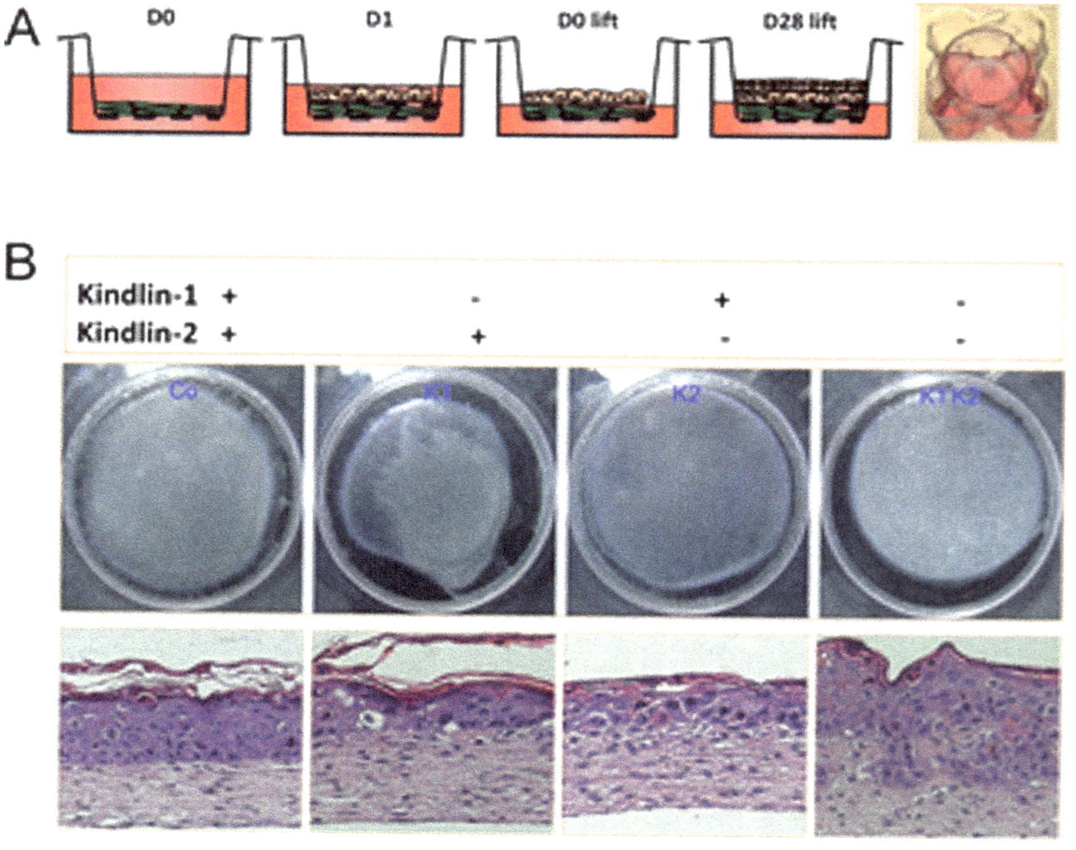

Fig. 3 (**a**) Schematic representation of the method for in vitro skin reconstruction; green, dermal equivalent, yellow epidermal equivalent; the right panel shows a picture of an organotypic skin coculture. (**b**) Reconstructed skin with normal fibroblasts and kindlin-1, or kindlin-2, or kindlin-1 and -2 deficient keratinocytes. upper panels: reconstructed skin in culture; lower: histological section of human reconstructed skin after 2 weeks. Note anomalies in the epidermal structure when the kindlin-deficient keratinocytes were used

3. Add 30 mL of OTC medium containing 50 μg/mL ascorbic acid (*see* **Note 8**). Incubate at 37 °C, 5% CO_2. Change culture medium three times a week.
4. After 2 weeks, reconstructed skin is then ready for various types of analysis after removal from the insert filters.
5. The reconstructed skin tissues will be snap frozen or formalin fixed for the further analysis. According to the need of the experiment, the reconstructed skin can be cultivated at the air- liquid interface for more than 28 days. However, a well-organized basement membrane is already obtained after 14 days of culture at the air–liquid interface (Fig. 3b).

4 Notes

1. Ascorbic acid stock solution must be prepared immediately before use and protected from light.
2. The incubation times are depending on tissue type, for the blister roof we usually use 20–35 min, for a full thickness skin at 60 min and for the foreskin at 60–90 min.
3. Pre-incubate KGM medium in the cell culture incubator for about 1 h, the keratinocytes will attach and spread more efficiently.
4. Coating the cell-culture flasks with different feeder layer sometime improve the attachment and spreading of the EB keratinocytes.
5. For the skin model construction usually will be performed in 6-well plates. However, they have already performed this model using 12-well plates according to the need of the experiment. The size of the cell culture dish to be used is thus not limited and is to be chosen according to the desired type of reconstruction and experiment to be performed. Therefore, filter inserts and deep well culture plates and glass rings should be adjusted for that particular type of cell culture dish.
6. Take care not to produce any air bubbles, pre-incubate the pipet used for this step in minus 20 degree helps for the manipulation.
7. Place one end of plate higher so that you can see the epithelium within the ring. Gently turn the glass ring using two sterile forceps to remove glass ring, take care not to disturb the epithelium.
8. The lower surface of the reconstructed skin must be in direct contact with culture medium.

Acknowledgments

The authors would like to thank current and former members of the laboratory. A special thank is said to Juna Leppert team members who have contributed to adapt the protocols to obtain this for her valuable suggestions on the manuscript.

References

1. Germain L, Rouabhia M, Guignard R et al (1993) Improvement of human keratinocyte isolation and culture using thermolysin. Burns 19:99–104
2. Bruckner-Tuderman L, Has C (2014) Disorders of the cutaneous basement membrane zone–the paradigm of epidermolysis bullosa. Matrix Biol 33:29–34

3. Has C, He Y (2017) Focal adhesions in the skin: lessons learned from skin fragility disorders. Eur J Dermatol 27:8–11
4. Has C (2018) Advances in understanding the molecular basis of skin fragility. F1000Res 7: 279
5. Has C, Nyström A, Saeidian AH et al (2018) Epidermolysis bullosa: molecular pathology of connective tissue components in the cutaneous basement membrane zone. Matrix Biol 71–72: 313–329
6. Bardhan A, Bruckner-Tuderman L, Chapple ILC et al (2020) Epidermolysis bullosa. Nat Rev Dis Primers 6(1):78
7. Gagnoux-Palacios L, Vailly J, Durand-Clement M et al (1996) Functional Re-expression of laminin-5 in laminin-gamma2-deficient human keratinocytes modifies cell morphology, motility, and adhesion. J Biol Chem 271: 18437–18444
8. Herz C, Aumailley M, Schulte C et al (2006) Kindlin-1 is a phosphoprotein involved in regulation of polarity, proliferation, and motility of epidermal keratinocytes. J Biol Chem 281:36082–36090
9. Piccinni E, Di Zenzo G, Maurelli R et al (2013) Induction of senescence pathways in Kindler syndrome primary keratinocytes. Br J Dermatol 168:1019–1026
10. He Y, Maier K, Leppert J et al (2016) Monoallelic mutations in the translation initiation codon of KLHL24 cause skin fragility. Am J Hum Genet 99:1395–1404
11. Maier K, He Y, Esser PR et al (2016) Single amino acid deletion in kindlin-1 Results in partial protein degradation which can be rescued by chaperone treatment. J Invest Dermatol 136:920–929
12. Stark H-J, Szabowski A, Fusenig NE et al (2004) Organotypic cocultures as skin equivalents: A complex and sophisticated in vitro system. Biol Proced Online 6:55–60

Chapter 12

Engineering a Multilayered Skin Equivalent: The Importance of Endogenous Extracellular Matrix and Model Flexibility for Nuanced and Accurate Skin Studies

Lucy Smith, Victoria Maltman, Paola De Los Santos Gomez, Kirsty Goncalves, and Stefan Przyborski

Abstract

Human skin equivalents (HSEs) are a valuable tool for both academic and industrial laboratories to further the understanding of skin physiology in a range of health and disease situations. There have been many advances in the development of HSEs that successfully recapitulate the structure of human skin in vitro; however a main limitation is variability due to the use of complex protocols and exogenous extracellular matrix (ECM) proteins. Previously, we have developed a robust, reproducible, and unique full thickness HSE, using a consistent scaffold, commercially available cells and defined low-serum media. Alvetex® scaffold technology allows fibroblasts to produce their own endogenous ECM proteins, alleviating the need for exogenous collagen and supporting epidermal differentiation and stratification. This model is highly flexible, with the potential to use other cell populations for more complex HSEs, alter the culture regime to experimental needs, and perform preclinical testing using the same tools as in vivo studies. Our full thickness skin equivalent is generated using a detailed step-by-step protocol, which sequentially forms the multi-layered structure of human skin in vitro and is detailed below.

Key words Skin equivalent, Organotypic culture, Alvetex® scaffold, Skin tissue engineering, Endogenous extracellular matrix deposition, Clinical measures, Complex models

1 Introduction

1.1 Structure of Human Skin

Human skin has a complex structure and is composed of three morphologically and functionally distinct layers: the epidermis, dermis, and hypodermis (*see* Fig. 1). The epidermis acts as a multifunctional physical barrier to withstand mechanical stress, protect against percutaneous pathogen invasion, and prevent unregulated water and electrolyte loss [1]. Populated by a multitude of cell types, including mechanosensory Merkel cells, immunoregulatory

Authors Lucy Smith and Victoria Maltman have equally contributed to this chapter.

Thomas Biedermann and Sophie Böttcher-Haberzeth (eds.), *Skin Tissue Engineering: Methods and Protocols*, Methods in Molecular Biology, vol. 2922, https://doi.org/10.1007/978-1-0716-4510-9_12,

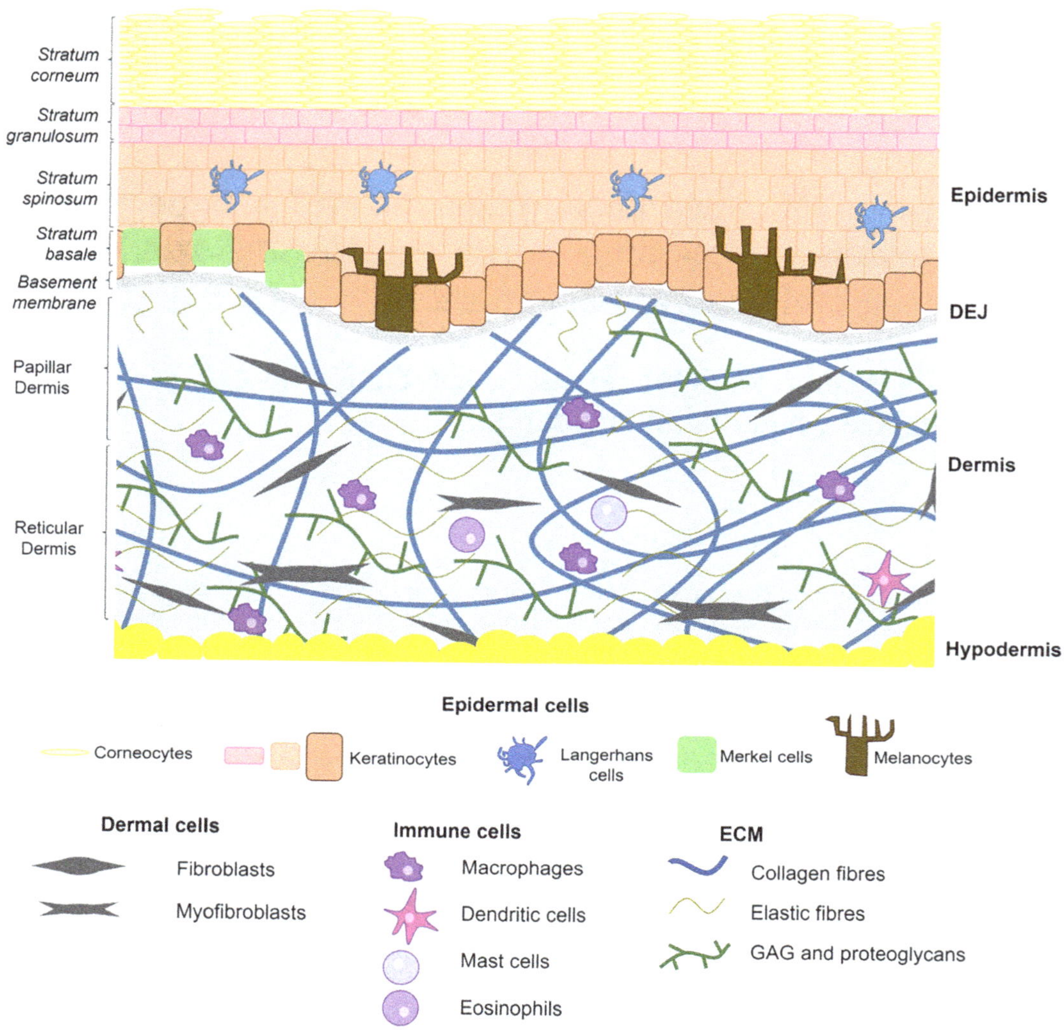

Fig. 1 Schematic of human skin. Skin contains three morphologically distinct layers: the epidermis, dermis, and hypodermis. The epidermis is a stratified epithelial compartment composed of the *stratum basale (S.b.)*, *stratum spinosum* (*S.s.*), *stratum granulosum* (*S.g.*), and *stratum corneum* (*S.c.*). The epidermis is supported by the dermis, which is comprised of the papillary (*P.d.*) and reticular (*R.d.*) layers. The dermis is perfused with blood vessels, and contains a multitude of cell types such as fibroblasts, which synthesize the extracellular matrix, and immune cells such as macrophages and dendritic cells. Dermal appendages include hair follicles, sebaceous glands, and sweat glands. The hypodermis is composed of subcutaneous adipose tissue, which provides insulation

Langerhans cells, and photoprotective melanocytes, the epidermis is predominantly comprised of keratinocytes, around 95%, organized in a stratified squamous manner [2–4]. Proliferative keratinocytes arise from asymmetrically dividing stem cell populations in the *stratum basale,* undergoing profound morphological and structural changes during programmed sequential differentiation to form the suprabasal *stratum spinosum* and *stratum granulosum* layers [5]. During terminal differentiation, keratinocytes undergo cornification, becoming anuclear, flattened corneocytes embedded in the lipid-rich, proteinaceous *stratum corneum*. Covalent cross-linking of proteins and lipids within this layer produces the water-resistant, protective barrier function of skin against the external environment [6].

The epidermis is tightly adhered to the dermis via the interdigitating dermo-epidermal junction, which facilitates substance exchange and the polarization of basal keratinocytes [7]. Mechanical strength and elastic recoil are key properties provided by the dermis, which is composed of two distinct layers: the superficial papillary dermis containing sparsely arranged collagen and elaunin fibers, and the deep reticular dermis containing abundant collagen fibers and a dense elastic network [8]. These protein networks are embedded in an amorphous, non-fibrous matrix composed of proteoglycans and glycosaminoglycans such as hyaluronic acid, chondroitin sulfate, and dermatan sulfate, which are synthesized by dermal fibroblasts. The hypodermis is a layer of subcutaneous fat that provides insulation and storage.

1.2 Bioengineering Multilayered Skin Models In Vitro to Bridge the Gap Between Two-Dimensional Tissue Culture and Animal Models

Traditionally, skin research has been conducted by culturing homogeneous keratinocyte or fibroblast populations on two-dimensional plastic substrates. Although widely accepted, high-throughput, and cost-effective, this method fails to represent the multi-layered complexity of human skin. Animal models, particularly mouse models, are considered an accurate alternative; as well as providing an in vivo skin sample to perform experiments, they also allow systemic effects of active compounds to be observed. However, despite some biological similarities, there are fundamental genetic and structural discrepancies between murine and human skin. Only 30% of human skin-associated genes are expressed in murine skin [9], and architectural differences include hair follicle distribution patterns, a lack of rête ridges, decreased epidermal thickness, and the presence of the panniculus carnosus in mouse skin, a feature absent in humans [10]. Furthermore, a general shift in scientific practices towards reducing the use of animals in research coupled with European Union legislation which prohibits animal testing for active compounds in cosmetics [11] means that animals models are now not considered to be a suitable option for many scientific studies.

Organotypic HSEs overcome the limitations of in vivo animal models and in vitro two-dimensional cell culture, providing an indispensable platform technology for skin research. Over the last few decades, significant progress has been made in the development of physiologically relevant HSEs that successfully recapitulate the complexity of human skin, and these tools are now used globally, in both academic and industrial research laboratories, for a range of applications. Engineering HSEs in vitro is technically challenging due to their multi-layered, multicellular structure, and therefore, detailed step-by-step protocols are required. In the first step, fibroblasts are seeded into the chosen support matrix to form the dermal compartment. Formation of the epidermis is more complex due to the presence of sequentially differentiated layers; the accepted method involves initially seeding keratinocytes on to the dermis in submerged culture in low calcium conditions to promote proliferation and the formation of the *stratum basale*. To achieve epidermal differentiation, HSEs are then raised to the air-liquid interface in high calcium conditions, a process known as the calcium switch; this change in calcium concentration mediates the characteristic basal-to-suprabasal switch of keratin 14 to keratin 10 [12]. Due to their highly complex structure, HSEs must be extensively analyzed, characterized, and validated to prove successful recapitulation of skin architecture in vitro before use in applications.

Previously, we have described the use of a porous, polystyrene scaffold to develop reproducible HSEs using commercially available primary cell populations and a low-serum medium [13]. Using commercially available products removes sources of variability and the nature of the scaffold, which includes no exogenous proteins, allows fibroblasts to synthesize and deposit endogenous extracellular matrix as the model matures. Diverse, abundant, endogenous ECM can be increasingly observed in the dermal compartment over time (*see* Fig. 2), forming a robust stroma on which the epidermis can be successfully supported without keratinocyte infiltration. The nature of this model allows for its use as a novel assay to monitor ECM production in response to various stimuli and stressors without interference from exogenous proteins.

The resultant fully human tissue model closely resembles the architecture of native human skin (*see* Fig. 3), as cells in the epidermis display the characteristic organized stratified nature of squamous epithelia, with columnar cells in the *stratum basale* progressively differentiating into the flattened anuclear corneocytes of the *stratum corneum*.

1.3 Developing Diverse Human Skin Equivalents for Accurate Biological Research

HSEs have multifaceted applications in dermatological research, and there are many commercially available HSEs including Epiderm FT™ (MatTek Corporation, Ashland, MA, USA), T-skin (Episkin, L'Oréal, France) and Phenion (AG&Co, KGaA, Dusseldorf, Germany). Models such as these are widely used as they are standardized, quality controlled and closely resemble the structure of human skin, offering quick, reliable turnarounds for studies and outputs. However, the inherent inflexibly of these products means that many biological parameters of the models cannot be altered, such as incorporating other cell types or performing interventions during tissue formation. In addition, commercial HSEs must often be cultured in proprietary media according to the manufacturer's recommendations to guarantee viability for a specific timescale [14].

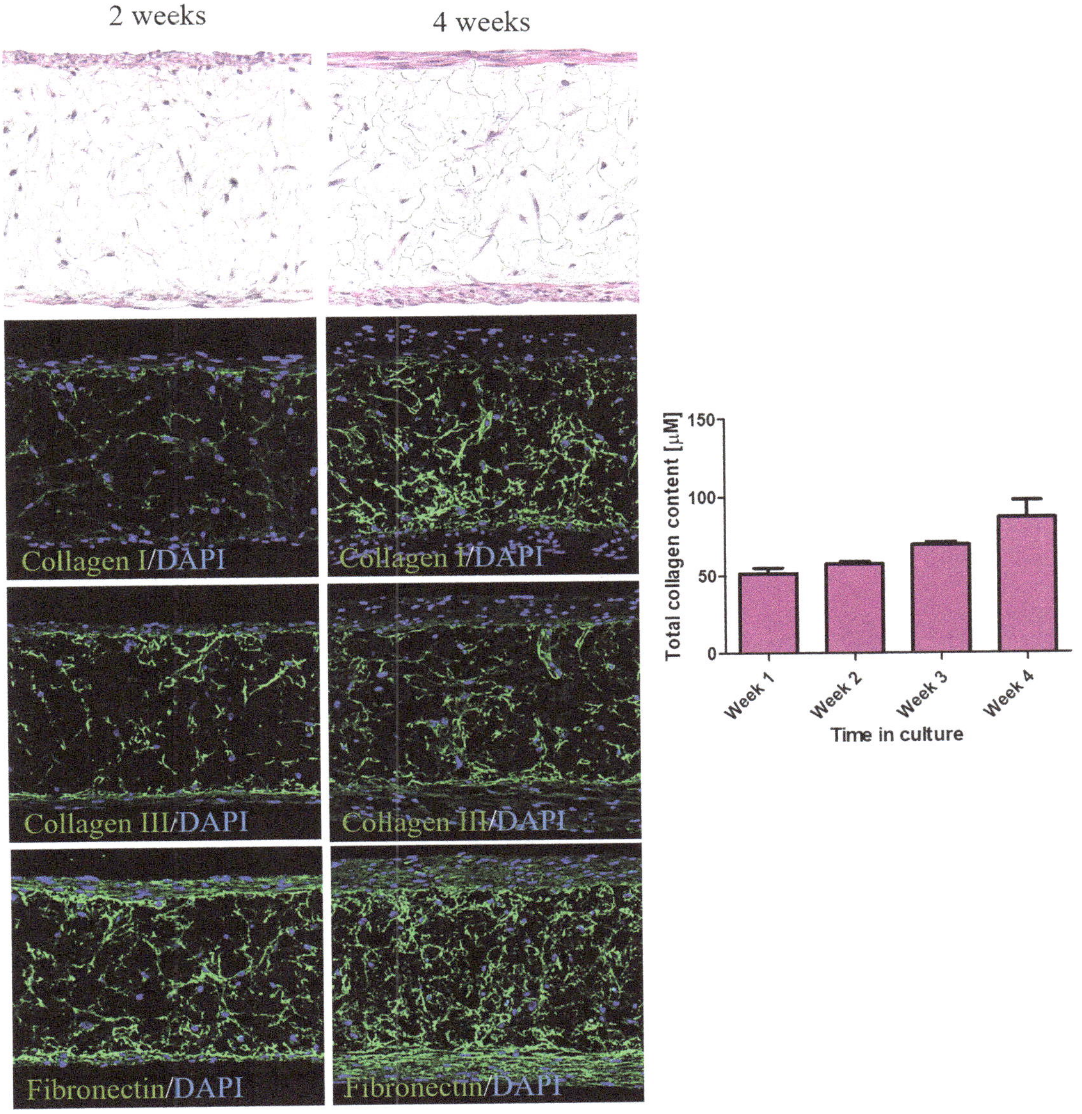

Fig. 2 Deposition of endogenous ECM over time. Representative H&E micrographs of fibroblasts cultured within the Alvetex® scaffold for 2 and 4 weeks to generate a dermal equivalent. Immunofluorescence analysis of collagen I, collagen III, and fibronectin deposited within the scaffold demonstrates increased ECM protein production over time, as does quantification of total collagen content using a hydroxyproline assay

The ability to adapt a 3D tissue model to experimental needs is a critical and essential feature, in order to tailor experimental set ups to answer specific biological questions and ensure that models accurately represent the complex, multicellular nature of in vivo counterparts. In addition, flexibility in culture practices and model format can also offer greater efficiency in experiments and maximize limited or costly resources. The original Alvetex® full

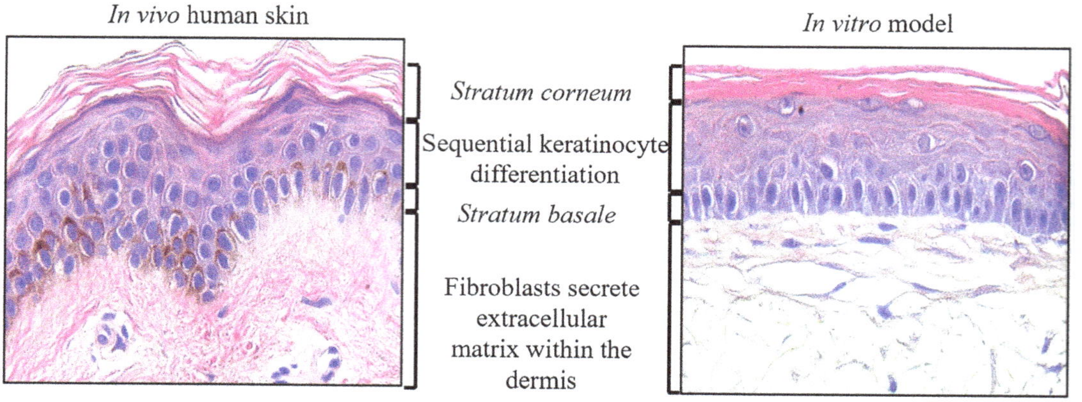

Fig. 3 Histological analysis of the full thickness skin equivalent closely resembles human skin. Representative H&E micrographs of the skin equivalent, in comparison to healthy human skin. The skin equivalent resembles the morphology of human skin, and the self-assembled dermal compartment provides support for the successful differentiation and stratification of the epidermis. The skin sample is from the buttock site of a 23-year-old female, and provided by Procter and Gamble, Cincinnati, USA

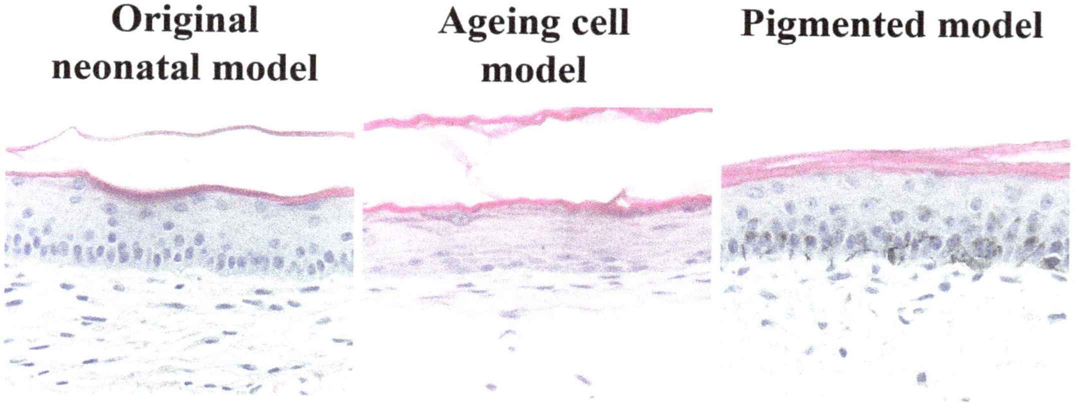

Fig. 4 Use of additional cell populations to create complex HSEs. Since the development of the original Alvetex® model, other cell populations have been used to create HSEs suitable for studying diverse topics including cellular ageing and skin pigmentation, all using commercially available primary cells. Note the differences in cellular structure and tissue morphology

thickness HSE protocol utilizes commercially available neonatal populations of the two most common cell types found in skin and a low-serum growth medium, but this methodology offers significant scope for modification to enable nuanced studies investigating the functions and responses of human skin in health and disease (*see* Fig. 4).

1.4 Alternative and Additional Cell Populations

The multifaceted nature of skin is enacted through highly specialized cell populations resident within it, and their importance is reflected in skin involvement in complex, diverse processes such as protection from the environment, sensation, and immune function. These cells are essential to skin anatomy and function, yet previously many HSEs have utilized only the most common cell populations in their construction: dermal fibroblasts and epidermal keratinocytes. Within these populations alone, significant differences in phenotype and function can be observed when comparing cells from difference sources, notwithstanding additional variations due to disease state. In turn, the co-culture of skin cell populations results in notable impact on their behavior, especially in the context of responses to environmental stressors [15]. As in many tissue engineering applications, there has been a shift in the skin biology field to develop more complex, physiologically representative HSEs, the most common types of which are briefly discussed here.

Ageing The study of cellular and tissue ageing is of paramount importance in the face of an ageing global demographic [16], and this too applies to skin, a tissue which not only experiences ageing, but is a visible sign of the effects of ageing on the human body. Multiple groups have developed HSEs to investigate aspects of skin ageing using a range of methods, including cell populations from aged donors to study intrinsic (chronological) ageing [17] or inducing cellular damage to mimic extrinsic ageing by exogenous stressors on skin cell populations [18]. The incorporation of one or both of these into the Alvetex® full thickness HSE is highly feasible and has been used successfully to develop models to study the impact of senescent fibroblasts on the structure and function of the epidermis [19, 20] and the importance of cellular communication between the dermal and epidermal compartments [21].

Melanocytes and Skin Pigmentation Skin pigmentation is another highly studied area of skin biology for both health-based and cosmetic applications, with topics of interest including the impact of UV exposure, pigmentation dysregulation during disease/ageing and differential responses of different human skin tones to common environmental stimuli. Processes such as these which heavily rely on the activities of epidermal melanocytes resident in the basal layer cannot be accurately investigated in HSEs lacking pigmented cell populations, yet many UV based studies lack melanocyte protection [22–27], leading to unreliable conclusions regarding the impact of UV radiation on skin. Additionally, the impact of melanocytic crosstalk on the function of both keratinocytes and fibroblasts cannot be underestimated considering that melanocytes account for around 5–10% of the human epidermis

[28]. Diverse melanocyte populations have been successfully incorporated into the Alvetex® full thickness HSE to study the impact of UV and cosmetic agents [29, 30] and more critically, to closely investigate proposed mechanisms of melanogenesis, with a more detailed discussion and methodology presented in Chap. 9.

1.5 Methodologies and Model Foundations

In vitro model flexibility should also be considered in the more practical sense of experimental constraints and downstream analytical needs. Restrictive culture regimes or limitations on model size can be highly problematic, giving little room to account for higher throughput or lower yields when using less proliferative cell populations.

Fibroblasts are known to grow successfully in many media types, with HSEs using diverse media compositions for the creation of a dermal compartment. Additional work on the Alvetex® HSE has demonstrated the ability to culture the dermal compartment in other media, resulting in reduced culture time [31] and alterations in ECM production, providing an alternative culture option which also uses fewer resources/labor and is useful if the stringent requirements of industrial collaborators, such as animal free systems, are of less importance.

Constraints on tissue model size can be challenging to deal with, particularly when purchasing commercial products, which makes consideration and choice of methodology even more important before embarking on experimental work. Cell growth/yields and intended analyses are the most critical aspects for determining model size, alongside factors such as study size, media requirements and labor involved. Many commercial products are available in various sizes, enabling the user to design and implement more efficient and effective experimental set ups, and the same can be said for Alvetex® inserts. While the standard format is considered to be the 12-well Alvetex® Scaffold insert, varied formats can be used successfully by simply scaling the seeding densities according to the size of the insert. Smaller formats are especially useful with cell populations which return less favorable yields, such as ageing cells, whereas larger formats provide more material for treatment and analysis.

1.6 Human Skin Equivalents as a Pre-clinical Tool

The ability to adapt in vitro study models to align with real world applications and clinical work by using the same techniques and measurements is a key strength, making them a powerful pre-clinical tool. By being able to directly compare the same measurements between in vitro and in vivo work, in vitro models can be used both as an informative screening tool for preclinical studies and to assess the impact of in vivo practices and stressors on human tissues (*see* Fig. 5).

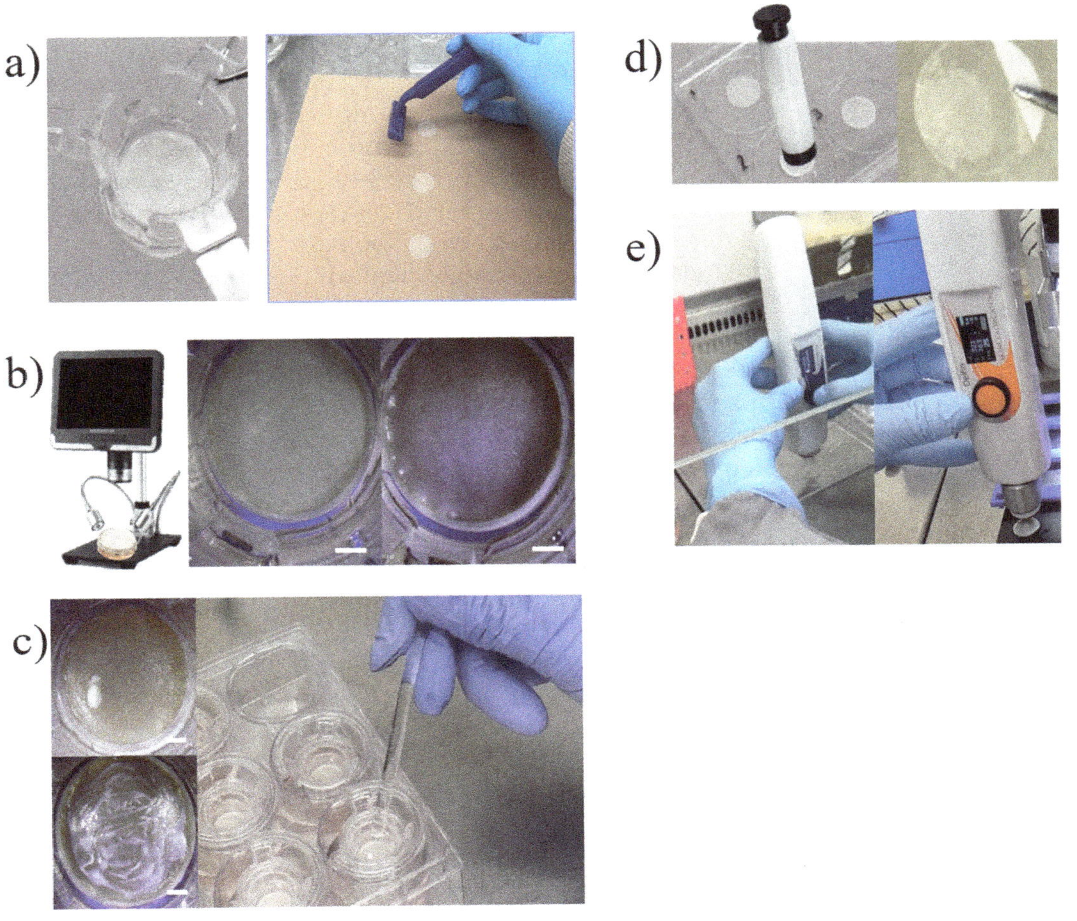

Fig. 5 Use of clinical tools and methods with the Alvetex® full thickness skin model. A range of clinical tools and measures can be used to analyze HSEs: (**a**) Passing a razor across the HSE to simulate shaving; (**b**) Detailed photography of the HSE surface to identify changes in texture or monitor damage and repair; (**c**) Application of topical moisturisers to HSEs; (**d**) Tape stripping HSEs for analysis and simulation of mechanical damage; (**e**) Measurement of transepidermal water loss (TEWL) using a handheld vapometer and analysis of model pigmentation using the handheld SkinColorCatch device

Tape stripping is a commonplace dermatological technique usually used in the clinic to assess multiple skin parameters, including barrier function and stratum corneum composition and organization. While this technique has been successfully used in its own right to analyze these parameters in in vitro HSEs, tape stripping skin has also been characterized in an ex vivo study as a proxy for mechanical exfoliation, particularly razor damage insults [32]. When investigated in the Alvetex® HSE, repeated tape stripping induced barrier damage and inflammation similar to that seen in a dry razor pass [31], correlating with the aforementioned study and providing an alternative strategy to investigate mechanical damage on skin. Shaving is an almost ubiquitous grooming practice; razor passes can also be successfully performed on full thickness HSEs and introducing additional parameters such as shaving foam and moisturisers via topical application mimics a standard shaving regime, as seen in Costello et al. [30], as well as opening up the possibility for a range of topical application studies. Achieving such similarity with normal skin-related practices allows for in-depth investigation into their molecular and cellular effects and future innovations based on biological insights.

Undertaking increasingly applied techniques such as these requires the parallel uptake of clinical measures and tools with HSEs, to allow for comparison. Even simple photography of HSE model surfaces can provide a wealth of information, particularly in studies involving damage. Providing a barrier to the external environment is a key role of skin, thus barrier function is a parameter often assessed in both in vivo and in vitro studies. Transepidermal water loss (TEWL) is an objective measurement of diffusing water vapor from the surface of the stratum corneum and is considered one of the most important measures of skin integrity [33]; thus it is often used to assess the impact of interventions on skin health in academic and clinical work due to its ease of measurement with hand-held technologies. TEWL is known to change in relation to various parameters, so the ability to isolate specific exogenous stressors in the laboratory provides an opportunity to investigate their direct impact in a controlled environment and correlate these with molecular and cellular changes. Similarly, the use of technology to objectively assess skin tone and erythema is rapidly replacing measures such as the Fitzpatrick scale which is limited in its options/descriptors, failing to accurately represent the nuanced nature of skin behavior [34]. Common measures in dermatology studies include Individual Typology Angle and Melanin Index. These tools can also be used with in vitro HSEs to reliably measure the effect of stressors known to impact on skin pigmentation, again allowing for detailed investigation [29, 30].

1.7 Protocols for Creating and Analyzing Full Thickness Skin Equivalents

Below we present protocols for full thickness human skin model generation which utilizes commercially available primary keratinocyte and fibroblast cell stocks. Provided are two protocols for the creation of the dermal compartment of the full thickness models, followed by a generic epidermal compartment protocol, which is the same for both methods. The original dermal protocol utilizes a *low-serum growth medium*. This method is best followed if a protocol with no animal-derived products is paramount, i.e., for example, studies closely linked to cosmetic pre-clinical trials. Additionally, we include an adapted dermal protocol with an alternative medium which benefits from a *reduced culture time* and may be preferred by academic-linked research projects. Also provided are protocols/methods for down-stream analysis of the skin models utilizing clinically relevant tools and techniques.

2 Materials

2.1 Bioengineering Human Skin In Vitro

1. Human neonatal dermal fibroblasts (HDFn).
2. Human neonatal keratinocytes (HEKn).
3. T175 flasks.
4. 12-well Alvetex® insert.
5. 6-well plate.
6. Complete Medium 106 (*Low-Serum Method*): Human Fibroblast Basal Expansion medium (also known previously as Medium 106) supplemented with 10 mL LSGS, 10 μg/mL gentamicin and 0.25 μg/mL amphotericin B.
7. Low Serum Growth Supplement (LSGS) (*Low-Serum Method*): final concentrations of the components in 500 mL supplemented medium are:

   2% v/v fetal bovine serum, 1 μg/mL hydrocortisone, 10 ng/mL human epidermal growth factor, 3 ng/mL basic fibroblast growth factor, 10 μg/mL heparin.
8. Complete Dulbecco's Modified Eagle Medium (DMEM) (*Reduced Culture Time Method*): 500 mL DMEM supplemented with 50 mL Fetal Bovine Serum, 5 mL Penicillin/Streptomycin and 5 mL 200 mM L-glutamine.
9. Fetal Bovine Serum (FBS) (*Reduced Culture Time Method*).
10. EpiLife®.
11. Human keratinocyte growth supplement (HKGS): final concentrations of the components in 500 mL supplemented medium are 0.2% v/v Bovine pituitary extract (BPE), 0.01 μg/mL Recombinant human insulin-like growth factor-I, 0.18 μg/mL Hydrocortisone, 5 μg/mL Bovine transferrin, 0.2 ng/mL Human epidermal growth factor.
12. Penicillin/Streptomycin.
13. Ascorbic acid.
14. Keratinocyte growth factor (KGF).
15. $CaCl_2$ Solution.
16. TGFβ1 Recombinant Human Protein.
17. Trypsin EDTA.
18. Trypsin Neutralizer.
19. Sterile Dulbecco's phosphate buffered saline (DPBS).
20. Phosphate-buffered saline (PBS).
21. Trypan blue solution.
22. Sterile forceps.

2.2 Processing the Skin Equivalent for Paraffin Wax Embedding

1. Paraformaldehyde (PFA).
2. 30, 50, 70, 80, 90, 95, and 100% ethanols.
3. Histoclear.
4. Histoclear II.
5. Paraffin Wax.
6. Tissue-processing cassette.
7. Plastic molds.
8. Mayers Hematoxylin.
9. Eosin.

3 Method

The full thickness skin equivalent is developed using a step-by-step protocol, which involves the generation of a robust dermal equivalent by fibroblasts producing endogenous extracellular matrix proteins within the Alvetex® scaffold, followed by the generation of a well-organized stratified epidermal compartment (Fig. 6). Whether the low-serum or reduced culture time method is followed, the same generation of endogenous extra-cellular matrix is formed, allowing a fully robust and reproducible epidermis to develop above.

All steps are to be carried out in a sterile, category 2 laminar flow hood at room temperature. When disposing of waste materials, follow your respective guidelines and COSHH (Control of Substances Hazardous to Health) forms (*see* **Note 1**).

3.1 Preparation of Alvetex® Scaffold

1. The Alvetex® scaffold must be treated with ethanol to render it hydrophilic before use (*see* **Note 2**). Pre-treat 12-well Alvetex® scaffolds with:
   (a) 70% ethanol
   (b) 1x sterile PBS

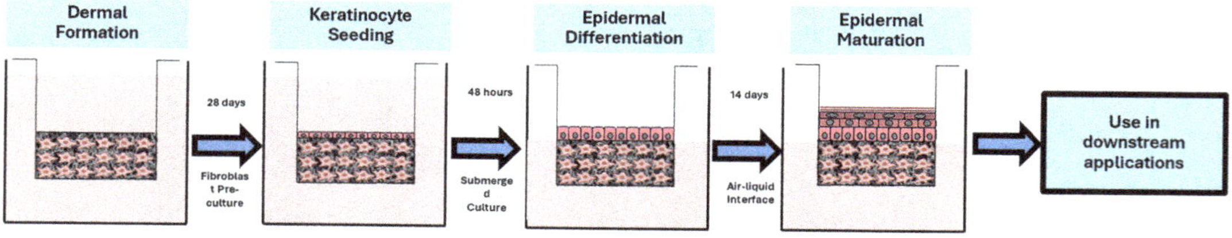

Fig. 6 Methodology used in the generation of the full thickness skin equivalent. In the first phase, a dermal component is generated by seeding dermal neonatal fibroblasts into Alvetex® scaffold and allowing them to establish and generate endogenous extracellular matrix proteins such as collagen and elastin. In the second phase, HEKn are seeded onto the dermal component, and the full thickness skin equivalent is sequentially cultured in submerged and air-liquid interface conditions

(c) Complete Fibroblast Medium (either low-serum complete medium or reduced culture time complete medium, depending on which method is being followed below).

(d) Leave scaffold submerged in complete medium until required.

3.2 Low-Serum Dermal Compartment Method

3.2.1 Revival of Fibroblasts

1. Prepare a bottle of complete Medium 106.
2. Revive and culture human dermal neonatal fibroblasts in complete Medium 106, as per manufacturer's instructions.
3. Incubate HDFn at 37 °C and 5% CO_2 in a humidified environment and change the medium the day after revival to remove the cryopreservative and every 2–3 days subsequently until 80% confluency is reached.

3.2.2 Seeding of Fibroblasts on to Alvetex® Scaffold

1. Prepare trypsin-EDTA by diluting to 1:10 in PBS.
2. Discard medium from the HDFn flask and wash with 5 mL sterile PBS.
3. Trypsinize fibroblasts using 5 mL trypsin-EDTA dilution at 37 °C until cells have detached.
4. Neutralize the Trypsin-EDTA with Trypsin Neutralizer, as per manufacturer's instructions.
5. Centrifuge the cells at 200× *g* for 5 min.
6. Perform cell counts using a trypan blue exclusion assay.
7. Adjust cell density according to number of dermal equivalents to be seeded.
8. Seed 0.5×10^6 HDFn in 100 μL complete Medium 106 onto 12-well Alvetex® scaffold inserts and incubate at 37 °C in a 5% CO_2 humidified incubator for 2 h (*see* **Note 3**).
9. Add 10 mL complete Medium 106 supplemented with 5 ng/mL TGF β1 and 100 μg/mL ascorbic acid.
10. Maintain dermal equivalents for 28 days with 2× media changes/week.

3.3 Reduced Time Dermal Compartment Method

3.3.1 Revival of Fibroblasts

1. Prepare a bottle of complete DMEM.
2. Revive and culture human dermal neonatal fibroblasts in complete DMEM, as per manufacturer's instructions.
3. Incubate HDFn at 37 °C and 5% CO_2 in a humidified environment and change the medium the day after revival to remove the cryopreservative and every 2–3 days subsequently until 80% confluency is reached.

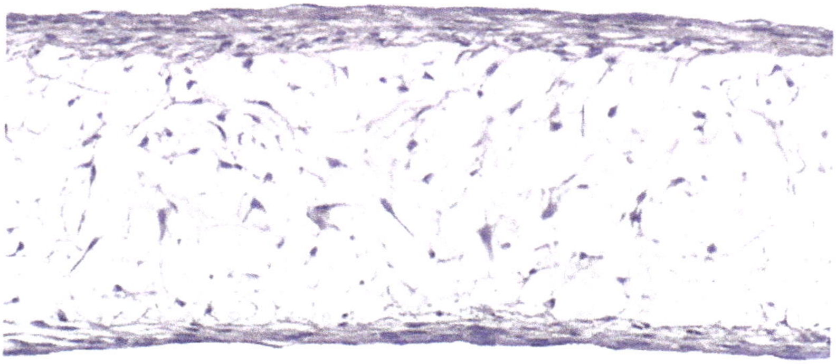

Fig. 7 Mature dermal equivalent. Representative H&E micrograph of a mature dermal equivalent

3.3.2 Seeding of Fibroblasts on to Alvetex® Scaffold

1. Prepare trypsin-EDTA by diluting to 1:10 in PBS.
2. Discard media from the HDFn flask and wash with 5 mL sterile PBS.
3. Trpsinize fibroblasts using 5 mL trypsin-EDTA dilution at 37 °C until cells have detached.
4. Neutralize the Trypsin-EDTA with Trypsin Neutralizer, as per manufacturer's instructions.
5. Centrifuge the cells at 200× *g* for 5 min.
6. Perform cell counts using a trypan blue exclusion assay.
7. Adjust cell density according to number of dermal equivalents to be seeded.
8. Seed 0.5×10^6 HDFn in 100 μL complete DMEM onto 12-well Alvetex® scaffold inserts and incubate at 37 °C in a 5% CO_2 humidified incubator for 2 h.
9. Add 10 mL complete DMEM supplemented with 5 ng/mL TGF β1 and 100 μg/mL ascorbic acid.
10. Maintain dermal equivalents for 28 days with 2× media changes/week (Fig. 7).

3.4 Revival of Keratinocytes

1. Prepare a bottle of EpiLife® medium (500 mL EpiLife® medium supplemented with 5 mL HKGS). HEKn are to be revived 6–8 days before seeding onto dermal equivalents.
2. Revive and culture HEKn seeded at 2.5×10^3 cm^{-2} according to the supplier's instructions in 25 mL in complete EpiLife® medium in a T175 flask.
3. Incubate HEKn at 37 °C and 5% CO_2 in a humidified environment and change the medium the day after revival to remove the cryopreservative and every 2–3 days subsequently until 80% confluency is reached.

3.5 Seeding of Keratinocytes onto the Dermal Equivalent

1. Prepare a bottle of complete EpiLife® medium (EpiLife® supplemented with 5 mL HKGS and 10 μg/mL gentamicin and 0.25 μg/mL amphotericin B).
2. Trypsinize keratinocytes using Trypsin EDTA until cells have detached.
3. Neutralize the Trypsin EDTA with Trypsin Neutralizer, as per manufacturer's instructions.
4. Centrifuge the cells at 200× *g* for 5 min.
5. Perform cell counts using a trypan blue exclusion assay.
6. Seed 1.3×10^6 human epidermal neonatal keratinocytes in 200 ul Epilife medium on to the mature dermal equivalent and incubate at 37 °C in a 5% humidified incubator for 2 h.
7. Add 10 mL complete EpiLife medium supplemented with 10 ng/mL KGF, 140 uM $CaCl_2$ and 100 ug/mL ascorbic acid.
8. Incubate at 37 °C in a 5% CO_2 humidified incubator for 48 h in complete EpiLife® medium supplemented with 10 ng/mL KGF, 140 μM $CaCl_2$ and 100 μg/mL ascorbic acid.

   Raising Full Thickness Models to the Air-Liquid Interface.

9. After 48 h, raise the full thickness skin equivalents to the air-liquid interface.
10. Culture the full thickness skin equivalent in complete EpiLife® medium supplemented with 10 ng/mL KGF, 1.64 mM $CaCl_2$ and 100 μg/mL ascorbic acid.
11. After 7 days at the air-liquid interface, the epidermis is fully formed; however it can be maintained for longer for long-term experiments (Fig. 8).

3.6 Harvesting the Skin Equivalents for Paraffin Wax Embedding

1. Remove the skin equivalent from the plastic insert holders using a pair of flat-ended sterile forceps.
2. Wash the skin equivalent three times in 5 mL PBS and fix the skin equivalents in 4% PFA for 2 h at room temperature, or overnight at 4 °C.
3. Wash the skin equivalents three times in 5 mL PBS.
4. Sequentially dehydrate the skin equivalents in 30%, 50%, 70%, 80%, 90%, and 95% ethanol v/v for 10 min each.
5. Place the skin equivalents in 100% ethanol for 30 min.
6. Transfer the skin equivalents to tissue processing cassettes and place in Histoclear II for 30 min.
7. Incubate the skin equivalents in a 1:1 ratio of Histoclear II: molten paraffin wax in a convection oven at 65 °C for 30 min.

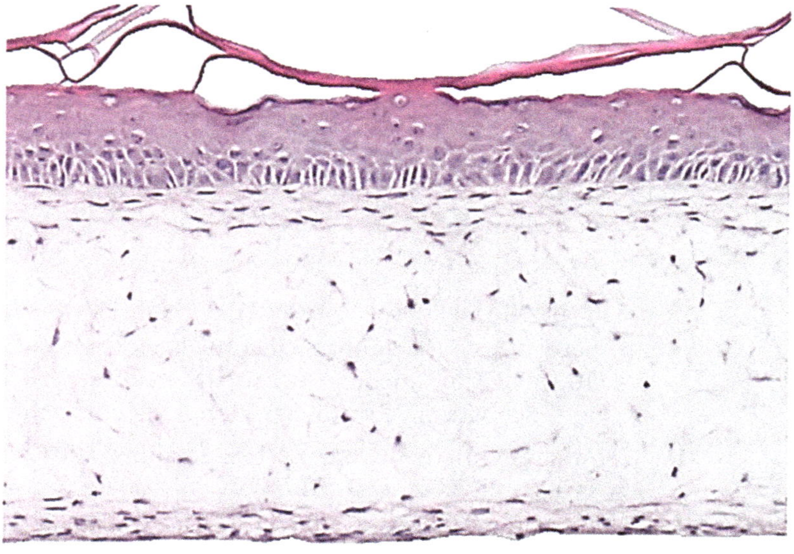

Fig. 8 Mature full thickness skin equivalent Representative H&E micrograph of a mature full thickness skin equivalent

8. Place the skin equivalents in 100% molten wax for at least 1 h at 65 °C, prior to cutting the models in half and embedding in tissue processing molds.
9. Allow the embedded samples to set overnight prior to downstream applications.

3.7 Processing Samples for H&E Analysis

1. Section the transverse skin equivalents at a thickness of 5 μm using a rotary manual microtome.
2. Place sections in a 37 °C water bath prior to mounting on to charged slides.
3. Allow slides to dry on a 30 °C heated slide rack before staining.
4. Deparaffinize the sections in Histoclear for 5 min.
5. Sequentially rehydrate the skin equivalents in 95% ethanol, 70% ethanol and distilled water for 1 min each.
6. Stain the slides with Mayers Hematoxylin for 5 min, which stains negatively charged components, such as nucleic acids, blue.
7. Wash away any excess Mayers Hematoxylin in distilled water for 30 s.
8. Transfer the samples to alkaline alcohol to ensure that the hematoxylin dye appears blue.
9. Sequentially dehydrate the samples in 70% ethanol and 95% ethanol for 30 s each.

10. Stain with Eosin for 30 s to stain positively charged components pink.
11. Further dehydrate the samples twice in 95% ethanol for 10 s, then twice in 100% ethanol for 15 and 30 s, respectively.
12. Place the skin equivalents in Histoclear twice for 3 min each.
13. Mount a clean coverslip onto the slide using Omnimount.
14. Leave to dry inside a fume hood at room temperature prior to imaging with a light microscope.

3.8 Processing Samples for Immunofluorescence Analysis

1. Section the transverse skin equivalents at a thickness of 5 μm using a rotary manual microtome.
2. Place sections in a 37 °C water bath prior to mounting on to charged slides.
3. Allow slides to dry on a 30 °C heated slide rack prior to use.
4. Deparaffinize the sections in Histoclear for 15 min.
5. Sequentially rehydrate the skin equivalents in 100% ethanol, 70% ethanol and PBS for 5 min each.
6. Perform antigen retrieval in pH 6 citrate buffer at 95 °C for 20 min.
7. After cooling, block and permeabilize the slides in 150 μL 20% neonatal calf serum in 0.4% Triton-X-100 in PBS for 1 h at room temperature.
8. Add 150 μL of the primary antibody diluted in blocking solution, and incubate with the samples overnight in a humidified chamber at 4 °C (Table 1).
9. Wash the samples three times in PBS for 10 min each.
10. Incubate the samples with 150 μL of the relevant secondary antibody diluted in blocking solution for 1 h at room temperature (Table 2).

Table 1
Examples of primary antibodies used for immunofluorescence staining with suppliers and working concentrations

Antibody	Supplier	Product code	Dilution
Keratin 10	Abcam	ab76318	1:100
Keratin 14	Abcam	ab7800	1:100
Filaggrin	Abcam	ab17808	1:100
Collagen I	Abcam	ab34710	1:100
Collagen III	Abcam	ab7778	1:100
Fibronectin	Abcam	ab23750	1:100

Table 2
Examples of secondary antibodies used for immunofluorescence staining with suppliers and working concentrations

Antibody	Supplier	Product code	Dilution
Alexa Fluor 594 Donkey Anti-Rabbit IgG (H + L)	ThermoFisher Scientific	10798994	1:1000
Alexa Fluor 488 Donkey Anti-Mouse IgG (H + L)	ThermoFisher Scientific	10544773	1:1000
Alexa Fluor 488 Donkey Anti-Rabbit IgG (H + L)	ThermoFisher Scientific	10424752	1:1000

11. Wash the samples three times in PBS for 10 min each.
12. Mount the coverslip onto the slides using Hardset Vectashield with DAPI.
13. Allow the Hardset Vectashield with DAPI to set for 15 min at room temperature, prior to imaging with a confocal microscope.

4 Notes

1. All solutions and equipment that comes into contact with the cells must be sterile and used within a Category 2 laminar flow hood. Aseptic technique should also be used accordingly.
2. Alvetex® scaffold is hydrophobic; therefore, the initial ethanol wash is important to render it hydrophilic prior to the seeding of cells.
3. Alvetex® scaffold is commercially available in a range of formats, and additional insert sizes can be used to generate the full thickness model, if cell seeding densities are adjusted according to surface area.

References

1. Proksch E, Brandner JM, Jensen JM (2008) The skin: an indispensable barrier. Exp Dermatol 17:1063–1072. https://doi.org/10.1111/j.1600-0625.2008.00786.x
2. Brenner M, Hearing VJ (2008) The protective role of melanin against UV damage in human skin. Photochem Photobiol 84:539–549. https://doi.org/10.1111/j.1751-1097.2007.00226.x
3. Merkel Fr (1875) Tastzellen und Tastkörperchen bei den Hausthieren und beim Menschen. Archiv f mikrosk Anat 11:636–652. https://doi.org/10.1007/BF02933819
4. Streilein JW, Bergstresser PR (1984) Langerhans cells: antigen presenting cells of the epidermis. Immunobiology 168:285–300. https://doi.org/10.1016/S0171-2985(84)80117-5
5. Eckert RL, Rorke EA (1989) Molecular biology of keratinocyte differentiation. Environ Health Perspect 80:109–116. https://doi.org/10.1289/ehp.8980109
6. Rice RH, Green H (1979) Presence in human epidermal cells of a soluble protein precursor of the cross-linked envelope: activation of the cross-linking by calcium ions. Cell 18:681–694. https://doi.org/10.1016/0092-8674(79)90123-5
7. Briggaman RA, Wheeler CE (1975) The epidermal-dermal junction. J Invest Dermatol

65:71–84. https://doi.org/10.1111/1523-1747.ep12598050

8. Cotta-Pereira G, Guerra Rodrigo F, Bittencourt-Sampaio S (1976) Oxytalan, elaunin, and elastic fibers in the human skin. J Invest Dermatol 66:143–148. https://doi.org/10.1111/1523-1747.ep12481882
9. Gerber PA, Buhren BA, Schrumpf H et al (2014) The top skin-associated genes: a comparative analysis of human and mouse skin transcriptomes. Biol Chem 395:577–591. https://doi.org/10.1515/hsz-2013-0279
10. Wong VW, Sorkin M, Glotzbach JP et al (2011) Surgical approaches to create murine models of human wound healing. J Biomed Biotechnol 2011:969618. https://doi.org/10.1155/2011/969618
11. European Commission (2009) Regulation (EC) No 1223/2009 of the European Parliament and of the Council of 30 November 2009 on cosmetic products (recast)
12. Bikle DD, Xie Z, Tu C-L (2012) Calcium regulation of keratinocyte differentiation. Expert Rev Endocrinol Metab 7:461–472. https://doi.org/10.1586/eem.12.34
13. Costello L, Fullard N, Roger M et al (2019) Engineering a multilayered skin equivalent: the importance of endogenous extracellular matrix maturation to provide robustness and reproducibility. Methods Mol Biol 1993:107–122. https://doi.org/10.1007/978-1-4939-9473-1_9
14. EpiDermFT in vitro 3D Tissue. In: MatTek Life Sciences. https://www.mattek.com/mattekproduct/epidermft/. Accessed 13 Mar 2024
15. Duval C, Régnier M, Schmidt R (2001) Distinct melanogenic response of human melanocytes in mono-culture, in co-culture with keratinocytes and in reconstructed epidermis, to UV exposure. Pigment Cell Res 14:348–355. https://doi.org/10.1034/j.1600-0749.2001.140506.x
16. United Nations (2019) Ageing. In: United Nations. https://www.un.org/en/global-issues/ageing. Accessed 17 Sept 2021
17. Dos Santos M, Metral E, Boher A et al (2015) *In vitro* 3-D model based on extending time of culture for studying chronological epidermis aging. Matrix Biol 47:85–97. https://doi.org/10.1016/j.matbio.2015.03.009
18. Weinmüllner R, Zbiral B, Becirovic A et al (2020) Organotypic human skin culture models constructed with senescent fibroblasts show hallmarks of skin aging. npj Aging Mech Dis 6:1–7. https://doi.org/10.1038/s41514-020-0042-x
19. Smith L, Costello L, Low E et al (2022) 121 Using in vitro human skin equivalents to understand the complexities of skin ageing and identify therapeutic interventions to restore skin health. J Invest Dermatol 142:S21. https://doi.org/10.1016/j.jid.2022.05.127
20. Smith L, Low E, Miwa S et al (2023) 357 bioengineered human ageing skin tissue: a new in vitro platform to investigate the biology of ageing skin and restore skin health. J Invest Dermatol 143:S61. https://doi.org/10.1016/j.jid.2023.03.362
21. Costello L (2020) Bioengineering novel ageing skin equivalents and their application to study the importance of dermal-epidermal interactions during skin ageing. Doctoral, Durham University
22. Huang XX, Bernerd F, Halliday GM (2009) Ultraviolet a within sunlight induces mutations in the epidermal basal layer of engineered human skin. Am J Pathol 174:1534–1543. https://doi.org/10.2353/ajpath.2009.080318
23. Lejeune F, Christiaens F, Bernerd F (2008) Evaluation of sunscreen products using a reconstructed skin model exposed to simulated daily ultraviolet radiation: relevance of filtration profile and SPF value for daily photoprotection. Photodermatol Photoimmunol Photomed 24:249–255. https://doi.org/10.1111/j.1600-0781.2008.00370.x
24. Marionnet C, Pierrard C, Lejeune F et al (2010) Different oxidative stress response in keratinocytes and fibroblasts of reconstructed skin exposed to non extreme daily-ultraviolet radiation. PLoS One 5:e12059. https://doi.org/10.1371/journal.pone.0012059
25. Meloni M, Farina A, de Servi B (2010) Molecular modifications of dermal and epidermal biomarkers following UVA exposures on reconstructed full-thickness human skin. Photochem Photobiol Sci 9:439–447. https://doi.org/10.1039/b9pp00164f
26. Mieremet A, van Dijk R, Gooris G et al (2019) Shedding light on the effects of 1,25-dihydroxyvitamin D3 on epidermal lipid barrier formation in three-dimensional human skin equivalents. J Steroid Biochem Mol Biol 189:19–27. https://doi.org/10.1016/j.jsbmb.2019.01.022
27. Vioux-Chagnoleau C, Lejeune F, Sok J et al (2006) Reconstructed human skin: from photodamage to sunscreen photoprotection and anti-aging molecules. J Dermatol Sci Suppl 2:S1–S12. https://doi.org/10.1016/j.descs.2006.08.001

28. Thingnes J, Lavelle TJ, Hovig E et al (2012) Understanding the melanocyte distribution in human epidermis: an agent-based computational model approach. PLoS One 7:e40377. https://doi.org/10.1371/journal.pone.0040377
29. De Los Santos Gomez P, Costello L, Goncalves K et al (2024) Comparison of photodamage in non-pigmented and pigmented human skin equivalents exposed to repeated ultraviolet radiation to investigate the role of melanocytes in skin photoprotection. Front Med 11. https://doi.org/10.3389/fmed.2024.1355799
30. Goncalves K, De Los Santos Gomez P, Costello L et al (2023) Investigation into the effect of skin tone modulators and exogenous stress on skin pigmentation utilizing a novel bioengineered skin equivalent. Bioeng Transl Med 8: e10415. https://doi.org/10.1002/btm2.10415
31. Costello L, Goncalves K, Maltman V et al (2023) Development of a novel in vitro strategy to understand the impact of shaving on skin health: combining tape strip exfoliation and human skin equivalent technology. Front Med (Lausanne) 10:1236790. https://doi.org/10.3389/fmed.2023.1236790
32. Dabboue H, Builles N, Frouin É et al (2015) Assessing the impact of mechanical damage on full-thickness porcine and human skin using an in vitro approach. Biomed Res Int 2015: 434623. https://doi.org/10.1155/2015/434623
33. Akdeniz M, Gabriel S, Lichterfeld-Kottner A et al (2018) Transepidermal water loss in healthy adults: a systematic review and meta-analysis update. Br J Dermatol 179:1049–1055. https://doi.org/10.1111/bjd.17025
34. Harvey VM, Alexis A, Okeke CAV et al (2024) Integrating skin color assessments into clinical practice and research: a review of current approaches. J Am Acad Dermatol 91:1189. https://doi.org/10.1016/j.jaad.2024.01.067

Chapter 13

Bioprinting a Tri-layered Skin Analogue

Hilal Mete Gunaydin, Saskia M. Fogg, Richard J. A. Moakes, and Liam M. Grover

Abstract

Although it is now possible to use tissue engineering approaches to regenerate the dermis and epidermis of the skin, replicating the cellular framework and functions of the skin's three primary layers—epidermis, dermis, and hypodermis remains challenging. Here, we describe the use of a bioprinting method known as suspended layer additive manufacturing (SLAM) to produce a structure with a gradient of mechanical properties, supporting the growth of various cell types, to create a continuous tri-layered skin substitute that closely mimics human skin. SLAM utilizes fluid gels as a temporary support matrix to facilitate the extrusion bioprinting of soft hydrogel-based bioinks into complex, functional, three-dimensional (3D) tissue structures. The printed structures are initially designed using computer-aided design (CAD) and may be printed using a variety of bioinks that can replicate the three primary skin layers: epidermal, dermal, and hypodermal. To accurately reproduce the cellular framework and functions of human skin, it is crucial to use materials that can duplicate the natural extracellular matrix (ECM) structure. This structure provides structural support and appropriate biochemical cues to the embedded cells. Here we describe a method that uses pectin to mimic ECM polysaccharides, along with collagen, which is a major component of human skin ECM. The resulting structure is supported within an agarose support-bed and ionically cross-linked prior to removal. To accurately replicate the cellular framework of human skin, human epidermal keratinocytes (hEKs), human dermal fibroblasts (HDFs), and adipose-derived stem cells (ADSCs) were integrated into the scaffold. A 21-day culture of the skin construct showed that the cellular components were crucial in remodelling the printed structure into architectures resembling those of healthy skin. Integration of the implant, demonstrated by the mobilization of adipose tissue from the surrounding area into the construct, occurred within 7 days post-ex vivo implantation into a simulated porcine wound.

Key words Tissue engineering, 3D bioprinting, Suspended layer additive manufacturing (SLAM), Tri-layered skin implant, Hydrogel-based bioinks

1 Introduction

Human skin is both an important component and the largest organ in the human body [1–4]. The human skin is a multi-layered structure which predominantly consists of the epidermis, dermis, and hypodermis [1, 2, 5]. The epidermis contains densely packed

Thomas Biedermann and Sophie Böttcher-Haberzeth (eds.), *Skin Tissue Engineering: Methods and Protocols*, Methods in Molecular Biology, vol. 2922, https://doi.org/10.1007/978-1-0716-4510-9_13,

keratinocytes, forming the initial barrier against external substances. It also includes keratin and collagen type IV/VII [2]. The dermis is abundant in blood vessels, nerve endings, and various glands, and is primarily made up of fibroblasts that produce collagen for ECM composed primarily of collagen type I and elastin. The dermis provides structural support, nutrition, and sensation through its blood vessels and nerves [1, 2, 6]. The hypodermis, also known as subcutaneous tissue, is the innermost layer of human skin. It is primarily composed of adipose tissue, blood vessels, nerves, and hair follicles, with collagen and elastin as essential ECM components [2]. This layer supplies nerves and blood vessels to the upper layers, playing a crucial role in re-epithelialization, wound healing, and angiogenesis [7] .

While superficial skin injuries can heal through immunological responses [1–3], large-scale or deep injuries cannot fully restore the original cellular environment, hindering the regeneration of hair follicles and sweat glands [1, 6, 8]. Incorporating the hypodermal layer in skin replacements is crucial for treatments involving full-thickness skin loss. Despite its vital role in wound healing, recent advances in tissue engineering have overlooked the significance of the subcutaneous tissue [2, 9, 10]. Indeed, studies suggest that adipose-derived stem cells (ADSCs) from the subcutaneous layer support epidermal morphogenesis, promote neovascularization, and accelerate wound healing [11, 12]. Therefore, accurately replicating natural skin necessitates including all layers, particularly the subcutaneous layer.

Hydrogel bioinks have been used as supports for the growth a range of tissues in vitro. They are widely used because they are highly hydrated and permeable meaning that it is possible to embed cells within them while maintaining high viability [3]. Additionally, they can be designed to exhibit the necessary rheological characteristics to enable extrusion including low viscosity and yield stress while avoiding the high extrusion pressures that can jeopardize cell viability during printing. However, these characteristics also compromise the manufacturability of these bioinks, resulting in their collapse and a loss of print fidelity following deposition onto a hard surface [13–15]. This property-level dichotomy is resolved by embedded printing techniques, where bioinks are deposited into a temporary support matrix which suspends the bioink, preventing the collapse of the structure until it can be stabilized via crosslinking [13, 16–19]. SLAM utilizes fluid gels as a temporary support matrix. Composed of micrometer-sized gel particles, fluid gels behave in bulk as viscoelastic fluids with shear-thinning and self-healing properties. This enables the support matrix to act as a liquid as the cartridge needle passes through it, yet rapidly recover solid-like properties to support the deposited bioink.

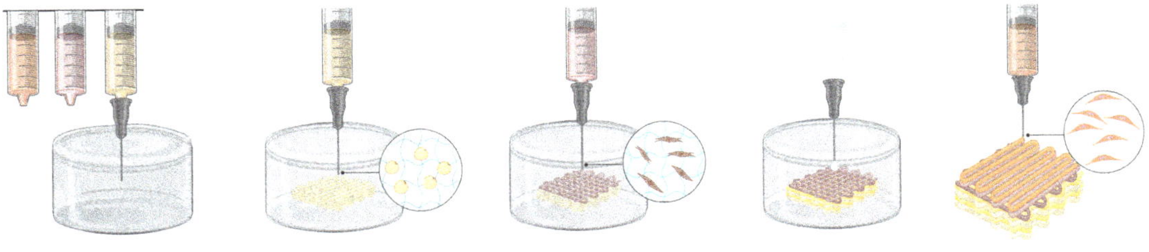

Fig. 1 A schematic illustrating the bioprinting of a tri-layered skin analogue using SLAM with multiple cartridges for depositing different hydrogel layers to create a multilayered construct. **a** Agarose fluidgel acts as a support bed, providing a temporary matrix to suspend the bioinks. **b** Bioink, prepared from a collagen/pectin (1:1) blend and adipose-derived stem cells (ADSCs), is extruded into the suspension medium to replicate the hypodermal layer ($15 \times 15 \times 7$ mm^3). **c** Bioink composed of collagen/pectin (2:1) and human dermal fibroblasts (HDFs) is extruded to mimic the dermal layer, including both reticular and papillary layers ($15 \times 15 \times 1$ mm^3). **d** Post-fabrication, constructs are solidified by local injection of $CaCl_2$ (200 mM), inducing gelation. **e** Low shear washing with deionized water releases the construct from the supporting agarose gel. Human epidermal keratinocytes (hEKs) are then seeded on top of the construct and cultured to form the epidermal layer

Here we report a method for the manufacture of a tri-layered skin substitute using the SLAM methodology. The method includes the formulation of a bioink formed from a blend of collagen and pectin that can be tailored to exhibit the distinctive mechanical properties found in each skin layer. Importantly, this material allows for the long-term maintenance of the embedded cells and the development of a mature tri-layered tissue engineered structure.

The entire fabrication process, including the sequential deposition of different hydrogel layers using SLAM and the creation of the epidermal, is schematically illustrated in Fig. 1.

2 Materials

Prepare all solutions with de-ionized sterile water and analytical-grade reagents.

2.1 Agarose Solution (for the Support Bed)

1. Agarose (type 1 low EEO).
2. Heat 100 mL deionized water to approximately 60 °C (*see* **Note 1**). Disperse 500 mg agarose powder into the heated water while continuously stirring. Place the solution in an autoclave and heat to 120 °C for 15 min. Remove the solution from the autoclave and allow to cool to 100 °C with constant agitation in a Class II biological safety cabinet. Immediately transfer the solution to the rheometer as specified in Subheading 3.1.

2.2 Pectin Solution

1. Low methoxy (LM) pectin powder (*see* **Note 2**).
2. Measure out 80 mL de-ionized water and add it to a Duran bottle. Weigh 5 g LM pectin powder and transfer it to the bottle, activating the magnetic stirrer. Begin dissolving the pectin powder under constant agitation to create a colloidal suspension. Top up the solution to 100 mL (*see* **Note 3**).
3. Autoclave the colloidal suspension at 121 °C for 15 min (*see* **Note 4**). Allow the solution to cool to 25 °C while continuously mixing at 700 rpm (*see* **Note 5**). Store the solution at 4 °C (*see* **Note 6**).

2.3 Collagen Solution

1. Collagen (PureCol EZ Gel).
2. 5 mg/mL collagen solution: prepare under sterile conditions (*see* **Note 7**) by dissolving collagen powder in Dulbecco's Modified Eagle Medium (DMEM) to achieve a final concentration of 5 mg/mL. This preparation method is similar to the procedure described in Subheading 2.2. Store the solution at 4 °C (*see* **Note 6**).

2.4 Cell Culture

This process was carried out in a Class II biological safety cabinet to maintain sterility.

1. Human skin tissue samples including subcutaneous adipose tissue (*see* **Note 8**). Transport and store samples under appropriate conditions (*see* **Note 9**).
2. Dulbecco's Modified Eagle Medium (DMEM).
3. Dispase II solution (*see* **Note 10**).
4. Use a sterile surgical scalpel blade (size 24 is recommended for easier cutting) along with a blade holder [20].
5. 0.5 mg/mL Collagenase D (from *Clostridium histolyticum)*: dissolve Collagenase D powder in sterile phosphate-buffered saline (PBS) (*see* **Note 11**).
6. 1× Trypsin-EDTA solution: mix 0.5 g porcine trypsin, 0.2 g EDTA, 4 Na per liter Hank's Balanced Salt Solution with phenol red, filter sterile (*see* **Note 11**).
7. Basal media: Measure 90 mL DMEM and pour it into a sterile container. Add 10 mL Fetal Bovine Serum (FBS) to the DMEM. Add 2.5 mL HEPES buffer to the mixture, followed by 1 mL penicillin/streptomycin solution. Mix the solution thoroughly to ensure that all ingredients are evenly distributed. Filter the media through a 100 μm cell strainer (*see* **Note 12**). Store the prepared media in a sealed bottle at 4 °C until use (*see* **Note 6**).
8. Cell strainer (100 μm, 70 μm, and 40 μm) (*see* **Note 12**).
9. T25 flask.

10. Keratinocyte growth medium.
11. Mitomycin-C treated 3 T3-J2 murine fibroblast feeder layers.
12. Sieve, mesh size 100.
13. Collagenase type I.
14. Erythrocyte lysis buffer.
15. PBS (phosphate buffered saline).
16. Adipogenic media (*see* **Note 13**): supplement the basal media with the following additives: 500 μM isobutyl-methylxanthine (IBMX), 50 μM indomethacin, and 1 μM dexamethasone. Store the prepared media at 4 °C (*see* **Note 6**). Ensure that the media is kept sterile.

2.5 Fabrication of Tri-layered Structure

1. Blends of collagen and pectin in ratios of 2:1 and 1:1.
2. Human dermal fibroblasts (HDFs).
3. Adipose-derived stem cells (ADSCs).
4. 3D bioprinter (INKREDIBLE™).
5. Petri dish.
6. Agarose fluid gel (serving as the support bed): place upon the z-stage in a petri dish.
7. 200 mM $CaCl_2$ (*see* **Note 14**).
9. Needles with internal diameters of 150, 200, 250, 410, and 610 μm.
10. Adipogenic media (*see* **Note 13**).
11. Human epidermal keratinocytes (hEKs).

2.6 Live/Dead Cell Viability Test

1. Calcein AM.
2. Propidium iodide (PI).
3. Fluorescent microscope.

2.7 Ex Vivo Integration Study

1. Porcine skin.
2. Q-Tracker cell labelling kit (kit 488).

2.8 Embedding 3D Printed Skin Equivalents

1. 10% neutral buffered formalin solution.
2. Paraffin.
3. 70, 95, 100% w/v ethanol solutions.
4. Xylene (*see* **Note 19**).
5. Tissue Embedder (Leica EG 1150C).

2.9 Hematoxylin and Eosin (H&E) Staining

1. Paraffin-embedded samples.
2. Microtome.
3. Adhesion-coated slides.

4. Xylene.
5. Absolute, 70%, and 30% w/v ethanol solutions.
6. Hematoxylin solution.
7. Tap water.
8. Eosin solution.
9. Mounting solution.
10. Coverslips.
11. Confocal microscope (Leica CTR 6500).

3 Methods

Carry out all procedures at room temperature unless otherwise specified.

3.1 Agarose Fluid Gel Support Matrix

1. Place a rheometer in a Class II biological safety cabinet. Sterilize the rheometer and safety cabinet to maintain the sterility of the agarose solution. Heat the rheometer, equipped with cup and bob geometry, to 100 °C.
2. Transfer the 100 °C agarose solution to the rheometer (*see* **Note 15**). Use the rheometer to cool the agarose fluid from 100 to 20 °C at a controlled rate of 1 °C per minute, maintaining a constant shear rate of 1200/s throughout (*see* **Note 16**).
3. After preparation, allow agarose fluid gels to age for 24–48 h before use, as they undergo an ageing process due to their granular nature (*see* **Note 17**). Store agarose fluid gels at 4 °C to maintain their integrity and properties over time (*see* **Note 16**).

3.2 Collagen: Pectin Bioinks

1. Mix the collagen solution with the pectin solution (5 mg/mL in DMEM) in a 2:1 or 1:1 ratio as follows (*see* **Note 7**).
2. To create a bioink for the papillary layer, use a 2:1 blend of collagen to pectin to resuspend 3.0×10^6 cells/mL of human dermal fibroblasts (HDFs) at passage 7.
3. To create a bioink for the reticular layer, use a 2:1 blend of collagen to pectin to resuspend 1.5×10^6 cells/mL of HDFs at passage 7.
4. To create a bioink for the hypodermal layer, use a 1:1 blend of collagen to pectin to resuspend 5.0×10^5 cells/mL of adipose-derived stem cells (ADSCs) at passage 5.

3.3 Isolation of Human Cells from Skin

1. Section the dermo-epidermal tissue from the subcutaneous adipose tissue and submerge in dispase II solution at 4 °C overnight before separating the dermal and epidermal counterparts.
2. Isolate the human dermal fibroblasts (HDFs) by incubating the dermis in collagenase D solution overnight, followed by trypsin treatment and mechanical agitation using a scalpel. Centrifuge the cells and re-suspend in the basal media before expanding in a T25 flask.
3. Isolate the keratinocytes by treating the epidermis with trypsin followed by mechanical mincing. Quench enzymatic reactions using keratinocyte growth medium (KGM). Centrifuge, resuspend, and filter the cells through 70 μm cell strainer before plating cells onto mitomycin-C treated 3 T3-J2 murine fibroblast feeder layers.
4. Isolate stromal vascular fraction (SVF) from adipose tissue by mechanical mincing followed by sieving (dissociation sieve size 100 mesh) and incubating in collagenase type I. Use media to quench the reaction of collagenase and sequentially filter the SVF through 100 and 40 μm filters. Centrifuge the solutions, and discard the lipid, mature adipocyte, and aqueous layers. Suspend the pellet in erythrocyte lysis buffer followed by further centrifugation before plating in the basal media.
5. Sub-culture extracted SVF as per standard trypsinization protocols for 21 days to produce an ADSCs population.
6. Culture ADSCs in adipogenic media.
7. Maintain the culture of HDFs, HEKs, and ADSCs for 21 days.

3.4 Fabrication of Implant Constructs

1. Create agarose fluid gel and bioinks according to the specifications outlined in Subheading 3.1 and 3.2, respectively.
2. Align and calibrate the bioprinter to the optimized printing parameters (Optimal printing conditions were obtained to be: flow rate—3.8 $\mu L\, s^{-1}$, pressure—5 kPa, and ϕID 410 μm) and place a petri dish containing the agarose fluid gel upon the z-stage.
3. Upload the G-code for the construct to the printing software to print a tri-layer consisting of the following dimensions: hypodermis layer (15 × 15 × 7 mm^3), reticular layer (15 × 15 × 1 mm^3), and papillary layer (15 × 15 × 1 mm^3).
4. Ensure the printer automatically switches inks upon each successive layer (*see* **Note 18**).
5. Following fabrication, gel constructs using 200 mM $CaCl_2$ through injection to the local area.

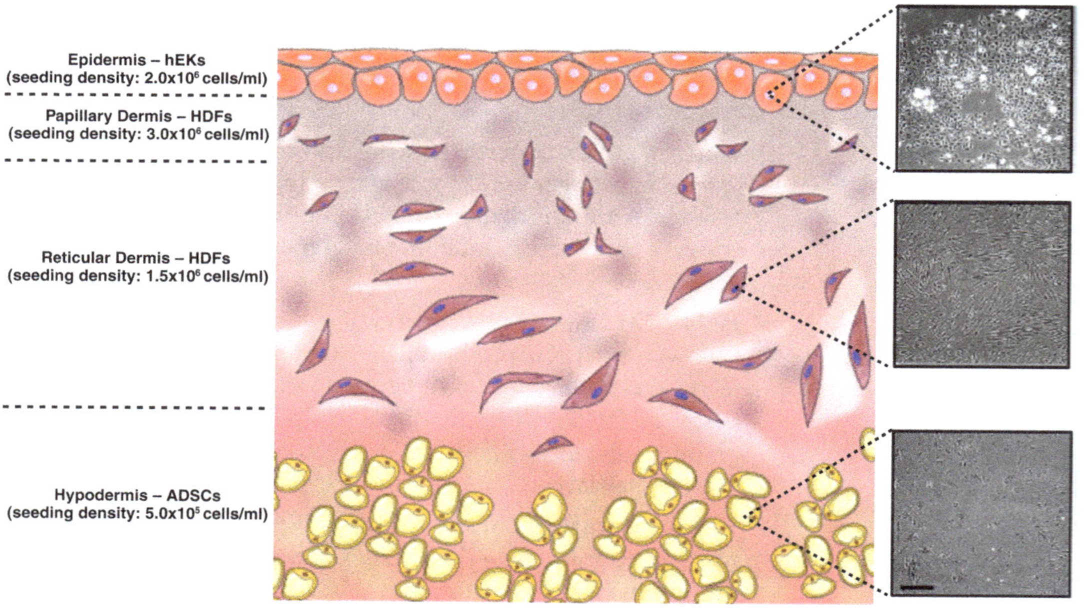

Fig. 2 Illustration of the skin structure, highlighting the skin layers, hypodermis, papillary and reticular dermis, and epidermis. The accompanying micrographs depict 2D-cultured adipocytes (ADSCs), human dermal fibroblasts (HDFs), and human epidermal keratinocytes (hEKs) before their integration into the structure at optimized densities to substitute the skin layers (scale bar represents 200 μm)

6. Immediately add adipogenic media to the suspension bath and incubate the construct (37 °C, 5% CO_2, and 95% air) to allow the collagen to undergo fibrillogenesis.
7. After 24 h, remove the construct from the support bath by gently lifting it out and rinsing it with deionized water. Culture in adipogenic media for the following 14 days.
8. Seed keratinocytes (2×10^6—passage 6) on the top of the construct and culture for a further 7 days; once confluent, raise the top of the construct to the air-liquid interface to create an epidermal layer.

The skin's physical, chemical, and biological architecture was successfully replicated by carefully designing the three distinct layers: the epidermal, dermal (including both papillary and reticular), and the hypodermal, as demonstrated in the schematic in Fig. 2.

3.5 Live/Dead Cell Viability

1. Section and stain 3D printed constructs, as shown in the photograph in Fig. 3 with calcein AM (staining live cells green) and propidium iodide (PI—staining dead cells red) reagents for the visualization of cells.
2. Visualize using a fluorescent microscope.

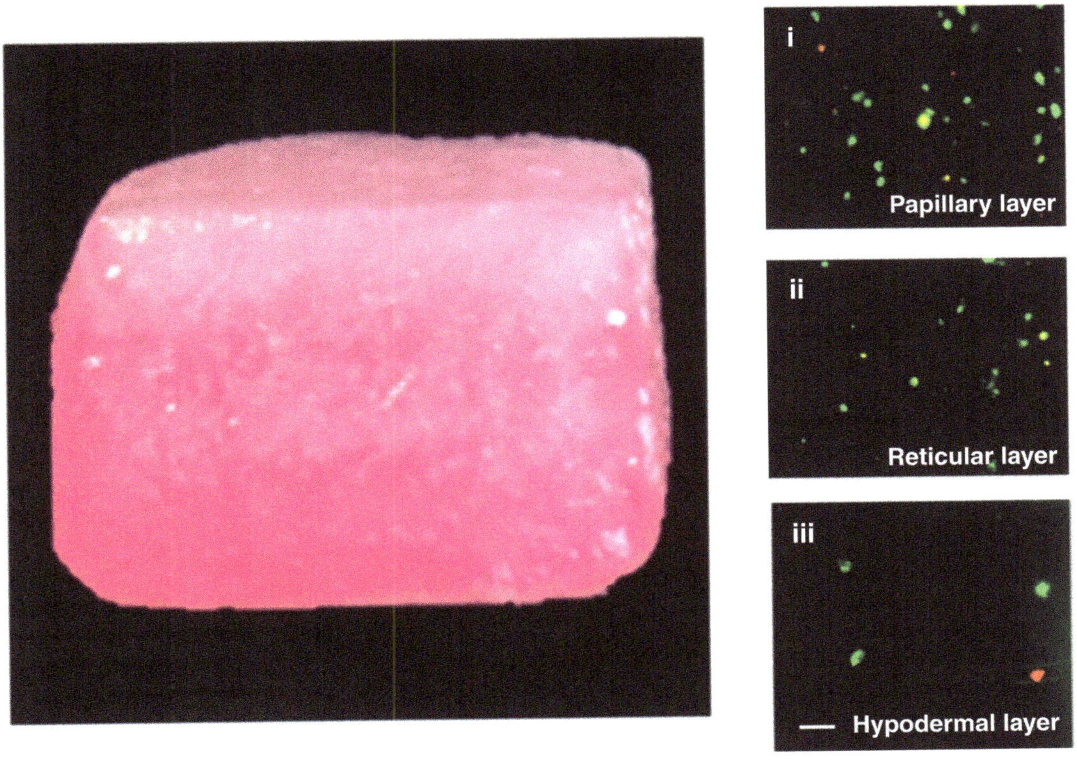

Fig. 3 Photograph of the 3D printed construct (dimensions: 15 × 15 × 9 mm^3). The increased collagen density is evident from the whitening of the construct at the top, with the purple/pink coloration resulting from the indicator in the culture media. Live/dead staining is presented for the (i) papillary, (ii) reticular, and (iii) hypodermal layers (scale bar represents 200 μm)

3.6 Ex Vivo Implant Integration Study

1. Wash freshly slaughtered porcine skin in absolute ethanol before forming a simulated wound throughout the epidermal, dermal, and hypodermal layers using a scalpel.
2. Label HDFs using a Q-Tracker cell labelling kit (kit 488), as per the kit's instructions prior to addition to the bioink.
3. Print a skin equivalent to fit the dimensions of the porcine wound model followed by solidification and culture.
4. Place the skin equivalent within the wound cavity and culture for 21 days before fixing, embedding, and sectioning.

3.7 Embedding of Implants in Paraffin

1. Fix 3D printed implants in 10% neutral buffered formalin overnight at 4 °C for a specified time to preserve the structure (*see* **Note 19**).
2. Dehydrate the samples by immersing them sequentially in increasing concentrations of ethanol: 70% ethanol for 30 min,

95% ethanol for 2 h, and 100% ethanol for 3 h, using fresh ethanol for each step and limiting exposure to 60 min per concentration (*see* **Note 20**).

3. Transfer the dehydrated samples into xylene for clearing (*see* **Note 21**).
4. Use a tissue embedder to immerse the samples in molten paraffin wax.
5. Quickly place paraffin-saturated samples into appropriate molds and ensure they are oriented correctly according to a protocol provided by Zhanmu et al. [21] (*see* **Note 22**).

3.8 Haematoxylin and Eosin (H&E) Staining

1. Cut paraffin-embedded samples into 7 μm thick sections.
2. Mount the sections onto adhesion-coated slides.
3. Immerse the slides in xylene to remove the paraffin (*see* **Note 21**).
4. Rehydrate the tissue sections by immersing them in a graded ethanol series (absolute, 70%, and 30%).
5. First, stain the slides with hematoxylin.
6. Rinse the slides under running tap water.
7. Counterstain with eosin.
8. Dehydrate the stained slides in a graded ethanol series (30%, 70%, and absolute) (*see* **Note 20**).
9. Submerge the slides in xylene to clear the tissue sections (*see* **Note 21**).
10. Add mounting solution to the slides. Place a coverslip over the sections to protect them.
11. Examine the prepared slides under a confocal microscope.

The similarity of the implant to human skin was assessed after 21 days of culture through histological staining with H&E histological staining, as shown in Fig. 4.

4 Notes

1. It has been observed that dispersing the agarose powder when the water temperature is between 50 and 80 °C facilitates better dispersion.
2. Pectin is a complex polysaccharide primarily composed of methoxy-esterified α, d-1, 4-galacturonic acid units. Pectins are classified based on their methoxy content and their gel formation speed. Generally, pectins can be divided into high methoxy pectin (more than 50% esterified) and low methoxy (LM) pectin (less than 50% esterified) [22]. LM pectins

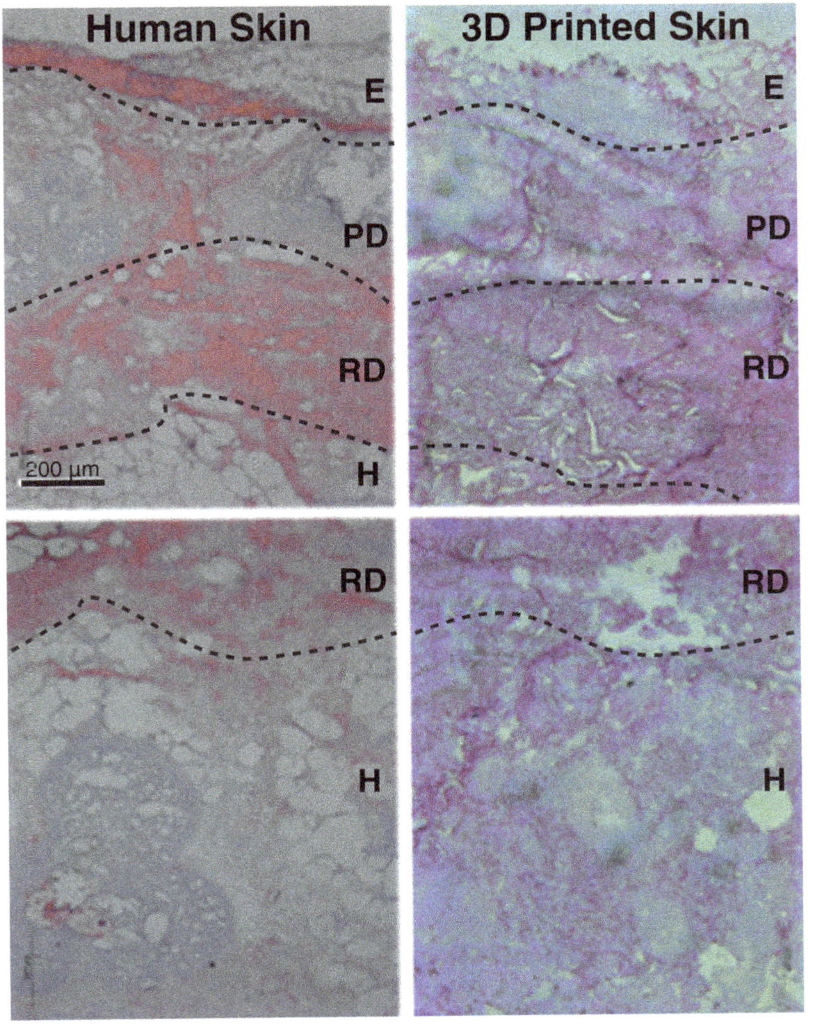

Fig. 4 Histological analysis of (left) human skin, showcasing the epidermis (E), papillary dermis (PD), reticular dermis (RD), and hypodermis (H) layers stained with hematoxylin and eosin; (right) 3D printed skin following 21 days of culture (scale bar represents 200 μm)

typically have a slower gel formation time. However, when combined with collagen, they demonstrated a shorter gelation time compared to environments containing only collagen [12].

3. Adding less amount of water initially and topping up to total volume after adding the volume of powder ensures more precise experimental process.
4. Autoclaving at 121 °C provides sterilization.
5. This gradual cooling procedure preserves the stability and integrity of the prepared solution, ensuring that it remains effective and reliable for subsequent use.

6. Stock solutions were kept sealed at 4 °C to maintain sterility until further use.
7. This procedure was conducted in a Class II biological safety cabinet to ensure sterility.
8. Human skin tissue samples were obtained from Genoskin (France) under French law L. 1245. The tissue samples were sourced from an elective human abdominoplasty with the patient's consent; thus, no additional ethics approval was required.
9. Samples were transported in phosphate buffered saline (PBS) in a sterile container at 4 °C within 4 h post-excision.
10. Dispase II, a protease enzyme used at 50 U/L, facilitates the gentle dissociation of the epithelium by enzymatically breaking down the connections between dermal and epidermal layers [23]. The dermo-epidermal tissue was incubated overnight at 4 °C in dispase II solution.
11. Storage temperature is 2–8 °C, while treatment at 37 °C is used to facilitate the isolation of cells.
12. Filter the cell suspension through 40, 70, and 100 μm cell strainers to remove small cellular debris, large cellular debris, and cell aggregates, ensuring a homogeneous cell suspension [20].
13. Adipogenic differentiation medium, along with its constituent components, serves as a crucial tool for imparting subcutaneous tissue functions to cells and for monitoring these functionalities.
14. After fabrication, the constructs were stabilized by injecting 200 mM $CaCl_2$ into the local area.
15. Agarose fluid gels can be prepared using a magnetic or overhead stirrer; however, when using these methods, it is difficult to precisely control the cooling rate and the shear rate will depend on the type of stirrer used, limiting reproducibility.
16. It is possible to reduce the temperature of the agarose fluid to 80 °C before transferring it to the rheometer to save time, without affecting the properties of the fluid gel.
17. It is common practice to wait 24–48 h after producing agarose fluid gels before using them. This is due to agarose fluid gels undergoing an aging process caused by their granular matter nature [24].
18. Optimal printing conditions were obtained to be: flow rate—3.8 μL/s, pressure—5 kPa, and ϕID 410 μm.
19. Ensure that fixation time in 10% formalin is optimized, typically ranging from 24 to 48 h, adjusted according to the sample's thickness and type.

20. Increase the ethanol concentration gradually to avoid causing tissue shrinkage and damage [21].
21. Use xylene under a fume hood because of its harmful fumes.
22. Since paraffin solidifies below 56 °C, transfer the molds quickly to melted paraffin at 58–60 °C. Once embedded in paraffin, the blocks can be stored at room temperature for extended periods. However, be cautious, as paraffin blocks can become brittle if they cool too much [21].

References

1. Koch L, Michael S, Reimers K et al (2015) Chapter 13 - Bioprinting for skin. In: Zhang LG, Fisher JP, Leong KW (eds) 3D bioprinting and nanotechnology in tissue engineering and regenerative medicine. Academic Press, pp 281–306., ISBN 9780128005477. https://doi.org/10.1016/B978-0-12-800547-7.00013-8
2. Chaudhry MS, Czekanski A (2023) In-situ bioprinting of skin – a review. Bioprinting 31:e00271. https://doi.org/10.1016/j.bprint.2023.e00271
3. Kang MS, Kwon M, Lee SH et al (2022) 3D printing of skin equivalents with hair follicle structures and epidermal-papillary-dermal layers using gelatin/hyaluronic acid hydrogels. Chem Asian J 17:e202200620. https://doi.org/10.1002/asia.202200620
4. Yang Y, Xu R, Wang C et al (2022) Recombinant human collagen-based bioinks for the 3D bioprinting of full-thickness human skin equivalent. Int J Bioprint 8:145–160. https://doi.org/10.18063/ijb.v8i4.611
5. Shen Z, Cao Y, Li M et al (2021) Construction of tissue-engineered skin with rete ridges using co-network hydrogels of gelatin methacrylated and poly(ethylene glycol) diacrylate. Mater Sci Eng C 129:112360. https://doi.org/10.1016/j.msec.2021.112360
6. Weng T, Zhang W, Xia Y et al (2021) 3D bioprinting for skin tissue engineering: current status and perspectives. J Tissue Eng 12. https://doi.org/10.1177/20417314211028574
7. Yu JR, Navarro J, Coburn JC et al (2019) Current and future perspectives on skin tissue engineering: key features of biomedical research, translational assessment, and clinical application. Adv Healthc Mater 8:e1801471. https://doi.org/10.1002/adhm.201801471
8. Jin R, Cui Y, Chen H et al (2021) Three-dimensional bioprinting of a full-thickness functional skin model using acellular dermal matrix and gelatin methacrylamide bioink. Acta Biomater 131:248–261. https://doi.org/10.1016/j.actbio.2021.07.012
9. Sun W, Liu Z, Xu J et al (2023) 3D skin models along with skin-on-a-chip systems: a critical review. Chin Chem Lett 34:107819. https://doi.org/10.1016/j.cclet.2022.107819
10. Schmidt FF, Nowakowski S, Kluger PJ (2020) Improvement of a three-layered in vitro skin model for topical application of irritating substances. Front Bioeng Biotechnol 8. https://doi.org/10.3389/fbioe.2020.00388
11. Moakes RJA, Senior JJ, Robinson TE et al (2021) A suspended layer additive manufacturing approach to the bioprinting of tri-layered skin equivalents. APL Bioeng 5:046103. https://doi.org/10.1063/5.0061361
12. Zimoch J, Zielinska D, Michalak-Micka K et al (2021) Bio-engineering a prevascularized human tri-layered skin substitute containing a hypodermis. Acta Biomater 134:215–227. https://doi.org/10.1016/j.actbio.2021.07.033
13. Colly A, Marquette C, Frances JM et al (2022) Classification of the emerging freeform three-dimensional printing techniques. MRS Bull 48(1):69–92. https://doi.org/10.1557/S43577-022-00348-9
14. Zeng X, Meng Z, He J et al (2022) Embedded bioprinting for designer 3D tissue constructs with complex structural organization. Acta Biomater 140:1–22. https://doi.org/10.1016/J.ACTBIO.2021.11.048
15. Fonseca AC, Melchels FPW, Ferreira MJS et al (2020) Emulating human tissues and organs: a bioprinting perspective toward personalized medicine. Chem Rev 120:11093. https://doi.org/10.1021/ACS.CHEMREV.0C00342
16. Chen S, Tan WS, Bin Juhari MA et al (2020) Freeform 3D printing of soft matters: recent advances in technology for biomedical engineering. Biomed Eng Lett 10:453–479. https://doi.org/10.1007/S13534-020-00171-8

17. Hinton TJ, Jallerat Q, Palchesko RN et al (2015) Three-dimensional printing of complex biological structures by freeform reversible embedding of suspended hydrogels. Sci Adv 1:e1500758. https://doi.org/10.1126/SCIADV.1500758
18. Moxon SR, Cooke ME, Cox SC et al (2017) Suspended manufacture of biological structures. Adv Mater 29:1605594. https://doi.org/10.1002/ADMA.201605594
19. Yang B, Liu T, Gao G et al (2022) Fabrication of 3D GelMA scaffolds using agarose microgel embedded printing. Micromachines 13:469. https://doi.org/10.3390/MI13030469
20. Henrot P, Laurent P, Levionnois E et al (2020) A method for isolating and culturing skin cells: application to endothelial cells, fibroblasts, keratinocytes, and melanocytes from punch biopsies in systemic sclerosis skin. Front Immunol 11:566607. https://doi.org/10.3389/fimmu.2020.566607
21. Zhanmu O, Yang X, Gong H et al (2020) Paraffin-embedding for large volume bio-tissue. Sci Rep 10:12639. https://doi.org/10.1038/s41598-020-68876-5
22. Lochhead RY (2017) The use of polymers in cosmetic products. In: Cosmetic science and technology: theoretical principles and applications. Elsevier Inc, pp 171–221. https://doi.org/10.1016/B978-0-12-802005-0.00013-6
23. Sarmin AM, Connelly JT (2022) Fabrication of human skin equivalents using decellularized extracellular matrix. Curr Protoc 2:e393. https://doi.org/10.1002/cpz1.393
24. D'Oria G, Gunes DZ, Lequeux F et al (2023) Fluid gels' dual behaviour as granular matter and colloidal glass. Food Hydrocoll 137:108401. https://doi.org/10.1016/J.FOODHYD.2022.108401

Chapter 14

Transdifferentiation of Human Dermis Derived Fibroblasts into iNeurons and Cultivation in a Collagen Type 1 Hydrogel

Nathalie A. Senn and Thomas Biedermann

Abstract

Skin injuries pose a significant medical challenge, particularly deep, extensive full thickness skin wounds that do not heal properly. These wounds often lack sensory nerves, which are crucial for skin sensitivity. The development of pre-innervated skin substitutes containing nerve cells offers a promising solution to enhance skin sensitivity in transplanted areas. For the production of such skin substitutes, direct transdifferentiation of fibroblasts into iNeurons using small molecules offers a fast and efficient approach.

In this chapter, we provide a method of human dermis derived fibroblast transdifferentiation into iNeurons, and culture of iNeurons in a collagen type 1 hydrogel with a detailed protocol establishing a 3D model of pre-innervated skin substitutes. This hydrogel can further be used for development of pre-innervated dermal skin substitutes for clinical application, for investigation of disease mechanisms and drug screening.

Key words Transdifferentiation, Small molecules, Fibroblasts, iNeurons, Collagen type 1 hydrogel

1 Introduction

Skin defects represent a significant challenge in modern medicine, particularly deep and extensive wounds that do not heal without intervention. Tissue engineering offers a viable solution to treat such skin wounds by utilizing autologous dermo-epidermal skin substitutes. Enriching these substitutes with neurons presents an innovative approach to enhance skin sensitivity in the transplanted area [1]. To minimize the rejection risk of the skin substitute, autologous somatic cells can be differentiated into neurons [2]. For this purpose, fibroblasts stand out as a preferred choice for inducing neuron differentiation, primarily due to their abundance, accessibility, resilience, and well-defined identity [3].

Induced neurons, commonly known as iNeurons, mark a groundbreaking innovation in neuroscience and regenerative medicine. Unlike neurons derived from neuronal stem cells, iNeurons

Thomas Biedermann and Sophie Böttcher-Haberzeth (eds.), *Skin Tissue Engineering: Methods and Protocols*, Methods in Molecular Biology, vol. 2922, https://doi.org/10.1007/978-1-0716-4510-9_14,

are produced through cellular reprogramming or induction, forcing somatic cells to adopt a neuronal fate [4]. In regard to induction, the generation of iNeurons can be achieved by small molecules. These small molecules enter cells and target signaling pathways, epigenetic modifications, and metabolic processes. Apart from being cost-effective, easy to produce and store, and standardized, small molecules have a rapid and reversible effect [4–6].

We transdifferentiated human dermis derived fibroblasts with small molecules into mature glutamatergic iNeurons based on the method of Labarrade et al. [7].

Upon successful generation, iNeurons can be used to produce pre-innervated autologous skin substitutes, promising a novel approach to addressing skin defects.

2 Materials

2.1 Labware

1. 0.22 μm filter top 500 mL bottle filter units.
2. 15 mL Falcon tubes.
3. 25 cm^2 culture flasks.
4. 1.5 mL Eppendorf tubes.
5. 12-well cell culture inserts.
6. 12-well companion plates.
7. Cell counter.
8. Humified cell culture incubator.
9. Cell culture centrifuge.
10. Cell culture microscope.

2.2 Cell Culture

1. Neuronal Induction Medium (NIM).

   NIM: one part neurobasal medium and one part DMEM/F12 (Dulbecco's Modified Eagle Medium/Nutrient Mixture F-12), 1% N-2 supplement (100×), 2% B-27 Plus Supplement (50×), 40 ng/mL BDNF (diluted in distilled water), 20 ng/mL GDNF (diluted in distilled water), 20 ng/mL IGF-1 (diluted in distilled water), and 0.1% gentamycin. Sterilize before use by filtration through a bottle filter unit and store at 4 °C.

2. Cocktail A.

   Cocktail A: NIM, 3 μM CHIR99021 (diluted in DMSO), 0.25 μM LDN193189 (diluted in distilled water), 10 μM RG108 (diluted in DMSO), 1 μM dorsomorphin (diluted in DMSO), 5 μM ISX9 (diluted in DMSO), 3 μM P7C3-A20 (diluted in DMSO), and 0.5 μM A83–01 (diluted in DMSO). Sterilize before use by filtration through a bottle filter unit and store at 4 °C.

3. Cocktail B.

   Cocktail B: NIM, 3 μM P7C3-A20 (diluted in DMSO), 10 μM forskolin (diluted in DMSO), 5 μM Y27632 (diluted in DMSO), 1 μM DAPT (diluted in DMSO), 1 μM PD0325901 (diluted in DMSO), 1 μM purmorphamine (diluted in DMSO), and 0.5 μM A83–01 (diluted in DMSO). Sterilize before use by filtration through a bottle filter unit and store at 4 °C.

4. Neuronal Maturation Medium (NMM).

   NMM: NIM, 20 ng/mL NT-3 (diluted in distilled water), and 2 μM Y27632 (diluted in DMSO). Sterilize before use by filtration through a bottle filter unit and store at 4 °C.

5. Fibroblast medium (DMEM+++).

   DMEM +++: Dulbecco's Modified Eagle Medium (DMEM), 5.6 mM D-glucose, 1 mM sodium pyruvate, 4 mM Gluta-MAX, 10% fetal calf serum, 5 μg/mL gentamycin, and 10 mM HEPES.

6. Cells.

   Human fibroblasts: isolated from healthy human skin sample according to Pontiggia et al. [8] and seeded to 25 cm^2 culture flasks in 5 mL DMEM+++/25 cm^2 flask. Medium must be changed every 2–3 days until flask is confluent.

2.3 Other Media or Solutions

1. Phosphate-buffered saline (PBS).
2. 0.5% Trypsin/EDTA.
3. 0.5% Soybean trypsin inhibitor (SBTI).
4. 0.4% Trypan Blue solution (in PBS).
5. Neutralization buffer (sterile): 0.31 M $NaHCO_3$, 0.15 M NaOH, 1 M HEPES in dH_2O.
6. Bovine collagen type 1: 5 mg/mL diluted to 4 mg/mL with 0.189 M acetic acid.
7. 4% paraformaldehyde.
8. Acetone/methanol: mix acetone/methanol 1:1, store at −20 °C.
9. PBS-T: 0.3% Triton-X in PBS.
10. 2% BSA: 2% bovine serum albumin in PBS.
11. 10% BSA/0.05% NaN_3: 10% bovine serum albumin/0.05% NaN_3 in PBS.

2.4 Immunofluorescence

1. PGP9.5 antibody (clone 13C4/J3C4).
2. Secondary fluorescence labelled antibody.
3. Hoechst 33342: 20 mM in PBS.

4. 100% glycerol.
5. Mounting medium for immunofluorescence.
6. Cover slips.

3 Methods

The whole procedure must be performed under a sterile laminar flow cabinet.

3.1 Trans-differentiation Process

1. After fibroblasts reached confluency, split fibroblasts and seed fibroblasts into collagen type 1 pre-coated 25 cm^2 culture flasks. Change medium every 2–3 days by adding 5 mL DMEM+++. Once reached 70% confluency, use fibroblasts for transdifferentiation.
2. Aspirate DMEM+++, wash cells with PBS and add 5 mL NIM/25 cm^2 culture flask. Incubate for 3 days at 37 °C.
3. Replace every following 3 days 50% of the medium with either Cocktail A or Cocktail B as shown in Fig. 1.
4. On day 18, replace 30% of the medium with NMM (*see* **Note 1**).
5. Analyze transdifferentiated cells by immunofluorescence staining for neuronal marker expression (Fig. 2) (*see* **Note 2**).

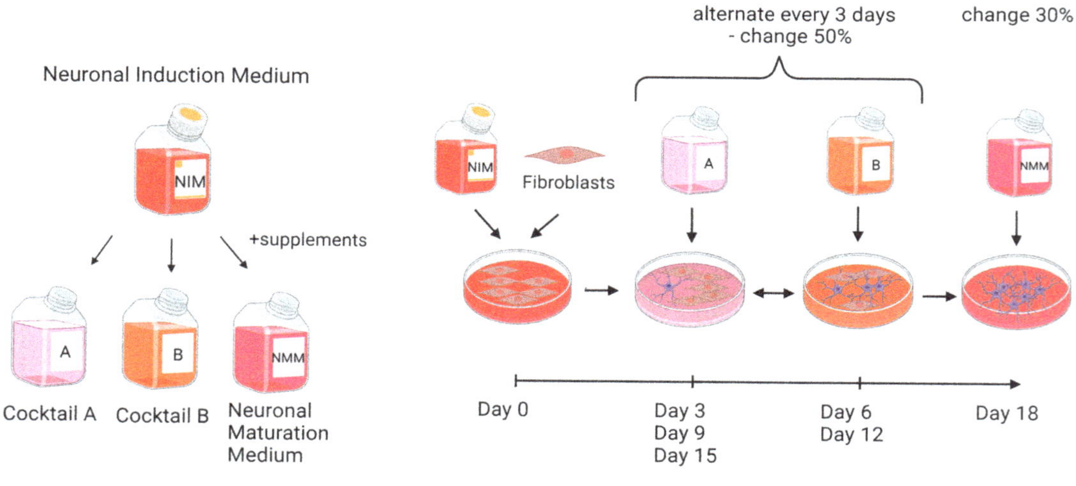

Fig. 1 Transdifferentiation process over time. To the neuronal induction media (NIM) different small molecules are added to produce Cocktail A, Cocktail B and the neuronal maturation media (NMM). The changes of fibroblasts into neurons during the transdifferentiation are shown in petri dishes. The arrow indicates the days on which media changes occur. (Created in BioRender. Biedermann, T. (2024) BioRender.com/g80z394)

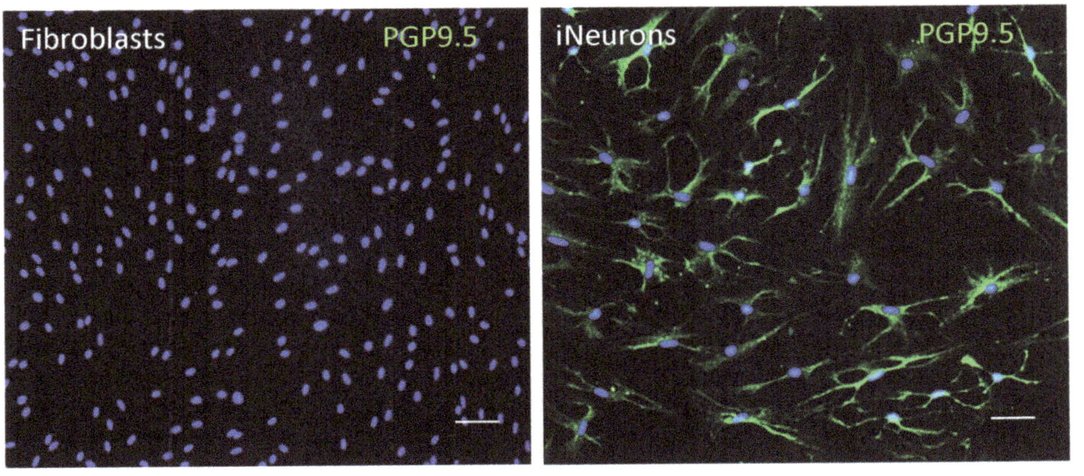

Fig. 2 Immunofluorescence staining of fibroblasts and iNeurons. Fibroblasts displayed no expression for the neuronal marker PGP9.5, whereas iNeurons were PGP9.5 positive. Hoechst (blue) was used for cell nuclei staining. Scale bar = 100 μm

3.2 Collagen Type 1 Hydrogel Preparation Containing iNeurons

1. Wash the cells three times with PBS for 5 min each.
2. Trypsinize iNeurons for 4 min at 37 °C using 3 mL 0.5% trypsin/EDTA/25 cm^2 culture flask until cells have detached from the flask.
3. Neutralize trypsin/EDTA with 100 μL 0.5% SBTI/25 cm^2 culture flask.
4. Transfer the cell solution into a 15 mL Falcon tube and centrifuge 3 min at 300 g at room temperature.
5. Remove the supernatant and resuspend the cell pellet with 1 mL NMM. Keep cells on ice.
6. Count the cells (e.g., using a cell counter) by trypan blue exclusion assay.
7. Dilute the cells to a concentration of 0.8×10^6 cells/mL.
8. Prepare for every collagen hydrogel an 1.5 mL Eppendorf tube containing 30 μL neutralization buffer, 20 μL cell suspension and 100 μL collagen type 1, carefully mix the solution (*see* **Notes 3** and **4**).
9. Distribute the solution evenly into a 12-well cell culture insert, placed in the well of a 12-well companion plate (*see* **Note 5**).
10. Incubate for 1 h at 37 °C to finalize collagen polymerization.
11. Add 1 mL NMM at the top of the hydrogel in the 12-well insert and 0.5 mL under the insert into the well.
12. Keep the hydrogel in an incubator for 14 days at 37 °C/5% CO_2 and change the medium every 2 days.

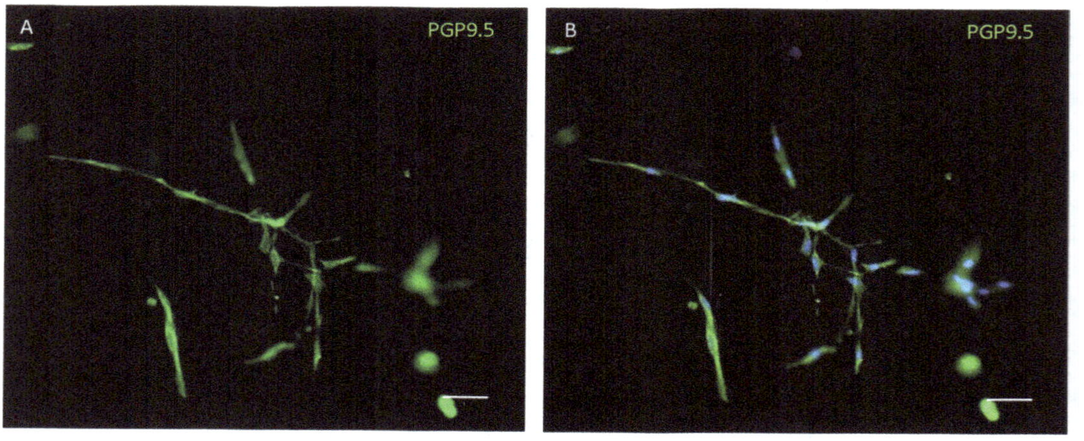

Fig. 3 Visualization of iNeurons in dermal skin substitutes. Whole mount staining was performed using the neuronal marker PGP9.5 (green, **a**) to investigate the behaviour of iNeurons in a 3D environment. **b** Merged image with Hoechst (blue), staining the cell nuclei. Scale bar = 100 μm

13. The hydrogel can now be used to create a dermo-epidermal skin substitute by adding keratinocytes on the top. Another possibility is to fix the hydrogel containing the iNeurons and perform a whole mount staining with a neuronal marker like PGP9.5, as shown in Fig. 3. For this, incubate the hydrogel with 4% paraformaldehyde for 1 h at 4 °C. Remove the PFA and permeabilize the hydrogel for 2 h with PBS-T at room temperature. Remove the PBS-T and add a mix of 10% BSA/0.05% NaN_3 for 2 h at room temperature. Remove the mix and add PGP9.5 antibodies, incubate overnight at 4 °C. Remove the antibody solution, wash with PBS for 6 h (change PBS every hour) at 4 °C. Remove the PBS and add a mix of 10% BSA/0.05% NaN_3 for 2 h at 4 °C. Remove the mix, add secondary antibodies and incubate overnight at 4 °C. Remove the antibody solution, wash for 6 h with PBS at 4 °C, change PBS every hour. Remove the PBS, add Hoechst 33342 for 20 min at 4 °C. Remove Hoechst and mount the hydrogel with 100% glycerol, store at 4 °C until microscopy (Fig. 3).

4 Notes

1. NMM can be used to further culture iNeurons. Media must be changed every 3 days.
2. For staining, fix cells with aceton/methanol for 5 min at −20 °C. Then wash cells with PBS and incubate cells for 30 min with 2% BSA at room temperature. Then add primary antibodies in 2% BSA to the cells and incubate for 1 h at room temperature, preferably in a moist dark chamber. Then wash cells three times

with PBS for 5 min and incubate again with 2% BSA in PBS at room temperature for 30 min. Then add secondary antibodies and incubated in a moist dark chamber for 1 h at room temperature. Wash cells twice with PBS for 5 min and then stain cell nuclei with Hoechst 33342, wash cells twice with PBS for 5 min, mount cells with mounting medium and cover them with a cover slip.

3. The neutralization buffer, collagen type 1 and cells have to be kept on ice until seeded to the hydrogel insert as otherwise the collagen type 1 starts to polymerize.

4. The order should be first neutralization buffer, then the cells, and then collagen type 1. The mixture should be done right before adding it onto the hydrogel insert. No pre-preparation is possible.

5. Take a 200 μL pipette for the procedure and avoid bubbles.

References

1. Biedermann T, Böttcher-Haberzeth S, Klar AS et al (2013) Rebuild, restore, reinnervate: do human tissue engineered dermo-epidermal skin analogs attract host nerve fibers for innervation? Pediatr Surg Int 29:71–78
2. Schutte SC, Kadakia F, Davidson S (2021) Skin-nerve co-culture systems for disease modeling and drug discovery. Tissue Eng Part C Methods 27:89–99
3. Ichim TE, O'Heeron P, Kesari S (2018) Fibroblasts as a practical alternative to mesenchymal stem cells. J Transl Med 16:212
4. Qin H, Zhao A, Fu X (2017) Small molecules for reprogramming and transdifferentiation. Cell Mol Life Sci 74:3553–3575
5. Yang Y, Chen R, Wu X et al (2019) Rapid and efficient conversion of human fibroblasts into functional neurons by small molecules. Stem Cell Rep 13:862–876
6. Xu Z, Su S, Zhou S et al (2020) How to reprogram human fibroblasts to neurons. Cell Biosci 10:116
7. Labarrade F, Botto JM, Imbert I (2023) Co-culture of iNeurons with primary human skin cells provides a reliable model to examine intercellular communication. J Cosmet Dermatol 22:2090–2098
8. Pontiggia L, Biedermann T, Meuli M et al (2009) Markers to evaluate the quality and self-renewing potential of engineered human skin substitutes in vitro and after transplantation. J Invest Dermatol 129:480–490

Part III

In Vitro and Vivo Models Testing Skin, Skin Cells, and Tissue Engineered Skin

Chapter 15

Deep Proteome Profiling of Primary Skin Fibroblasts

Sibilla Sander, Alexandre Leytens, and Jörn Dengjel

Abstract

Skin-derived primary cells and skin tissue are prime model systems to study molecular disease mechanisms on cell and tissue level. Cells and tissue can be cultivated ex vivo and protocols for both, simple 2D as well as 3D in vitro cultures reflecting different levels of complexity exist. Mass spectrometry (MS)-based proteomics is a prime approach to link molecular mechanisms to observable disease phenotypes. In this chapter, we describe in detail the analysis of 2D and 3D primary skin fibroblast cultures by MS. We focus on automated sample processing to increase throughput and usage of limited cell numbers to reduce costs. The described workflow supports the study of proteome regulation in large scale screening approaches, be it for drug discovery or in clinical studies.

Key words Fibroblasts, Mass spectrometry, Proteomics, Skin, SP3, Spheroid

1 Introduction

The skin is the outermost organ of the human body, protecting it from microbial, chemical, and physical insults, from loss of water and proteins. Proper functioning of the skin is vital for human health. Skin diseases lead to reduced quality of life and contribute to morbidity and mortality [1]. The human skin consists of three layers, epidermis, dermis, and subcutis, each of the layers having a distinct architecture and function. Importantly, each layer is made up of a specific cellular composition which in turn produces a characteristic extracellular matrix (ECM). Skin ECM is highly complex and essential for skin function [2–4]. The epidermis, the outermost layer, consists mainly of keratinocytes, one important function being the formation of a barrier towards the environment. The underlying dermal-epidermal junction zone is a complex basement membrane critical for dermal epidermal anchorage. The dermis is mainly made up of fibroblasts and provides elasticity, tensile strength, and resilience to the skin. The subcutis consists of loose connective tissue and is important for fat storage, major cell types being fibroblasts and adipocytes. Skin and skin-derived cells are not

Thomas Biedermann and Sophie Böttcher-Haberzeth (eds.), *Skin Tissue Engineering: Methods and Protocols*, Methods in Molecular Biology, vol. 2922, https://doi.org/10.1007/978-1-0716-4510-9_15,

only used to study skin pathophysiology. Due to its easy accessibility, skin is also a basic model system to study human organ function and principles of cell-cell communication. This can be done by simple 2D culture systems, or by 3D models [5], ranging from single or mixed culture spheroids [6] to organotypic 3D skin equivalents [7].

Mass spectrometry (MS)-based proteomics plays a major role in deciphering the proteome composition of skin and in studying protein functions [8]. New acquisition methods like Data-Independent Acquisition (DIA) contribute to increased sensitivity and throughput and support large scale screening approaches on proteome level. In DIA, compared to the more classical data-dependent acquisition (DDA), ions are not selected for fragmentation. Instead, groups of ions are measured, generating a "digital map" of the entire sample [9, 10]. This leads to less missing values across samples of a given experiment and supports bioinformatic data analyses to determine, e.g., disease-relevant regulated pathways.

In the current protocol, we describe a semi-automated MS sample preparation workflow making use of a liquid handling platform. MS samples are prepared in a 96-well format increasing throughput and reproducibility. The coupling of this workflow to DIA-based quantitative MS, which is more sensitive than DDA approaches, supports the processing of single 3D spheroids made up of some thousand cells. Thus, this workflow can be used to study skin cell proteome regulation in large scale screening approaches, be it for drug discovery or in clinical studies.

2 Materials

The following instruments/equipment are used to perform this procedure: Bravo Automated Liquid Handling Platform and lid-heated PCR instrument.

2.1 2D and 3D Spheroid Fibroblast Cultures

1. Dulbecco's Modified Eagle Medium (DMEM), 4.5 g/L high glucose, stable glutamine, with sodium pyruvate.
2. 10% Fetal Bovine Serum (FBS).
3. Normocin.
4. 2-Phospho-L-ascorbic acid trisodium salt.
5. Fibroblast medium: DMEM supplemented with 10% FBS, 100 μg/mL normocin and 50 μg/mL 2-phospho-L-ascorbic acid.
6. Trypsin-EDTA solution (200 mg/L trypsin, 500 mg/L EDTA).
7. BioFloat 96-well plate.

8. 6-well plate.
9. Dulbecco's Phosphate-Buffered Saline 1× w/o Ca and Mg (DPBS).
10. Syringe Driven Filters, 0.2 μm, 30 mm diameter.
11. Syringe.
12. Multichannel pipet.
13. Primary normal human fibroblasts (NHF).

2.2 3D Spheroids Harvest

1. Greiner Bio-One 96-well microplate for PCR, skirted.
2. Ice cold DPBS.
3. Thermowell Sealing Tape for 96-well plates.

2.3 2D Cell Culture Harvest

1. Cell scrapers.
2. Ice cold DPBS.
3. Centrifuge.
4. 1.5 mL reaction tubes.

2.4 AutoSP3

1. Lysis buffer: 50 mM Tris–HCl pH 7.5, 150 mM NaCl, 1 mM EDTA, 1% NP-40, 1% SDS.
2. Benzonase (≥250 units/uL).
3. ABC buffer: 100 mM Ammonium bicarbonate, pH 8.
4. Sera-Mag™ Speedbeads (GE Healthcare, Cat. No. 45152105050250).
5. Sera-Mag™ Speedbeads (GE Healthcare, Cat. No. 65152105050250).
6. cOmplete™ Protease Inhibitor Cocktail.
7. 550 mM Chloroacetamide (CAA).
8. 100 mM Dithiothreitol (DTT).
9. 50 mM Acetic acid.
10. 5% TFA (MS grade) in ddH2O.
11. Sequencing grade modified trypsin.
12. 100% Ethanol.
13. 80% Ethanol.
14. 96-well magnet (Alpaqua, Cat. No. A000400).
15. Agilent 250 μL tips boxes.
16. Deep reservoir wells plates.
17. Pierce™ BCA Protein Assay Kit.
18. PCR 8-stripes.
19. Magnetic rack.

20. Silica HPLC-column tip (I.D. 75 μm, New Objective, self-packed with ReproSil-Pur 120 C18-AQ, 1.9 μm (Dr. Maisch).
21. Gradient A (0.1% formic acid in water).
22. Gradient B (0.1% formic acid in 80% acetonitrile in water).

2.5 Mass Spectrometer

1. Orbitrap Exploris 480 mass spectrometer coupled to Vanquish NEO UHPLC system.

2.6 Software

1. Perseus software platform.
2. Spectronaut software version 19.
3. STRING v12.

3 Methods

In this protocol, we compare sample processing of traditional 2D primary normal human fibroblasts (NHF) cultures with low-input 3D primary NHF cultures. Proteome dynamics are studied by quantitative MS-based proteomics using state-of-the-art single-shot, DIA-liquid chromatography (LC)-tandem MS (MS/MS) [11]. Compared to DDA, DIA leads to less missing values and support usage of limited starting material. In the 3D setup, spheroids of 5000 NHFs are cultured in an ultra-low attachment U-bottom 96-well plate. For harvesting, they are moved in a V-bottom plate, which can be used directly for MS-sample preparation by autoSP3 [12, 13], thereby reducing further sample transfers and minimizing potential losses due to non-specific binding of proteins to hydrophobic surfaces. For the 2D setup, 100,000 NHFs are cultured in a 6-well plate, harvested by scraping, lysed in a tube and then transferred to a 96-well plate for MS-sample preparation. The 3D method offers the advantages of fewer cells, higher throughput and less handling, while giving rise to similar protein numbers using DIA-based proteomics.

3.1 2D Cell Culture and Harvest

1. Seed 100,000 cells/well in a 6-well plate in 2 mL of DMEM, supplemented with 10% FBS, 100 μg/ml normocin and 50 μg/ml 2-phospho-L-ascorbic acid (*see* **Note 1**).
2. Change medium every other day.
3. On day 7, remove medium and wash twice with 1 mL of PBS.
4. Add 0.5 mL of PBS and scrape the cells with a cell scraper.
5. Move everything to a reaction tube and centrifuge at 1000 rcf for 5 min.
6. Remove supernatant and snap freeze the pellet in liquid nitrogen.

3.2 3D Spheroid Culture

1. To transfer cells from 2D to 3D, wash cells once with PBS, digest with trypsin and resuspend cells in DMEM, supplemented with 10% FBS,100 μg/mL normocin and 50 μg/mL 2-phospho-L-ascorbic acid (*see* **Note 1**).
2. In a BioFloat 96-well plate, seed 5000 cells in 200 μL per well (*see* **Note 2**).
3. Centrifuge the plate at 290 rcf for 3 min at RT.
4. Change the medium every other day by centrifuging at 100 rcf for 3 min, removing 100 μL and replacing it with 100 μL fresh medium (*see* **Note 3**).
5. Spheroids should form and be visible on day 4.

3.3 3D Spheroids Harvest

1. On day 7, spheroids are ready to be harvested (*see* **Note 4**). Centrifuge the plate at 100 rcf for 3 min to pellet spheroids (Fig. 1a).
2. Collect spheroids by pipetting 20 μL from the bottom of the plate (Fig. 1b).
3. Transfer spheroids to a V-bottom plate well (Fig. 1c—*see* **Note 5**).
4. Wash with 100 μL of ice-cold PBS.
5. Centrifuge the plate for 5 min at 2650 rcf at 4 °C (Fig. 1d).
6. Carefully and quickly aspirate the PBS (Fig. 1e—*see* **Note 6**).
7. Repeat the wash steps twice and aspirate PBS (Fig. 1f—*see* **Note 7**).
8. These steps can be carried out with a multichannel pipet.

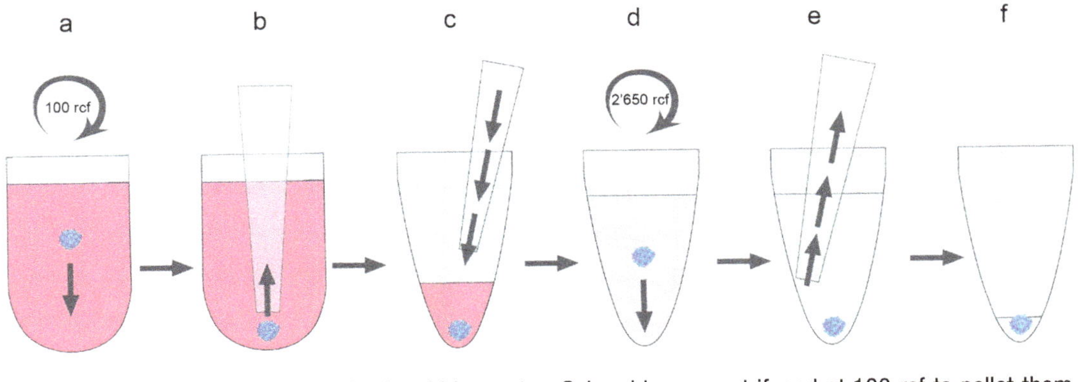

Fig. 1 Schematic representation of spheroid harvest. **a** Spheroids are centrifuged at 100 rcf to pellet them. **b** Spheroids are collected by aspirating 20 μL of medium from the bottom of each well. **c** Spheroids are moved to V-bottom plate and 100 μL of PBS are added to each well. **d** The plate is centrifuged 5 min at 2650 rcf to pellet spheroids. **e** The supernatant is carefully and quickly removed without touching the bottom of the well in order to not aspirate/loose spheroids. Steps **d** and **e** are repeated twice. **f** At the end of the process, supernatants are removed, and spheroids are snap frozen, ready to be processed for MS

3.4 SP3 Beads Preparation

1. Gently shake the Sera-Mag containers to resuspend beads (Fig. 2a).
2. Prepare a MasterMix of SpeedBeads: for each sample, use 100 μg (2 μL) of each SpeedBeads stock solution and combine these in a reaction tube (Fig. 2a—*see* **Note 8**).
3. With the help of the magnetic rack, remove the supernatant.
4. Wash beads by resuspending them in 500 μL of ddH_2O.
5. With the help of the magnetic rack, remove the supernatant and resuspend beads in ddH_2O to a final concentration of 40 μg/μL. For each sample, you will use 5 μL of this mix.
6. The beads are stable at 4 °C for 1 month.

3.5 Cell Lysis

1. Supplement the lysis buffer with proteases inhibitors (manufacturer's concentration) and benzonase (1/1000 v/v).
2. For the 2D cell culture, add lysis buffer (twice the volume of cell pellet) and incubate 30 min at RT.
3. Quantify the concentration of the 2D samples by BCA (manufacturer's protocol) and adjust protein concentration to 1 μg/μL.
4. Transfer 10 μL of sample in the Greiner 96-wells PCR Sapphire plates for the autoSP3 protocol.
5. For the frozen spheroids (already in the Greiner 96-well plate), add 5 μL of lysis buffer per well and incubate 30 min at RT.

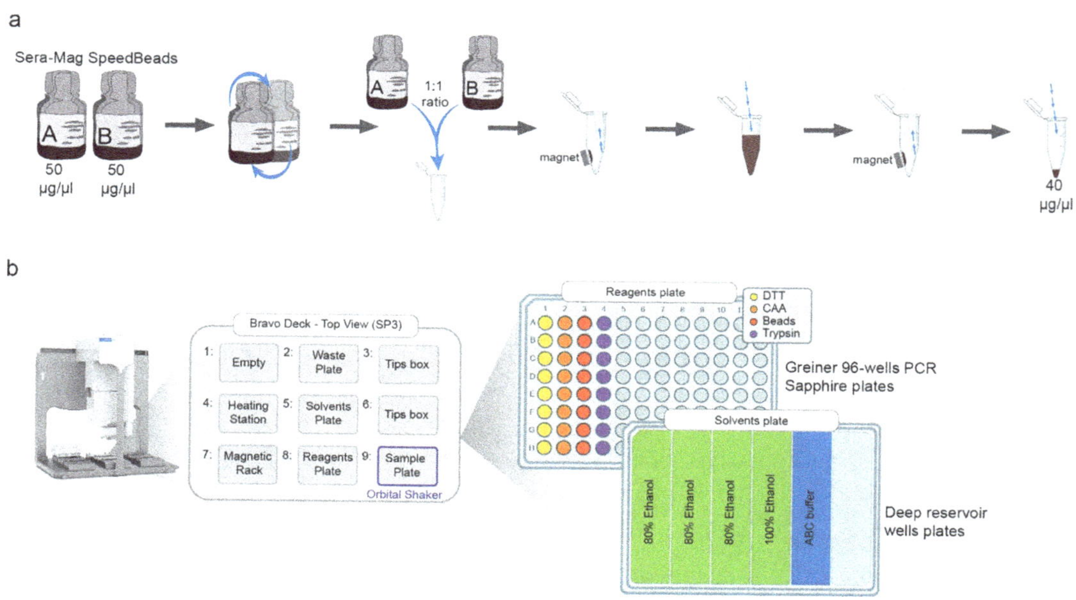

Fig. 2 Schematic representation of AutoSP3 setup. **a** SeraMag stocks are at an initial concentration of 50 μg/μL. Before use, they should be mixed by gentle shaking, combined into a reaction tube at a ratio of 1:1, washed with water and finally resuspended at a final concentration of 40 μg/μL. **b** Schematic representation of the plate deck of the Bravo Automated Liquid Handling Platform, reagent and solvent plates setup

3.6 AutoSP3

This protocol is adapted from Müller et al. [13].

1. Dilute CAA and DTT stocks 100-fold.
2. Resuspend trypsin in 50 mM acetic acid to a concentration of 8 ng/μL.
3. Prepare the reagent plate following the Bravo Automated Liquid Handling Platform settings and the given volumes (Fig. 2b—*see* **Note 9**).
4. Fill the necessary volumes of 100% ethanol, 80% ethanol and ABC buffer in the deep well solvent plate (Fig. 2b).

The steps from 5 to 17 (~2 h 30) are carried out by the Bravo Automated Liquid Handling Platform (Agilent).

1. Perform reduction: Add 5 μL of 1 mM DTT in each well, shake at 1500 rpm for 30 s and incubate 30 min at RT.
2. Perform alkylation: Add 5 μL of 5.5 mM CAA in each well, shake at 1500 rpm for 30 s and incubate 30 min at RT.
3. Add beads: Add 5 μL of beads and shake at 1500 rpm for 20 s.
4. Induce binding by adding 25 μL of 100% ethanol to each well.
5. Shake in 9 cycles of 30 and 90 s at 1500 rpm and 100 rpm, respectively.
6. Place plate on the magnet and wait 3 min.
7. Remove supernatant.
8. Wash with 180 μL of 80% ethanol and shake at 250 rpm for 60 s.
9. Place plate on the magnet and wait 3 min.
10. Remove supernatant.
11. Repeat wash steps [12–14] for a total of three times.
12. Add 5 μL of ABC buffer and 5 μL of trypsin.
13. Shake 30 s at 1500 rpm for 30 s.
14. Recover the plate with the samples, spin it for 5 s and seal it with a Thermowell sealing tape for 96-well plates.
15. Carefully vortex on low rpm to resuspend beads in the ABC buffer–trypsin solution (*see* **Note 10**).
16. Load the plate in a lid-heated PCR Machine and digest at 37 ° C for 16 h (*see* **Note 11**).
17. After digestion, remove the lid, acidify the samples with 1 μL of 50% TFA and gently shake.
18. Place the plate on the magnet and incubate 5–10 min.
19. Carefully transfer the supernatant to PCR tubes and store at −80 °C for LC-MS/MS analysis (*see* **Note 12**).

3.7 LC-MS/MS Measurements and Data Analysis

1. Perform LC-MS/MS analyses on a Orbitrap Exploris 480 mass spectrometer coupled to Vanquish NEO UHPLC system. Separate peptides on a fused silica HPLC-column tip (I.D. 75 μm, New Objective, self-packed with ReproSil-Pur 120 C18-AQ, 1.9 μm (Dr. Maisch) to a length of 20 cm) using a gradient of A and B: load samples with 0% B with a flow rate of 600 nl/min; separate peptides by 5%–30% B within 85 min with a flow rate of 250 nl/min. Set spray voltage to 2.3 kV and the ion-transfer tube temperature to 250 °C; use no sheath and auxiliary gas.
2. Operate the mass spectrometer in data independent mode, after each MS scan (m/z = 400–1200; resolution: 120,000, maximum injection time of 60 ms), 33 MS/MS scans with an isolation width of 24 m/z with 1 Da overlap, covering a range from 400 to 1200 m/z. Set the normalized AGC target value to 300%, resolution to 30,000 and normalize stepped collision energy to 25.5%, 27%, 30%.
3. Analyze the MS raw files using Spectronaut software version 19 and directDIA+ workflow against a UniProt full length Homo sapiens database (January 2023) and common contaminants, such as keratins, as reference.
4. Perform bioinformatics interpretation of MS data and statistical analyses with the freely available Perseus software platform [14].
5. Perform a log2 transformation of intra-group normalized protein intensities.
6. Filter out missing values and median-normalize results across groups.
7. With the current setup, expect between 4500 and 4800 protein identifications (IDs) per spheroid compared to the 5000 per 2D sample. Total protein IDs per experiment are similar (Fig. 3a—5028 protein IDs for 3D spheroid experiment compared to the 5085 proteins for the 2D experiment).
8. Perform Gene Ontology (GO)-term enrichment and cluster analyses with STRING v12 [15]. Wide coverage of cell compartments can be achieved with the 3D setup (Fig. 3b).
9. Analyze differential protein abundances in 3D spheroids and 2D samples by two-sample t-test with a permutation-based false discovery rate of 0.05 and a S0 value of 0.1 (Fig. 3c and d).

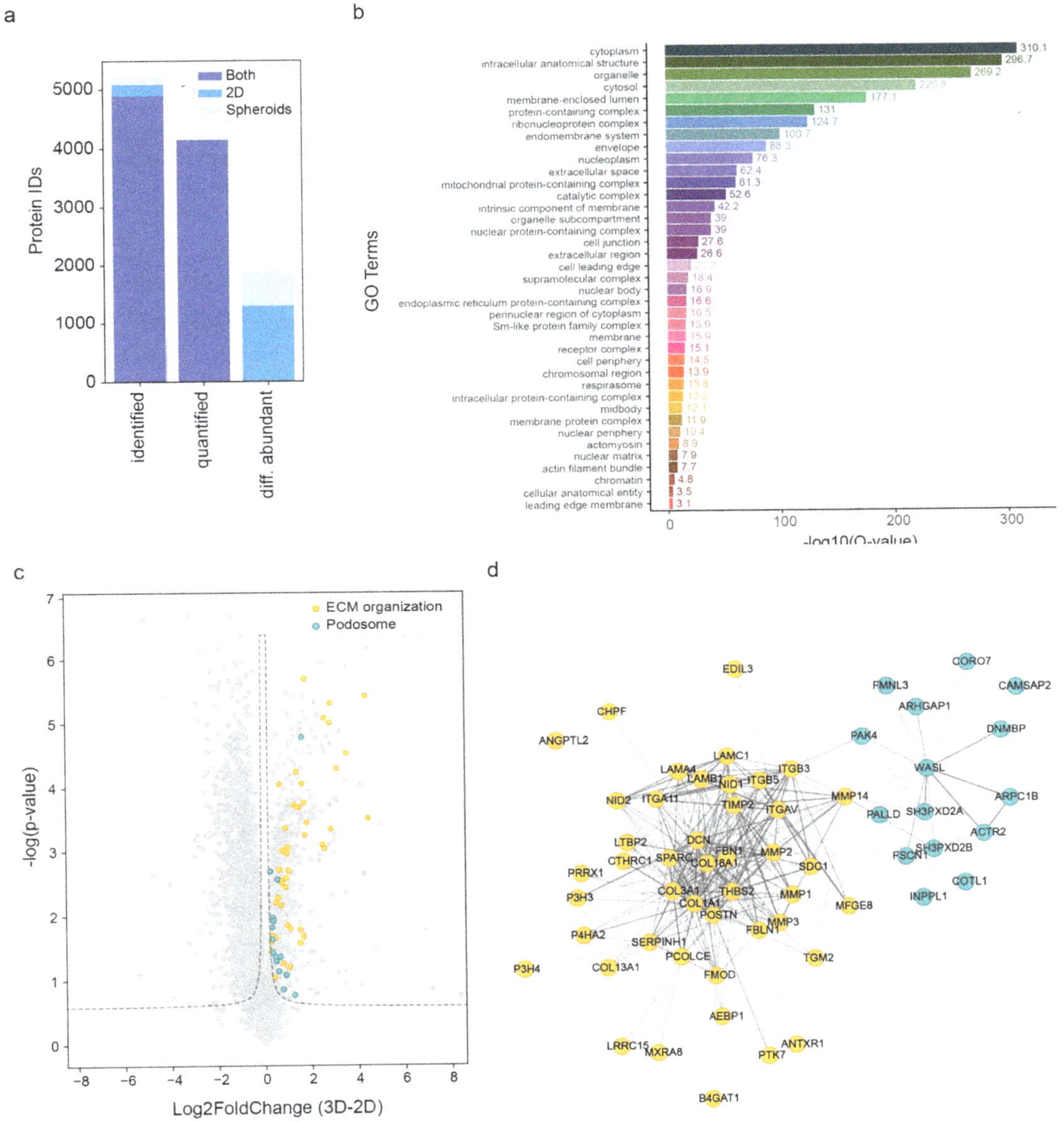

Fig. 3 DIA LC-MS/MS results from simple 2D and 3D in vitro NHFs cultures. **a** Proteins identified in at least one sample across the experiment, quantified in all samples and differentially abundant in 2D versus 3D. **b** GO cell component analysis indicates a comprehensive coverage of the cellular components, including the extracellular space. GO-Term enrichments are obtained with ClueGO v2.5.10 [18] by analyzing 5176 commonly identified proteins isolated from $n = 4$ spheroids. Respective q- values are noted (Two-sided hypergeometric test, $p < 0.001$, BH corrected). **c** Log2 fold change indicates the mean abundance level of each protein. Each dot represents a protein. Extracellular matrix and podosome proteins are enriched in a 3D compared to a 2D setup. **d** MCL cluster analysis (inflation parameter = 3) of significantly regulated proteins in spheroids (FDR: 0.05, s0 = 0.1). Clusters are analyzed for enriched GO-terms. Only cluster 1 (Proteins involved in the extracellular matrix) and cluster 9 (proteins involved in the podosome) are shown in orange and turquoise, respectively. Edges between proteins indicate interactions according to STRING DB.

4 Notes

1. 2-Phospho-L-ascorbic acid trisodium salt should be freshly supplemented to DMEM. It is a stable derivative of ascorbic acid, which is an essential co-factor of enzymes that catalyze proline and lysine hydroxylation [16], which in turn are critical for collagen maturation and stability. A stock solution of 5 mg/mL in ddH_2O can be prepared, sterile-filtered with a syringe driven filter (0.2 μm, 30 mm diameter), and stored at −20 °C.
2. Depending on the cell type, the number of cells might be adjusted. More than three biological replicates per cell type are advisable.
3. On day 1, it is possible to supplement the medium with collagen I at a final concentration of 3 μg/mL to support cells less likely to form stable spheroids. In that case, on day 0, 5000 cells should be seeded in 100 μL. On day 1, the plate should be centrifuged at 100 rcf for 3 min and 100 μL of medium supplemented with 10% FBS, 100 μg/mL normocin, 50 μg/mL 2-phospho-L-ascorbic acid, and 6 μg/mL collagen I rat tail should be added.
4. The spheroids are stable for more than 7 days.
5. The spheroids are barely visible by eye, checking the U-bottom well after the transfer is advisable.
6. A quick and careful movement is necessary to remove as much of PBS as possible without losing spheroids.
7. After removing the PBS for the second time (Fig. 1f), the plate can be sealed with a Thermowell sealing tape for 96-well plates and stored at −80 °C.
8. Take into consideration the extra volume necessary for the Bravo Automated Liquid Handling Platform to pipette pre-defined volumes ("over aspiration", ~15 μL).
9. Default settings are column 1: reduction; column 2: alkylation; column 3: beads; column 4: trypsin. The software calculates necessary volumes, which depend on the total number of samples.
10. Carefully resuspend the samples to avoid spillage.
11. The lid-heated PCR equipment reduces sample evaporation. As alternative: after 16 h at 37 °C, the plate can be kept at 10 °C.
12. If needed, the samples can be desalted by STAGE tips prior loading to the LC autosampler [17].

Acknowledgments

We thank D. Kressler and M. Stumpe for mass spectrometry support and Agilent Technologies for technical support regarding the Bravo Automated Liquid Handling Platform. This work was supported by the Swiss National Science Foundation as part of the NRP79 (SNF_206394_NFP79) and by the Canton and University of Fribourg as part of the SKINTEGRITY.CH research network.

References

1. Vandergriff T (2008) Anatomy and physiology, 4th edn. Dermatology
2. Nyström A, Bruckner-Tuderman L (2019) Matrix molecules and skin biology. Semin Cell Dev Biol 89:136–146
3. Nystrom A, Bernasconi R, Bornert O (2018) Therapies for genetic extracellular matrix diseases of the skin. Matrix Biol 71-72:330–347
4. Nystrom A, Bornert O, Kuhl T (2017) Cell therapy for basement membrane-linked diseases. Matrix Biol 57–58:124–139
5. Tolle RC, Gaggioli C, Dengjel J (2018) Three-Dimensional Cell Culture Conditions Affect the Proteome of Cancer-Associated Fibroblasts. J Proteome Res 17(8):2780–2789
6. Vu B, Souza GR, Dengjel J (2021) Scaffold-free 3D cell culture of primary skin fibroblasts induces profound changes of the matrisome. Matrix Biol Plus 11:100066
7. Berning M, Pratzel-Wunder S, Bickenbach JR et al (2015) Three-Dimensional In Vitro Skin and Skin Cancer Models Based on Human Fibroblast-Derived Matrix. Tissue Eng Part C Methods 21(9):958–970
8. Dengjel J, Bruckner-Tuderman L, Nystrom A (2020) Skin proteomics - analysis of the extracellular matrix in health and disease. Expert Rev Proteomics 17(5):377–391
9. Aebersold R, Mann M (2016) Mass-spectrometric exploration of proteome structure and function. Nature 537(7620):347–355
10. Gillet LC, Leitner A, Aebersold R (2016) Mass Spectrometry Applied to Bottom-Up Proteomics: Entering the High-Throughput Era for Hypothesis Testing. Annu Rev Anal Chem (Palo Alto, Calif) 9(1):449–472
11. Martinez-Val A, Guzman UH, Olsen JV (2022) Obtaining Complete Human Proteomes. Annu Rev Genomics Hum Genet 23:99–121
12. Gegner HM, Naake T, Aljakouch K et al (2024) A single-sample workflow for joint metabolomic and proteomic analysis of clinical specimens. Clin Proteomics 21(1):49
13. Muller T, Cremonini MA, Kliewer et al (2023) Automated Sample Preparation for Mass Spectrometry-Based Clinical Proteomics. Methods Mol Biol 2718:181–211
14. Tyanova S, Temu T, Sinitcyn P et al (2016) The Perseus computational platform for comprehensive analysis of (prote)omics data. Nat Methods 13(9):731–740
15. Szklarczyk D, Kirsch R, Koutrouli M et al (2023) The STRING database in 2023: protein-protein association networks and functional enrichment analyses for any sequenced genome of interest. Nucleic Acids Res 51(D1):D638–D646
16. Murad S, Grove D, Lindberg KA et al (1981) Regulation of collagen synthesis by ascorbic acid. Proc Natl Acad Sci USA 78(5):2879–2882
17. Rappsilber J, Mann M, Ishihama Y (2007) Protocol for micro-purification, enrichment, pre-fractionation and storage of peptides for proteomics using StageTips. Nat Protoc 2(8):1896–1906
18. Bindea G, Mlecnik B, Hackl H et al (2009) ClueGO: a Cytoscape plug-in to decipher functionally grouped gene ontology and pathway annotation networks. Bioinformatics 25(8):1091–1093

Chapter 16

Characterization of Human Skin Derived Cells by Raman Micro-Spectroscopy

Luca Pontiggia, Hesham Yosef, Thomas Biedermann, and Ueli Moehrlen

Abstract

Studying human skin biology can aid in understanding the pathophysiology of skin diseases and developing novel therapies, including tissue engineering approaches. Among various optical and spectroscopic techniques, Raman micro-spectroscopy stands out. This non-invasive method requires minimal sample preparation and provides molecular-level information on healthy and diseased cell types. The following chapter aims to provide a comprehensive guide to the basic methods of Raman micro-spectroscopy analysis of human epidermal samples from the perspective of their biochemical-molecular relationship, without the need for a deep and specialized knowledge in the field of spectroscopy. We do this based on two distinct examples: (1) Unstained cell subpopulations identified by hierarchical cluster analysis are compared with antibody treated sorted cell populations, and (2) skin cell type-specific molecular features are determined, and the composition of unknown cellular mixtures is revealed.

Key words Human skin cells, Raman micro-spectroscopy, Linear discriminant analysis classification training, Hierarchical cluster analysis of keratinocytes, Principal component analysis

1 Introduction

Raman micro-spectroscopy is a powerful analytical tool that uses the Raman effect to provide detailed chemical information about various biomolecules, body fluids, cells, and tissues. In biological and medical research, Raman micro-spectroscopy is gaining prominence because it is a non-destructive, label-free analytical technique, and provides near real-time data about the state on cellular level, i.e., whether cells are healthy and functional or start to decay. This makes Raman micro-spectroscopy invaluable in disease diagnosis and pathology studies [1, 2]. With the protocols presented in this chapter, we want to show how to use this tool in epidermal stem cell biology and skin tissue engineering, based on two examples.

Thomas Biedermann and Sophie Böttcher-Haberzeth (eds.), *Skin Tissue Engineering: Methods and Protocols*, Methods in Molecular Biology, vol. 2922, https://doi.org/10.1007/978-1-0716-4510-9_16,

Experiment A: Keratinocytes, comprising 90–95% of the human interfollicular epidermis (IFE) [3], undergo constant regeneration via self-renewal. The debate centers on the identity and location of these self-renewing keratinocytes [4]. While existing models, primarily based on mouse studies, suggest stem cells reside solely in the basal layer [5–14], our research revealed that both human basal and suprabasal keratinocytes possess self-renewing abilities, maintaining a fully stratified epidermis for 16 on the back of immuno-incompetent rats [15]. In this context, we analyzed the biochemical-molecular properties of basal and suprabasal keratinocytes by Raman micro-spectroscopy, just after isolation and after 7 days in culture. Basal and suprabasal keratinocytes lose their distinct Raman spectra in vitro. The biochemical (Raman fingerprint) properties of basal and suprabasal cells after culture in vitro became similar, which may reflect a similar regenerative potential [15]. In addition, the bias of the Raman measurements through the presence of the antibody needed for cell sorting could be excluded after subjecting a freshly isolated keratinocyte population (containing basal and suprabasal cells) to Hierarchical Cluster analysis (HCA).

Experiment B: Another pivotal possibility of Raman micro-spectroscopy is to analyze the spectral results and spectral band assignments and perform HCA of unknown cell populations. A classification model based on Linear Discriminant Analysis (LDA) can be implemented on the Raman spectra of reference cell populations to distinguish cell populations of unknown Raman spectra, meaning that the LDA classifier learns how to discriminate cell populations in an unknown mixture [16]. In this chapter, we present as an example the identification and counting of human primary keratinocytes, fibroblasts, and melanocytes in an unknown cell population derived from human skin.

2 Materials

2.1 Labware

1. 25 and 75 cm^2 tissue culture flasks.
2. Plastic 10 cm non-treated Petri dishes for tissue dissection.
3. Sterile scalpel, sterile forceps.
4. 50 mL and 15 mL sterile centrifugation tubes for cell culture.
5. Cell strainer with 100 μm pore size for 15 mL and 50 mL tubes.
6. Cell counter or Neubauer chamber for cell counting.
7. 40 μm cell strainer cap tube.
8. Microchannel chips with 0.17-μm-thick borosilicate cover glass bottom.

9. Flow Cytometer Sorter: e.g., FACS Aria III flow cytometer equipped with 405 nm (violet), 488 nm (blue), 561 nm (yellow/green) and 635 nm (red) lasers and 85 μm nozzle.
10. Biospex micro-Raman spectrometer with the following key features:
    (a) Inverted digitized light microscopic system.
    (b) Bright-field visualization camera.
    (c) 60×water immersion objective (1.1 NA, 0.2 WD) with correction collar set to 0.17 mm.
    (d) Motorized scanning stage with an x/y range of 180 × 120 mm. Repeatable relocation accuracy of 1 μm plus 0.01% of the running distance, z-focus range of 45 mm with a repeatable relocation accuracy of 0.5 μm plus 0.01% of the running distance.
    (e) Diode excitation laser of 785 nm wavelength, output power of 100 mW.
    (f) Highly sensitive Raman spectrograph with a frequency area of 100–3500 wavenumbers, optimized for the laser wavelength of 785 nm.
    (g) For detection of the Raman scattered photons: diffraction grating and Charge-Coupled Device Detector camera with 1024 × 127 pixels and a pixel size of 26 × 26 μm, operating with Peltier cooling at −80 °C.

2.2 Solutions

1. HBSS: Hank's balanced salt solution with 1.3 mM calcium, 0.9 mM magnesium, 4.2 mM sodium bicarbonate and 5.6 mM D-glucose.
2. Phosphate buffered saline (PBS) with 1.5 mM KH_2PO_4, 134 mM NaCl and 8.1 Na_2HPO_4-$7H_2O$.
3. Dulbecco's Phosphate buffered saline (DPBS): like PBS with 2.7 mM KCl.
4. Red blood cells lysis solution (ACK-Lysis buffer): 155 mM ammonium chloride, 10 mM potassium bicarbonate, 0.1 mM EDTA.
5. Bovine collagen I: dilute 5 mg/mL bovine collagen to 4 mg/mL with 0.189 M acetic acid.
6. 100% fetal calf serum.
7. 4% formaldehyde.

2.3 Skin Samples, Cells, and Complete Media

1. Human skin samples (*see* **Note 1**).
2. CT57: Epithelial Proliferation Medium supplemented according to the manufacturer and additionally supplemented with 5 μg/mL gentamycin.

3. CT40: CnT-40 Melanocyte Proliferation Medium supplemented according to the manufacturer and additionally supplemented with 5 μg/mL gentamycin.
4. DMEM: Dulbecco's Modified Eagle Medium containing 5.6 mM D-glucose, 1 mM sodium pyruvate, 4 mM GlutaMAX™, and supplemented with 10% fetal calf serum, 5 μg/mL gentamycin and 10 mM HEPES.
5. Skin biopsy medium: Dulbecco's Modified Eagle Medium supplemented with threefold concentration of Antibiotic-Antimycotic.

2.4 Enzymes

1. Trypsin/EDTA: 0.5% Trypsin / 0.2% EDTA Stock solution (10×). For some steps, dilute to 0.2× in DPBS (*see* Subheading 2.2.).
2. Dispase: Stock solution (50 caseolitic U/mL) diluted with HBSS (*see* Subheading 2.2.) (1:1) and supplemented with 5 μg/mL gentamycin.
3. 35 mg/mL collagenase (100× stock solution): 200 mg Collagenase G, 100 mg collagenase H in DPBS sterile filtered. 50 μL aliquots are stored at −20°C.
4. 350 μg/mL collagenase working solution: Dilute a 50 μL aliquot of 35 mg/mL collagenase (100× stock solution) with 5 mL HBSS.
5. Soybean trypsin inhibitor (SBTI): 1 g in 100 mL DPBS, filter sterile to obtain the 10 mg/mL stock solution. Store 15 mL aliquots at −20°C. Dilute an aliquot to 40 mL with DPBS for the 10× working solution.

2.5 Antibodies and Staining Compounds

1. iβ4-Alexa Fluor 488 (integrin β4 antibody, clone 422325).
2. Zombie Aqua Viability Dye.

2.6 Analysis of the Raman Data

1. The mpx-RamSES web-based software package built on the Python platform is utilized for the processing of Raman spectra. This includes:
   (a) **Cropping** to the spectral range of 480–1800/cm, which covers most of the biological-relevant spectral bands.
   (b) **Baseline** correction using an asymmetric least square fit.
   (c) Removal of the **cosmic** spikes.
   (d) **Smoothening** with a median filter.
   (e) **Interpolation** to continuous wave numbers.
   (f) **Normalization** by implementing a Band- or Unit-Vector-Normalization.
   (g) **Principal Component Analysis** (**PCA**): multivariate statistical analysis method used to reduce the dimensions of

large data sets by transforming the large data variables into smaller numbers of components called Principal components (PCs). The first principal component (PC1) explains for most important variances in the data set, the PC2 explains the second most important, etc. PCA is performed under Python 3, using the scikit-learn package. PCA Score plots illuminate differences and similarities among samples while the PCA Loadings display the Raman spectral differences that are used to compare the analyzed samples.

(h) **Linear Discriminant Analysis (LDA)**: linear transformations of the variables to the new discriminant functions (LDs) are computed, in which the directions of the LDs are determined in the spectral space.

(i) **Hierarchical cluster analysis (HCA)**: unsupervised statistical analysis and clustering method, to separate the Raman spectra based on the different spectral observations to different clusters.

(j) **Data visualization**.

3 Methods

As mentioned in the introduction, we want to show a way to use Raman micro-spectroscopy in epidermal stem cells biology and skin tissue engineering, based on two example experiments: In the following protocols, we refer to the experiments using the symbols A and B.

3.1 Isolation and Culture of Fibroblasts (B)

1. Cut the biopsy in fragments of approx. 0.2–0.4 cm^2 in a non-treated Petri dish by means of a sterile scalpel and transfer the fragments in a sterile 50 mL centrifugation tube.
2. Add 5 mL dispase solution and digest overnight at 4°C.
3. Transfer the digested skin fragments to a Petri dish. Separate epidermis and dermis using suitable sterile forceps. Transfer epidermis into a 15 mL and dermis into a 50 mL centrifugation tube.
4. For isolation of fibroblasts, digest the dermal tissue fragments in 5 mL collagenase working solution for 1 h in a water bath (37°C) by gentle shaking every 10 min. If available, a shaker located in an incubator is a convenient alternative (try to position the tube obliquely, so that the tissue fragments are separated from each other but remain covered with the digestive solution. If necessary, increase the volume of the solution).

In the meantime, the isolation of keratinocytes and melanocytes can proceed (see Subheadings 3.2 and 3.3*).*

5. To continue the isolation of fibroblasts, remove the digested dermal tissue from the water bath, pipet up and down with a 25 mL serological pipet for mechanical disruption.
6. Fill up to 14 mL, pass the solution through a 100 μm cell strainer and centrifuge for 3 min at 400 g, discard the supernatant.
7. Optional: Resuspend in 0.5 mL DPBS, add 4.5 mL ACK-Lysis buffer, incubate for 2 min at room temperature and centrifuge 3 min at 400 g.
8. Resuspend the cell pellet in 10 mL DMEM and plate in a 75 cm^2 cell culture flask (coating not necessary). Incubate the fibroblasts at 37°C, 5% CO_2.
9. Change the medium every 2–3 days and passage the cells at approximately 80–90% confluence (Fig. 1a).

3.2 Isolation and Culture of Keratinocytes (A), (B)

Proceed as in Subheading 3.1, **steps 1–3**.

1. For keratinocytes isolation, prepare first coated cell culture flasks: Dilute 120 μL bovine collagen I in 6 mL DPBS and add to a 75 cm^2 cell culture flask. Incubate at least 30 min at room temperature. Prepare one flask per each biopsy (approx. 1 g skin sample).
2. Add 2 mL pre-warmed 10× trypsin/EDTA to the 15 mL tube to start the digestion. Incubate for 3 min in a water bath (37° C) under manual continuous gentle shaking and brief vigorous shaking (Vortex-shaker) every minute.
3. Stop digestion by adding 2 mL 10× SBTI, and diluting with 10 mL DPBS.
4. Pass the cell suspension through a 100 μm cell strainer.
5. Centrifuge at 400 g for 3 min, remove supernatant, resuspend in 10 mL DPBS and centrifuge again at 400 g for 3 min.
6. Remove the supernatant, resuspend the cell pellet in 6 mL CT57 and plate at a cell density of approx. 100,000 cells/cm^2 into the collagen coated flasks after removal of the collagen coating solution. → *For FACS sorting, resuspend the keratinocytes first in 1 mL, count and proceed at 3.5.*
7. Incubate keratinocytes at 37°C, 5% CO_2. Change the medium every 2–3 days and passage the keratinocytes at approximately 80–90% confluence (Fig. 1b)

3.3 Isolation and Culture of Melanocytes (B)

Proceed as in Subheading 3.2, **steps 1–5**.

1. Remove the supernatant, resuspend the cell pellet in 6 mL CT40 and plate at a cell density of approx. 100,000 cells/cm^2 into a non-coated cell culture flask.

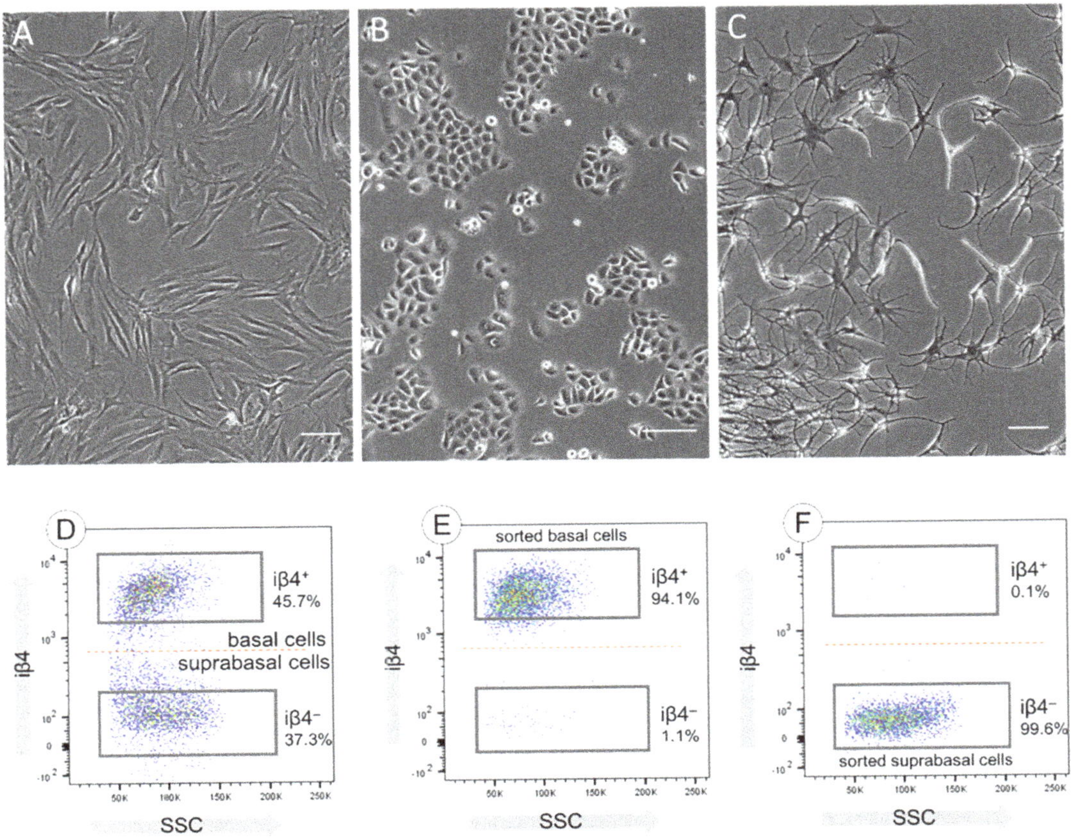

Fig. 1 Primary cells used for the tissue engineering of the human skin. a Fibroblasts at passage 1 cultured in DMEM. **b** Keratinocytes at passage 0 cultured in CT57. (**C**) Melanocytes at passage 0 cultured in CT40. **d–f** Discrimination of basal and suprabasal keratinocytes by integrin β4 expression: **d** FACS analysis of keratinocytes stained for integrin beta 4 (iβ4) only. Two distinct populations are visible: basal iβ4$^+$ and suprabasal iβ4$^-$ keratinocytes. Grey rectangles: sorting gates. **e** Sorted iβ4$^+$ population. **f** Sorted iβ4$^-$ population. (**d–f** are modified according to Pontiggia et al. [15])

2. Incubate at 37°C, 5% CO_2. Change the medium every 2–3 days and passage the melanocytes at approximately 80% confluence (Fig. 1c).

3.4 The "Unknown" Samples (B)

For experiment B, we "artificially" produced three unknown cell mixtures with different keratinocytes, melanocytes, and fibroblasts content: The cells were grown to 80% confluency in passage 0, harvested, counted, and set to the same cell concentration of one million per mL.

1. For the «unknown sample» X1 we mixed 75 μL of keratinocytes and 25 μL melanocytes.
2. For the «unknown sample» X2 we mixed 50 μL keratinocytes and 50 μL melanocytes and.

3. For the «unknown sample» X3 we mixed 50 μL of keratinocytes, 25 μL fibroblasts, and 25 μL melanocytes.
4. We proceeded to Subheading 3.6.6 for the preparation of the cells for Raman spectroscopy.

3.5 Fluorescence Activated Cell Sorting (FACS) of Epidermal Basal Cells (A)

This protocol is designed for the analysis and sorting of the freshly isolated keratinocytes from Subheading 3.2, **Step 6** *according to the integrin beta 4 expression.*

1. Prepare collagen-coated culture flasks: dilute 40 μL bovine collagen I in 2 mL DPBS and add to a 25 cm^2 cell culture flask.
2. Count the keratinocytes and set the cell concentration to 10 million/mL in a 15 mL tube.
3. Separate a portion of at least 100,000 cells as **the "unsorted" cell sample** to be directly analyzed by Raman spectroscopy. Continue with this sample at Subheading 3.6, **step 6**.
4. Add the integrin beta 4 (iβ4) Alexa Fluor 488 antibody and Zombie Aqua Viability Dye and incubate for 20 min on ice. Resuspend gently every 2 min.
5. Adjust the cell suspension to 4 mL with DPBS and underlay 1 mL FBS. Centrifuge at 400g for 3 min, discard supernatant.
6. Resuspend in 500 μL cell culture medium and pass through a 40 μm cell strainer cap tube.
7. Analyze the cells using a flow cytometer sorter and display the data in a iβ4 vs SSC dot plot (Fig. 1d, *see* **Note 2**).
8. Set the gates for the discrimination of $iβ4^+$ and $iβ4^-$ cells (grey rectangles, *see* **Note 3**).
9. Sort the cells and collect the sorted keratinocytes in 3 mL CT57 medium supplemented with 50% FCS.
10. Reanalyze the sorted samples to confirm the purity (Fig. 1e and f).
11. Centrifuge the collection tubes for 3 min at 400g, wash once by resuspending in 10 mL DPBS, centrifuge again and finally resuspend the sorted keratinocytes in 1 mL DPBS.
12. Separate portions of at least 100,000 cells as the fresh isolated, sorted "**$iβ4^+$ and $iβ4^-$ (day 0)**" cell samples to be analyzed by Raman spectroscopy. Continue with this sample at 3.6.6.
13. Centrifuge again the bulk of the cells and finally resuspend in 1 mL CT57.
14. Count and plate at a density of approx. 50,000 cells per cm^2 in the previously collagen-coated culture flasks.
15. Incubate at 37°C, 5% CO_2. Change the medium every 2–3 days and passage the cells at approximately 80% confluence.

16. After 7 days, the sorted cells are harvested as the "iβ4^{+} and iβ4^{-} (day 7)" according to Subheading 3.6.

3.6 Preparation of Cells for Raman Measurement (A), (B)

With the cell samples which are already in suspension proceed directly to Subheading 3.6, **step 6**.

1. Grow the cells to 80% confluency.
2. Separate cells from the substrate using the 0.2× Trypsin/EDTA solution.
3. Stop digestion by adding 0.2 mL of 10× SBTI, and diluting with 10 mL DPBS.
4. Centrifuge at 400 g for 3 min, remove supernatant, resuspend in 1 mL DPBS.
5. Count the cells and prepare a sample with 100,000 cells in 100 μL DPBS (1×10^6/mL).
6. Add 100 μL (one volume) of 4% PFA and incubate for 5 min at room temperature.
7. Adjust the volume to 1 mL with DPBS or PBS.
8. Centrifuge at 400 g for 3 min, discard the supernatant, resuspend in 1 mL PBS and centrifuge again.
9. Resuspend in 0.2 mL PBS and store at 4 °C until measurement.

3.7 Collection of Raman Spectra (A), (B)

This part of the protocol is modified from [16] *and is based on the Biospex System of MicrophononX* (Fig. 2a).

1. Initiate the Raman microscopic system and the control software (Fig. 2d, *see* **Note 4**). Open a new data sheet (Fig 2d(g)).
2. Select 60×water immersion objective. Carefully clean the lens using glasses cleaning wipes dipped in 70% ethanol.
3. Add 100 μL of deionized water on the top of the objective.
4. Add 25 μL of PFA-fixed cells in PBS buffer into a new channel of the microchannel chip.
5. Fix the microchannel chip on the microscope stage (Fig. 2b), assuring that water is captured between the lens of the objective and the bottom of the chip (Fig. 2c).
6. Change the camera mode to bright field and move the stage to localize cells (Fig. 2d(b)) and up in the Z axis to focus the cells (Fig. 2d(c)). Choose the best optical view of the cell (Fig. 2g).
7. Select a measurement point (e.g., nucleus or cytoplasma; red circle in Fig. 2d(b)).
8. From the best optical view move up 5–10 μm (Fig. 2h).
9. Switch off the bright field light and switch on laser and spectrometer continuous acquisition to adjust the best focus (Fig. 2j, *see* **Note 5**).

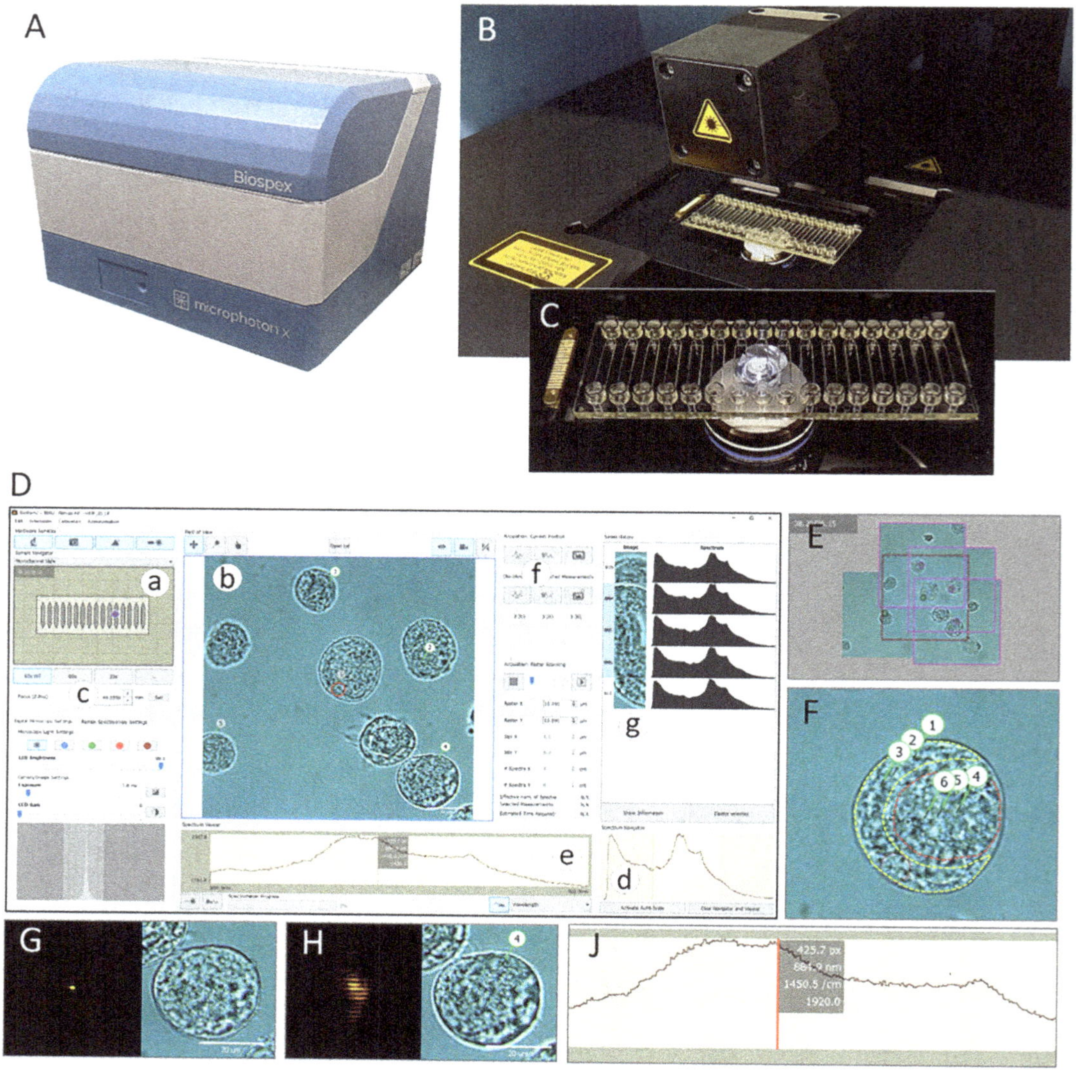

Fig. 2 The Raman microscopic system and the control software. a The Biospex microscope chamber, closed. **b** The stage with a micro channel slide is moved under the bright field light source. **c** A droplet water is captured between the objective lens and the bottom of the slide. **d** Control software window: (a) The sample navigator locates the laser beam on the specimen. Zooming in allows to visualize which areas of the sample have already been analyzed **e**. (b) The field of view allows to select the cells, to pin the region of interest, e.g., **f** the nucleus (red region) or the cytoplasma (yellow region), and to set the optimal focal point by changing (c) the Z-position of the stage. **g** The upper refraction point of the slide (left) corresponds approximatively to the best optical view of the cell (right); however, this focussing point is located under the cell at the contact point between cell membrane and glass. **h** Between 5 and 10 μm above is the most appropriate focussing point for the collection of the highest number of Raman photons. (d) The spectrum navigator and (e) the spectrum viewer report the whole or a portion of the last measured Raman spectrum. **j** The best focussing point with the highest number of photons is reflected by the highest contribution of proteins and DNA at the wavenumber 1450/cm (red vertical line) in the Raman spectrum. Once all the settings are performed, (f) the acquisition of brightfield images and Raman spectra can be started. (g) Brightfield images and Raman spectra are saved on the series history panel and can be exported

10. Pin the measurement point. Select and pin other similar points in other cells (Fig. 2f).
11. Set Laser Power to 120 mW, Exposure time to 10 s, accumulations number to 1.
12. Select the pinned measurement points on the data sheet (Fig. 2f) a start (Fig. 2d(f)) the acquisition of images and spectra.
13. Collect approx. at least 60 spectra per each cell type (Fig. 2d (g)) and export data (*see* **Note 6**).

The analysis of the mean spectrum, the performance of PCA, LDA and HCA is presented in the next paragraphs, separately for experiments A and B.

3.8 Mean Spectra, PCA and HCA (A)

Experiment A: *The Raman fingerprint of basal and suprabasal cells after culture* in vitro *becomes similar. See the workflow of experiment A in* Fig. 3.

1. Upload the data sets in two new groups on RamSES: **one group** for the fresh isolated cells (iβ4$^+$ and iβ4$^-$, day 0), a **second group** for the cultivated cells (iβ4$^+$ and iβ4$^-$, day 7) and a **third group** with the fresh isolated but unsorted cells (unsorted).
2. Under "Options" set the analysis range to 450–1800 nm (*see* **Note 7**).
3. Choose the Vector normalization method (*see* **Note 8**).
4. Enter the first group (day 0) and start the analysis to display the mean spectrum (Fig. 4a). Repeat the same with the second group (day 7, Fig. 4b).
5. Perform the PCA for both groups in 2D (Fig. 4c and d).

Note the ellipses describing the iβ4$^+$ and iβ4$^-$ cell populations: they only partially overlap at day 0, yet they completely overlap after 7 days of culture. Similar results in 3D and with LDA:

6. Merge the first (day 0) and second (day 7) group to a single new group by copy the datasets form a group to the other.
7. Enter the new group and perform the PCA in 3D (Fig. 4e).
8. To maximize the difference between the cell populations, perform LDA in 2D (Fig. 4f) or in 3D (Fig. 4g).

3.9 HCA (A)

To exclude the influence of the antibody staining needed for FACS, the fresh isolated keratinocytes population (containing basal and suprabasal cells) is subjected to Hierarchical Cluster analysis (HCA).

1. Enter the third group (unsorted, but also unstained) and start the analysis to display the mean spectrum (Fig. 4h).

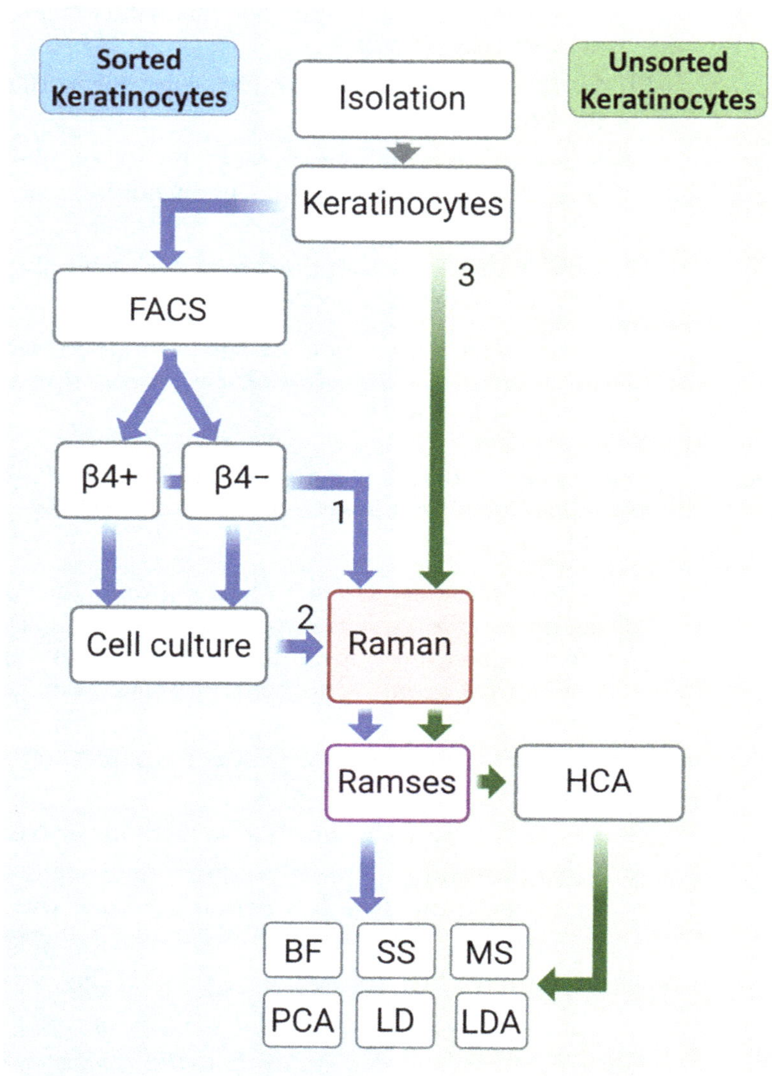

Fig. 3 Workflow Experiment A: The schema presents the various steps of Experiment A. On the left (blue part), the analysis of cells sorted by FACS is highlighted, on the right (green part), the measurement of cells without separation is emphasized. **Sorted cells**: After the isolation, keratinocytes are separated into β4-positive ($i\beta4^{+}$) and β4-negative ($i\beta4^{-}$) cells through FACS. Subsequently, the sorted cells are (1) directly measured with the Raman spectrometer or (2) first cultivated for 7 days and then analyzed. The data are passed to the RamSES platform for analysis and output of the results: Bright Field Images (BF), Single Spectra (SS), Mean Spectra (MS), Principal Component Analysis (PCA), PCA Loadings (LD), or Linear Discriminant Analysis (LDA). Similarly (right, green part), fresh isolated **unsorted cells** (3), are measured with the Raman spectrometer and the data are transferred to RamSES. Here, a Hierarchical Cluster Analysis (HCA, *see* also Fig. 4h) is carried out to obtain two groups of spectra (clusters) with the highest statistical variance. These are then compared with the data previously obtained with the separated cells (*see* Fig. 4l). (This figure was created in BioRender. Biedermann, T. (2024) BioRender.com/s71l636)

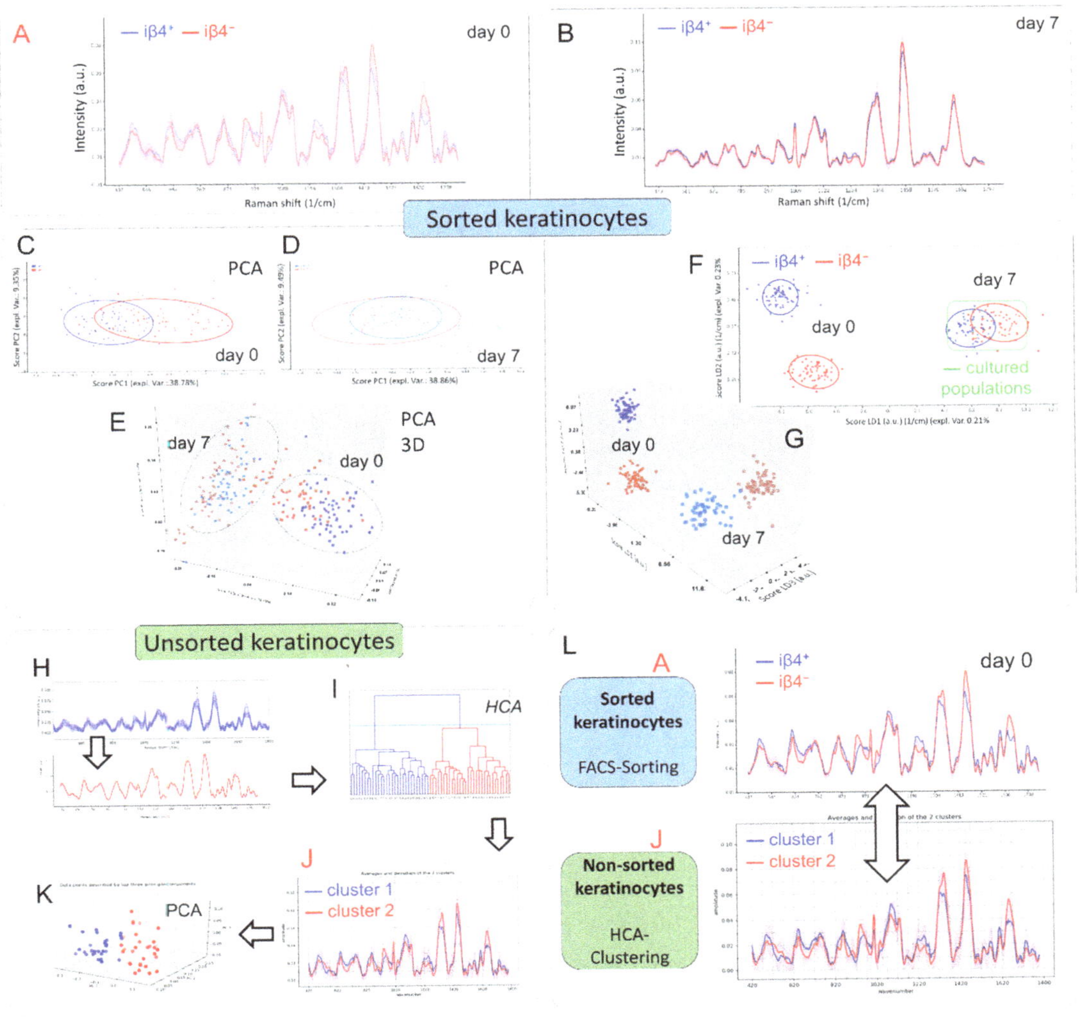

Fig. 4 Experiment A: Discrimination of freshly isolated basal iβ4⁺ and suprabasal iβ4⁻ keratinocytes by FACS sorting or by Hierarchical Cluster Analysis on the Raman data. Keratinocytes were isolated and directly measured by Raman micro-spectroscopy or first sorted by FACS and subsequently measured by Raman micro-spectroscopy. (**a** and **b**) Mean Raman spectra of iβ4⁺ (blue) and iβ4⁻ (red) cells (**a**) just after sorting or (**b**) after 7 days of culture. Shadows represent the standard deviation. **c** and **d** Score plot of the Principal Component Analysis (PCA:PC1xPC2) showing **c** two partially overlapping populations at day 0 and **d** two indistinguishable populations at day 7. Each dot represents a Raman measurement, and the ellipses describe 90% confidence interval. **e** 3D representation of the four cell populations (PC1xPC2xPC3): PC1 is the parameter which discriminates the best. The ellipses represent only a visual aid to locate the two groups day 0 and day 7, yet **f** the score plot of the Linear Discriminant Analysis (LDA) reveals better two distinct populations at day 0 (left) and two spectrally very similar 7 (right), which populations at day both completely differ from the original populations. The green region underlines the similarity of the cultured keratinocyte populations. **g** The 3D representation of LD1xLD2xLD3 shows, that the cultured populations (day7) are less distant from each other (and therefore differ less), than the fresh populations, (day 0). **h** The measured spectra of the untreated, unsorted population (containing basal and suprabasal cells) cells ($n = 60$) were first transformed into a mean spectrum prior to be subjected to **i** Hierarchical Cluster Analysis (HCA) and obtain the two clusters with the highest variance (blue and red). **j** The mean spectra of the obtained clusters are plotted and **k** Linear Discriminant Analysis (LDA) was performed to visualize two distinguishable cell

2. Apply the HCA clustering and divide the spectra in two groups (clusters) (Fig. 4i): cluster 1 (blue) and cluster 2 (red).
3. Display the mean spectra of the two clusters (Fig. 4j).
4. Perform PCA (Fig. 4k).
5. Compare the plots of Fig. 4a and j.

The striking similarity (Fig. 4l) *suggest that the presence of antibodies on the sorted cells does not influence the Raman Fingerprint of the cell populations, and that the discrimination of basal and suprabasal keratinocytes barely with Raman micro-spectroscopy is possible.*

3.10 Preliminary Analysis of the Cell Populations (B)

Experiment B: *We want to discriminate cell populations in an unknown mixture (see Workflow in* Fig. 5*). In a first step, the reference keratinocytes, fibroblasts, and melanocytes, and the unknown cell populations X1, X2, and X3 are analyzed.*

1. Bright field images of fibroblasts, keratinocytes, and melanocytes can be displayed for documentation purposes (Fig. 6a).
2. Upload the exported spectral data sets in two new groups on RamSES: one group containing the reference data (fibroblasts, keratinocytes, and melanocytes) and three groups, each containing a single unknown sample mixture X1, X2, or X3.
3. Under "Options" set the analysis range to 450–1800 nm (*see* **Note** 7).
4. Choose the Vector normalization method (*see* **Note 8**).
5. The single spectra for each population can be displayed to obtain a first impression of the internal variation of the data (Fig. 6b), (*see* **Note 9**).
6. Convert them to mean spectra to allow the discrimination of distinctive molecular features of keratinocytes, fibroblasts, and melanocytes (Fig. 6c): Carotenoids in fibroblasts (yellow regions) or lipids in keratinocytes and melanocytes.
7. Perform the PCA (Fig. 6d).
8. Perform the LDA in 2D (Fig. 6e) or in 3D (Fig. 6f), (*see* **Note 10**).
9. Bright field images of the unknown populations X1, X2, and X3 can be displayed for documentation purposes (Fig. 6g).
10. Display the single spectra of each unknown population (Fig. 6h). Observe the presence of the distinctive molecular

Fig. 4 (continued) populations (blue and red). However, **l** the mean spectra of the HCA clusters **j** were almost identical to the mean spectra of the stained and FACS-sorted populations **a**. This excludes the influence of the antibody staining needed for FACS on the Raman fingerprint of iβ4$^+$ and iβ4$^-$ cells. (**a**, **b**, and **f** are modified according to Pontiggia et al. [15])

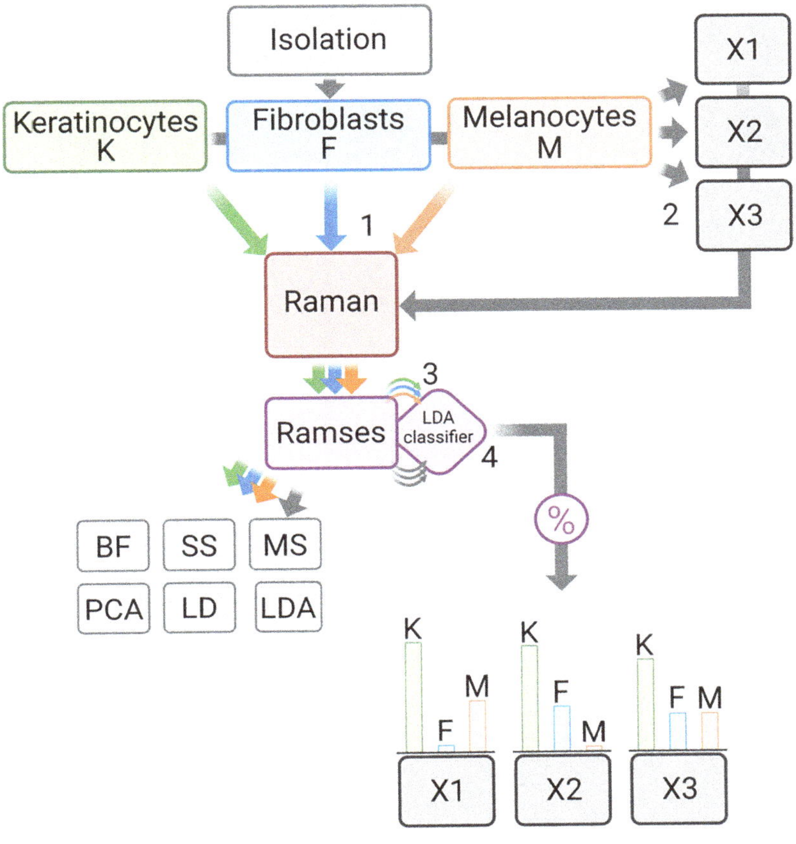

Fig. 5 Workflow Experiment B: After the isolation, keratinocytes (K), fibroblasts (F), and melanocytes (M) are (1) directly measured with the Raman spectrometer or (2) three different mixtures of them are produced (X1, X2, X3) and measured. The data are passed to the RamSES platform for analysis and output of the results: Bright Field Images (BF), Single Spectra (SS), Mean Spectra (MS), Principal Component Analysis (PCA), PCA Loadings (LD), Linear Discriminant Analysis (LDA). On RamSES the LDA classifier is (3) trained through the reference data set of K, F, and M (4). This model is used to identify the cell-types and percentages in the sample mixtures. (This figure was created in BioRender. Biedermann, T. (2024) BioRender.com/s71l636)

features of the references (e.g., carotenoids in fibroblasts, yellow regions).

11. Copy the unknown population X1 into the reference group and perform LDA in 2D (Fig. 6i). Observe the overlapping of the unknown population (black ellipse) with the reference populations.

The 2D plots give the impression of the overlapping of X1 with a single reference population, but this may not be true. In this case, the 3D view is the better option.

12. Perform LDA in 3D (Fig. 6j). Observe the overlapping of the unknown population (yellow dots) with the reference populations.

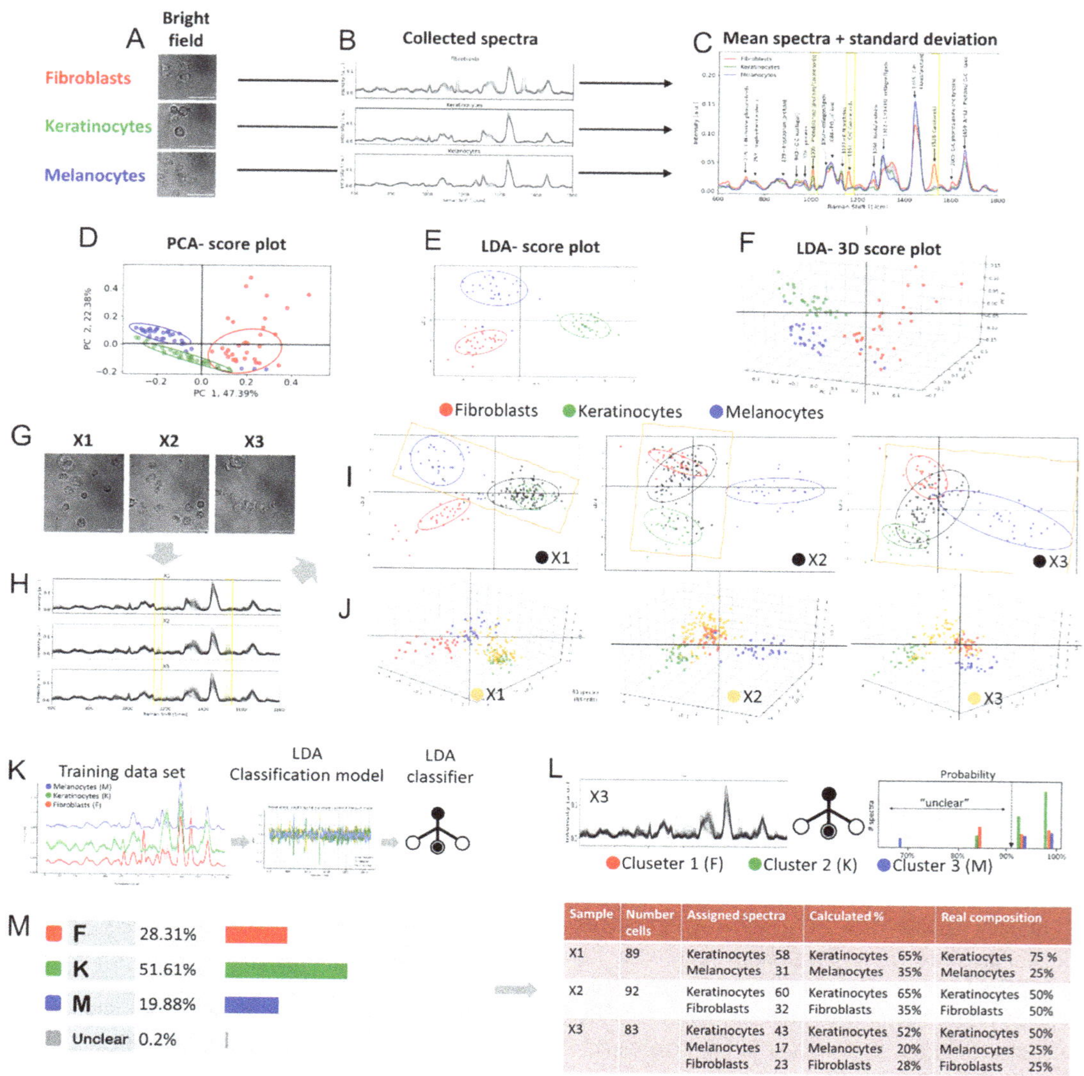

Sample	Number cells	Assigned spectra	Calculated %	Real composition
X1	89	Keratinocytes 58 Melanocytes 31	Keratinocytes 65% Melanocytes 35%	Keratiocytes 75 % Melanocytes 25%
X2	92	Keratinocytes 60 Fibroblasts 32	Keratinocytes 65% Fibroblasts 35%	Keratinocytes 50% Fibroblasts 50%
X3	83	Keratinocytes 43 Melanocytes 17 Fibroblasts 23	Keratinocytes 52% Melanocytes 20% Fibroblasts 28%	Keratinocytes 50% Melanocytes 25% Fibroblasts 25%

Fig. 6 Experiment B: The composition of unknown mixtures of fibroblasts, keratinocytes, and melanocytes is revealed by LDA classifier training. a Bright field images of fibroblasts, keratinocytes, and melanocytes. **b** Superimposed collected spectra of the three cell types. **c** Superimposed mean spectra with some band assignments: carotenoids are typical for the presence of fibroblasts (blue), some lipids assignments are more pronounced in melanocytes (green) and some other produced by DNA, tryptophan, and glycogen characterize keratinocytes (red). **d–f** PCA- and LDA- score plots can clearly differentiate between fibroblasts (red), keratinocytes (green), and melanocytes (blue). **d** PCA score plot of the mean spectra of **c**. **e** LDA score plot of the same data and **f** 3D representation of the LDA scores. **g–j** Analysis of the unknown cell mixtures X1, X2, and X3. **g** Bright field pictures. **h** Superimposed collected spectra of the three unknown mixtures X1, X2, X3. The presence of carotenoids (fibroblasts) is visible through the yellow region in X2 and X3, but not in X1. The distinctive features of keratinocytes and melanocytes are not identifiable here. However, **i** 2D-LDA analysis from sample X1, X2, X3 (black) compared with references samples (red, green, blue), gives only partial indication on the identity of the unknown mixtures. In contrast **j** in 3D-LDA the relationship of the unknown mixtures (yellow) to the references is clear: X1 overlaps with the keratinocyte "cloud" and to a lesser extent with the melanocytes data set. Similarly, X2 overlaps with fibroblasts but many data points also overlap with the keratinocyte population. X3 seems to be related with all three reference cell types. The yellow squares in the 2D plots **i** indicate the overlapping of the unknown population as revealed by the 3D view in **j**.

For better visualization, the yellow squares in the 2D plots indicate the overlapping of the unknown population in the 3D view.

13. Remove the X1 dataset from the reference group.
14. Repeat with the unknown data sets of X2 and X3.

3.11 LDA Classification of Three Unknown Cell Mixtures (B)

1. Each data set of an unknown population must be in a separate group.
2. Enter the reference data group and start the analysis (plot the mean spectra), which includes the training of the LDA classifier (Fig. 6k).
3. Apply the LDA cluster analysis (LDA clustering): choose which unknown cell population you want to analyze (Fig. 6l).
4. An optional step is to setup a threshold to identify spectra based on the percentage of the matching probability (optimally more than 90%). Spectra below will not be classified and will be considered "unclear" (Fig. 6l, right box).

   Read the results in the bar graph (Fig. 6l).

4 Notes

1. Biopsies (circumcisions) were obtained from the Department of Surgery of the University Children's Hospital Zurich and transported to the laboratory in a 50 mL tube with transport medium.
2. The choice of iβ4 vs FSC is also possible. Choose the variant which gives the better separation of $i\beta4^{+}$ and $i\beta4^{-}$ populations.
3. Set the largest possible distance between the gates to maximize the sorting efficiency and the purity of the collected populations. Contaminations below 0.5% are negligible, depending on the experimental aim. In our case the contamination of $i\beta4^{+}$ with $i\beta4^{-}$ cells (1.1%, Fig. 4b) could not be reduced further, yet did not compromise the results.
4. To ensure the accuracy of the Raman measurements, a calibration check should be done every week or month. This involves measuring the Raman band of the silicon substrate (at 521/cm) with the same method as the cell samples. If the results are

Fig. 6 (continued) k–**m** LDA classification of three unknown cell mixtures: **k** Training of the LDA classifier on three reference data sets of keratinocytes (K), melanocytes (M), and fibroblasts (F). **l** Analysis of the unknown mixture X3. The single spectra of X3 are shown (left). In the right panel, the set-up of the threshold for the matching probability of the single spectra against the references (K or M or F) is displayed. **m** Summary of the results of the LDA classification. The calculated composition is near to the real composition of the mixtures (table to the right)

consistent with previous data, the Raman measurements can continue. Otherwise, the spectrometer needs to be calibrated following the manufacturer's instructions.

5. Fresh and fixed tissues and cells show a common peak around 1450/cm which is the result of the vibration of CH functional groups in lipids, amino acid side chains, and carbohydrates [17]. The maximum signal at this point is indicative of an optimal signal across the whole spectrum. Generally this is between 5 and 10 micrometers above the bottom of the chip. To check this, set the exposure time to 3 s, switch on the laser and the spectrograph, switch off the bright field light. Move the stage in the Z-Axis to maximize the 1450/cm peak.
6. Data are exported as TIFF-images and CSV-Excel files.
7. In the Biospex System of MicrophononX, there is an optical filter that impacts the spectra in this region, necessitating the cropping around 450/cm. It is important to verify whether this behaviour differs in other Raman spectroscopy systems. Regardless, this region contains only very weak spectral information.
8. Normalization: The Uni-Vector normalization method is set by default. It can be applied when there are potential artefacts affecting the data, such as variations due to different measurement conditions or fluctuations in laser power. Also, it is particularly relevant when comparing old data with more recent measurements. Normalization can mitigate variations due to changes in instrumental conditions or experimental setups that can impact the results.
9. When displaying individual spectra, data consistency, standard error bars, and deviations between single spectra can be verified. Additionally, you can identify unusual bands in a spectrum that might not be related to the biological samples but rather stem from contaminants.
10. For an optimal classification process, it is necessary to collect large number of spectral data and compare the different cell types via LDA scores to check if that the reference cell types are well distinguished.

References

1. Gomes Da Costa S, Richter A, Schmidt U et al (2019) Confocal Raman microscopy in life sciences. Morphologie 103:11–16
2. Rangan S, Schulze HG, Vardaki MZ et al (2020) Applications of Raman spectroscopy in the development of cell therapies: state of the art and future perspectives. Analyst 145:2070–2105
3. Hoath SB, Leahy DG (2003) The organization of human epidermis: functional epidermal units and phi proportionality. J Invest Dermatol 121:1440–1446
4. Boehnke K, Falkowska-Hansen B, Stark HJ et al (2012) Stem cells of the human epidermis and their niche: composition and function in

epidermal regeneration and carcinogenesis. Carcinogenesis 33:1247–1258

5. Fuchs E (2008) Skin stem cells: rising to the surface. J Cell Biol 180:273–284
6. Clayton E, Doupé DP, Klein AM et al (2007) A single type of progenitor cell maintains normal epidermis. Nature 446:185–189
7. Lechler T, Fuchs E (2005) Asymmetric cell divisions promote stratification and differentiation of mammalian skin. Nature 437:275–280
8. Aragona M, Dekoninck S, Rulands S et al (2017) Defining stem cell dynamics and migration during wound healing in mouse skin epidermis. Nat Commun 8:14684
9. Jensen KB, Watt FM (2006) Single-cell expression profiling of human epidermal stem and transit-amplifying cells: Lrig1 is a regulator of stem cell quiescence. Proc Natl Acad Sci USA 103:11958–11963
10. Li A, Pouliot N, Redvers R et al (2004) Extensive tissue-regenerative capacity of neonatal human keratinocyte stem cells and their progeny. J Clin Invest 113:390–400
11. Pellegrini G, Dellambra E, Golisano O et al (2001) p63 identifies keratinocyte stem cells. Proc Natl Acad Sci USA 98:3156–3161
12. Li L, Clevers H (2010) Coexistence of quiescent and active adult stem cells in mammals. Science 327:542–545
13. Mascré G, Dekoninck S, Drogat B et al (2012) Distinct contribution of stem and progenitor cells to epidermal maintenance. Nature 489: 257–262
14. Senoo M (2013) Epidermal stem cells in homeostasis and wound repair of the skin. Adv Wound Care (New Rochelle) 2:273–282
15. Pontiggia L, Ahuja AK, Yosef HK et al (2022) Human basal and suprabasal keratinocytes are both able to generate and maintain dermo-epidermal skin substitutes in long-term in vivo experiments. Cells 11:2156
16. Yosef HK, Schutze K (2020) Raman trapping microscopy for non-invasive analysis of biological samples. Methods Mol Biol 2095: 303–317
17. Talari ACS, Movasaghi Z, Rehman S et al (2015) Raman spectroscopy of biological tissues. Appl Spectrosc Rev 50:46–111

Chapter 17

Standard Preparation Protocol of Human Skin and Skin Substitute Samples for Transmission Electron Microscopy

Gery Barmettler and Urs Ziegler

Abstract

This chapter is intended to be a guide to skin and skin substitutes studies dealing with the analysis of subcellular structures of cells and tissues by providing an exhaustive sample preparation protocol for Transmission Electron Microscopy (TEM). Critical steps to obtain good ultrastructure are mentioned as well as notes to improve the sample collection. New ways to automatize the imaging acquisition are discussed.

Key words TEM tissue preparation, Skin tissues, Skin substitutes, Bioengineered skin analogs, Tissue-engineered dermo-epidermal skin, Human-pigmented skin analog, Mesenchymal cells, Epidermal stratification and cornification tissue engineering

1 Introduction

The classical TEM analyses are still the standard methods for the representation of the ultrastructure of medical or biological preparations per se. In contrast to light microscopes, electron microscopes have a much higher resolution. This allows to resolve structural details and the morphology of subcellular components of cells, such as membranes, ribosomes, Golgi apparatus, cytoskeleton, mitochondria, vesicles, cell nuclei, and cell-cell contacts. Such structural detail can only be observed by electron microscopy in optimally prepared samples. This wealth of information in structural details is also the reason why TEM techniques remain essential today as well as in the future.

This method is routinely used in many areas of medicine and life sciences, so it is essential that the preservation of the ultrastructure is a prerequisite for the morphological evaluation of the data. One of the most crucial steps for analysis by TEM is the fixation of the native sample. Fixation must aim to preserve the ultrastructure as closely as possible to the native state. Except for cryo-electron

Thomas Biedermann and Sophie Böttcher-Haberzeth (eds.), *Skin Tissue Engineering: Methods and Protocols*, Methods in Molecular Biology, vol. 2922, https://doi.org/10.1007/978-1-0716-4510-9_17,

microscopy, fixation is done using chemical reactions to stabilize the structure. To obtain reproducible results the concentrations of buffers and fixatives, such as glutaraldehyde (GA) and osmium tetroxide (Os), the quality of reagents as well as the reaction time must be precisely controlled [1, 2].

An alternative to classical chemical fixation is fixation by high-pressure freezing (HPF). HPF results in a better ultra-structure preservation than the chemical fixation. However, one of the limitations of this method is the size of the specimen which should not be thicker than around 0.2 mm [3, 4].

The following protocol is based on standard preparations performed at the Center for Microscopy and Image Analysis of the University of Zurich [5–9]. Notes at the end give precise and further details on the preparation steps below. As classical method, TEM is still used to analyze skin and skin substitutes today [10–13].

2 Materials

2.1 Chemicals

1. Sodium-cacodylate powder.
2. Uranyl acetate (UAc) powder.
3. Lead(II) nitrate.
4. Lead(II) citrate.
5. Sodium hydroxide pellets.
6. Epon 812.
7. Durcupan ACM.
8. Dibutylphthalate.
9. Dodecenylsuccinic anhydride (DDSA).
10. Accelerator: Epoxy embedding medium (*2,4,6-Tris (dimethylaminomethyl)phenol*).
11. Molecular Sieve Dehydrate.
12. Ethanol absolute.
13. Propylene oxide.
14. 1 g Osmium tetroxide in ampoule.

2.2 Stock Solutions

1. 25% Glutaraldehyde (GA) solution, EM Grade.
2. 2% Osmium tetroxide (Os) aqueous solution.
3. 0.2 M Sodium-cacodylate buffer, pH 7.35 in water.
4. 3% Uranyl acetate in water.
5. Epon-Araldite stock solution (without catalysator, for approximately 150 mL): 70.89 g Epon, 92.35 g Durcupan, 8.68 g Dibutylphthalate.

2.3 Working Solutions

1. 1% Formvar solution in ethylene dichloride.
2. GA fixation buffer: 2.5% GA in 0.1 M sodium-cacodylate buffer.
3. Rinsing buffer: 0.1 M sodium-cacodylate buffer.
4. 1% Uranyl acetate in water.
5. Os fixation buffer: 1% (w/v) osmium tetroxide in 0.1 M sodium-cacodylate buffer.
6. UAc staining solution: 1% UAc in water.
7. Epon-Araldite working solution (approx. 20 mL): 11.7 g Epon-Araldite stock solution, 10.0 g DDSA, 620 mg accelerator.
8. 1% (w/v) Toluidine blue in water.
9. Reynold's lead solution [14]: (A) 1.33 g Lead(II) nitrate in 15 mL water; (B) 1.76 Lead(II) citrate in 15 mL water; (C) 8 mL 1 N NaOH. Mix A and B into a 50 mL volumetric flask, shake well until solution becomes homogeneously milky. Add 8 mL of solution C. Solution becomes clear. Fill up to 50 mL with water. The finished solution can be portioned and stored at 4 °C for several months.
10. 3% UAc in water.

2.4 Materials

1. Silicon nitride membrane, 50 nm membrane, 1 × 1 mm.
2. Copper Grid 2 × 1 mm.
3. Tabbed Copper Grid.
4. Single Edge Blades.
5. Double edge stainless steel razor blade.
6. Filter paper, hardened ashless (Whatman grade 542).
7. Parafilm®.

2.5 Instruments

1. Water system for 18.2 MΩ.cm ultrapure water.
2. Diamond knives for ultra-thin sectioning.
3. Diamond trimming knives.
4. Ultra-microtome.

3 Methods

1. Extract the sample and immerse it immediately in the GA fixation buffer (*see* **Notes 1–8**).
2. Subsequently, cut samples with razor blades to a size of 1 mm in one dimension. Keep the sample immersed always in GA fixation buffer (Fig. 1).

Fig. 1 Cutting tissue with two halves of a razor blade immersed in GA fixation buffer to reduce compression artifacts

3. Incubate 1 mm^3 sample in 1 mL GA fixation buffer for 2–4 h or overnight at 4 °C.
4. Wash 2× with 1 mL of rinsing buffer for 5 min each.
5. Postfix with 1 mL Os fixation buffer *on ice* (*see* **Note 9**) *for 1 h.*
6. Wash 3× with 1 mL water at room temperature (RT) for 5 min each.
7. Stain with UAc staining solution (*En bloc* staining) for 1 h at RT or overnight at 4 °C (*see* **Note 10**).
8. Wash 2× with water for 5 min each at RT.
9. Make a dehydration by an ascending ethanol series: 20 min 70% (v/v) ethanol, 20 min 80% (v/v) ethanol, 100% dried ethanol 2× for 20 min each (*see* **Notes 11–12**).
10. 2× propylene oxide for 20 min each (*see* **Note 13**)
11. 2× 1:1 propylene oxide/Epon-Araldite working solution for 1 h each, or 1× overnight.
12. 1× Epon-Araldite working solution for 1 h
13. Embed in fresh Epon-Araldite working solution in a silicon mold (*see* **Note 14**).
14. Polymerization for about 24 h at 60 °C.
15. Trimming, semi-thin and ultra-thin sectioning (*see* **Notes 15–19**).
16. Collection of the ultra-thin sections (*see* Fig. 2 and **Note 20**).
17. Post-staining (*see* **Notes 21–22**).
18. Imaging at transmission electron microscope (*see* **Note 23**).

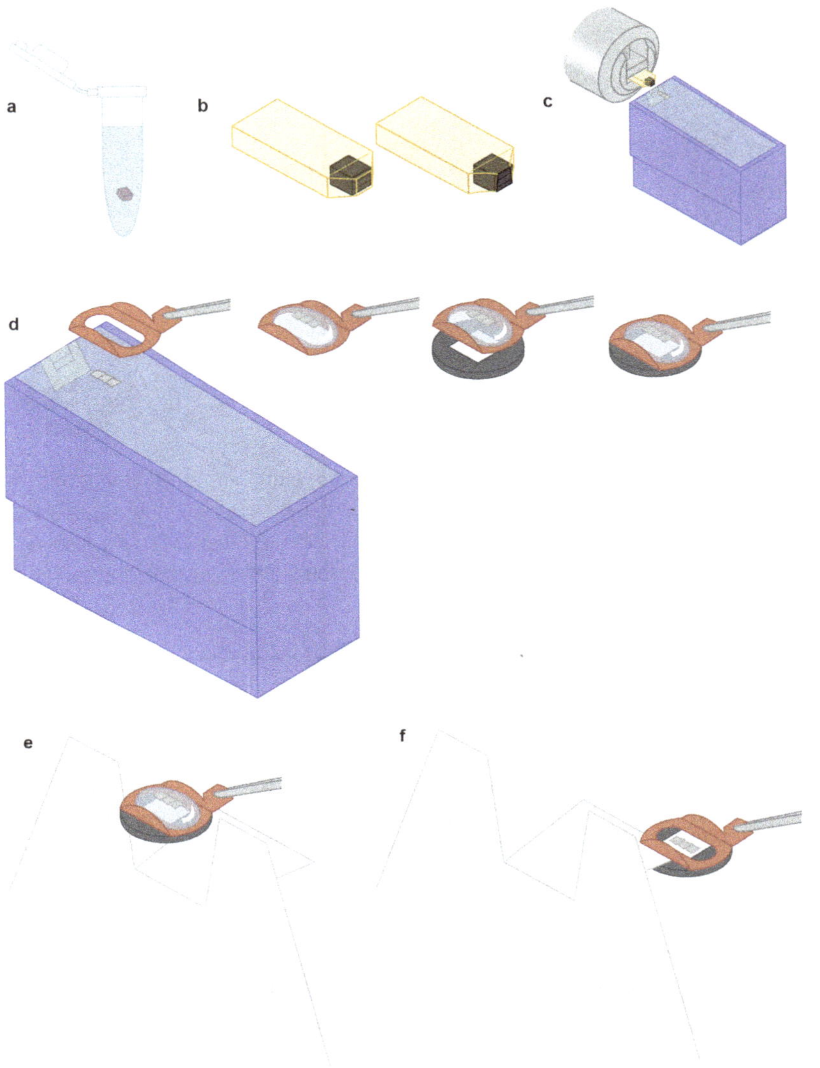

Fig. 2 Illustration showing the TEM sample preparation steps. **a** Process the tissue in an Eppendorf tube: first fixation, washing; second fixation, washing; *en bloc* staining, washing, dehydration, Epon-Araldite. In case of a high number of samples, a processing device (Leica EM TP) is recommended. **b** Embed the tissue in Epon-Araldite in a silicon mold and bring the tissue in the position with the help of a toothpick under a binocular as long the Epon-Araldite is liquid. After hardening in an oven at 60 °C for 24 h, first trim the block by hand with a razor blade, followed by a fine trimming using a glass or a diamond trimming knife (Diatome) in an ultra-sectioning device (Reichert or Leica). **c** Ultra-sectioning with a diamond knife (Diatome), 60–80 nm sections. **d** Pick up ultra-sections with a self-made modified slot grid (*see* Fig. 3) and mount them on a silicon nitride support film or a formvar-coated slot grid. **e** Under a binocular, suck away the water very slowly, selectively and alternately with the filter paper. **f** Separate the two with the filter paper

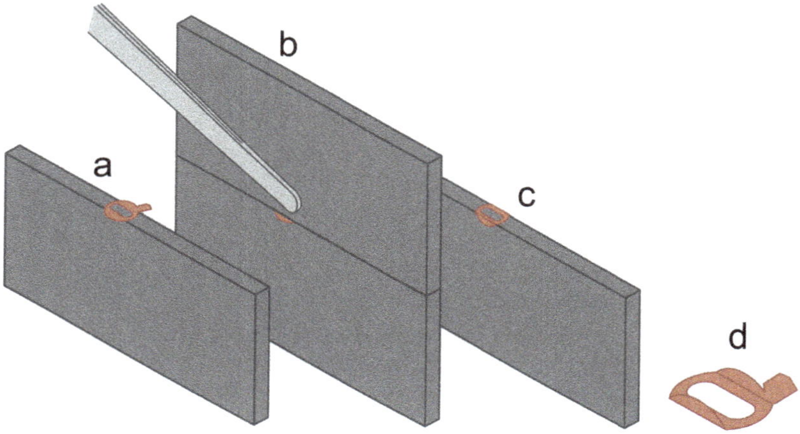

Fig. 3 Produce a perfect eyelet for section collection: a brass or aluminum plate with a thickness of 2 mm is used. **a** Place the tabbed copper slot grid with the tab on the edge. **b** Take a second, same plate and place it on top of it. Fasten the two plates with fingers or clamps. Form with the backside of a tweezer the protruding parts of the tabbed copper slot grid over the edges. **c** Release the plates **d** Remove the eyelet from the plate, it is now ready to use

4 Notes

1. The quality of the water for all the buffers and the solutions in electron microscopy technique is an essential factor. New water systems produce 18.2 MΩ.cm ultrapure water which is recommended to use.
2. The optimal concentration of the fixative and the best buffer must be determined if other buffers are employed or other samples than skin tissue is processed. We recommend for human skin tissue 2.5% glutaraldehyde (GA) in 0.1 M sodium-cacodylate buffer (pH 7.35). PBS has a very low buffer capacity and therefore it is not recommended as a buffer solution for fixation.
3. It is of the utmost importance to apply a freshly prepared fixative solution and at the same temperature as the tissue itself. Additionally, it must be applied immediately and without delay. Best results are achieved if the fixatives reacts with the biological material within seconds after a biopsy is taken [2].
4. To guarantee proper penetration of the fixation solution, the sample should not be thicker than 1 mm in one dimension [2]. If extracted specimen is larger than these dimensions, cut into smaller pieces either immediately or as quick as possible, ideally using razor blades as depicted in Fig. 1 while the sample is immersed in GA fixation buffer.
5. CAUTION: for the work with solutions containing hazardous chemicals like osmium, GA or ethylene dichloride, wear a lab

coat and gloves, work in a fume hood, and collect the waste in labeled containments. Chemicals like osmium, GA, and ethylene dichloride can irritate the respiratory tract, eyes and skin, and can occur chronic (long-term) health effects. Waste of all these substances must be collected and disposed according to the guidelines of Hazardous Waste Management of the University.

6. CAUTION: the sodium-cacodylate buffer contains arsenic which is carcinogen and may cause reproductive damage; always wear a lab coat and gloves, and collect and dispose of the waste in accordance with safety regulations.
7. Use fresh fixative solution with a surplus of approx. 10–20 times the volume of the sample.
8. Prevent any drying/pressing/squeezing or other damage of the sample as it may lead to degradation of the ultrastructure.
9. After washing of the sample with the buffer twice, it is optimal to make a second fixation with Os fixation solution in water. Osmium is a heavy metal and binds to the lipid membranes, this later provides a stronger contrast of these structures.
10. The *en bloc* staining technique with 1% UAc enhances the contrast of the tissue [15]. It is possible to omit this step and stain with a 2% uranyl acetate solution in water at the end of the procedure, onto the ultra-thin sections for 5–20 min, but the *en bloc* staining with 1% UAc generates a stronger contrast due to the diffusion of UAc through the tissue and the binding in the tissue is more powerful than the binding just on the surface of the ultra-thin sections.
11. Do not use denatured ethanol. Effects of denaturants on tissue are not known.
12. Use molecular sieve with a size of 3 Å for the 100% ethanol to produce water free ethanol. A high-water content in the sample can result in soft resin and thus difficult to section.
13. Propylene oxide has a high gas pressure. Using safe lock Eppendorf tubes or sealing the tube with a strip of Parafilm is recommended to prevent evaporation.
14. Once the samples are immersed in Epon-Araldite working solution use a toothpick under a binocular, to align the samples that they will later be correctly positioned for ultra-thin cutting (Fig. 2c).
15. CAUTION: Epon-Araldite contains hazardous substances. Embedding must be carried out in a fume hood and wearing gloves dense and resistant against solvents (Sempercare® nitrile skin^2 are recommended). The waste must be collected and disposed strictly in accordance with safety regulations.

16. Trimming to trapezoidal or rectangular pyramids is preferred. Sharp edges ensure better pressure distribution during cutting, which allows qualitatively better sections.
17. Semi-thin sections are optionally used to facilitate the correct localization of the target area in the tissue. For this purpose, 100–200 nm sections are made with the diamond knife (Diatome 35° or 45°). These sections are applied to a glass slide by a large, stainless-steel loop and dried on a heating plate at 50–60 °C. Drops of filtered toluidine blue 1% (w/v) in water are applied for 1–5 min, depending on tissue. Afterwards, the microscope slide is rinsed with water and air-dried. Locate region of interest using phase contrast light microscopy. Subsequently trim block around identified target area.
18. A critical step to perform automatic imaging is to mount the sections as flat as possible on support materials. Silicon nitride membranes are atomically flat, are chemically inert, do not deform in the TEM beam, have no type of impurities and are available in ready-to-use units. For many samples, the silicon nitride membrane (Ted Pella, Plano) is recommended. For good sections pick up, a self-made shaped unit is first produced in a very simple way with a tabbed copper slot grid with a tab. *See* Fig. 3.
19. The ultra-thin cut can now be carried out. The diamond knife is used to make 60–80 nm sections. The thickness of the sections can be estimated according to the optical interference color [16]. Gray 50–60 nm, silver 60–90 nm, gold 90–150 nm, purple 150–190 nm. According to this color scale, the ultra-microtome is adjusted. Once the sections are produced they float on the water surface and are manipulated with a hair on a stick (wooden or plastic) in the direction that the sections can be captured with an empty slot grid.
20. Wrinkle-free cutting on the grid is a goal that can be mastered with the following method: in addition to the usual cutting tools, a binocular, a Petri dish with filter paper and the self-formed slotted grid (Fig. 3d) are required. Cut the filter paper into a form (Fig. 2e) and glue it into the petri dish with double-sided adhesive tape. The sections float on the water surface of the diamond knife and are picked up with a tweezer using the self- formed slot grid. By pressing lightly on the water surface, the sections get stuck in the small drop. Then pull very close to the silicon nitride membrane and let only the small water droplet on the underside touch. Both grids now stick together. Under the binoculars, the water between them is sucked off very slowly, selectively and alternately at the prepared filter paper (Fig. 2e) until only a little water remains visible. The small water cushion reduces mechanical effects on the surface.

Then pull the two over the edge of the filter paper and pull the filter paper through between the two (Fig. 2f). When properly moved, the silicone membrane falls onto the filter paper in the Petri dish. The grid can now be placed in the grid-box with tweezers until further processing.

21. 1st post-staining: put grid on 1 drop of UAc 1% in water for 5–20 min, short wash on 3 drops of water, suck water away with a filter paper. Remarks: post staining is recommended to enhance contrast.
22. 2nd post-staining: put gird on 1 drop of Reynold's lead solution [14] for 5 min, short wash on 3 drops of water, suck water away with a filter paper. Store grid in a grid-box. Remarks: A few sodium hydroxide pellets are put around this drop. These prevent precipitates, by lead solution reacting with CO_2 from the air. Remarks: uranyl acetate and lead solution should be centrifuged before use. In an Eppendorf tube at 14,000 g for 3 min. Staining should be performed in a closed Petri dish on a Parafilm, which is covered for incubation. UAc and lead solutions must be carefully collected and disposed according to safety regulations.

    The sample is now ready for microscopy.
23. Once samples are processed according to the protocol described above ultrathin sections are examined with a transmission electron microscope (optionally, a short post-staining step can be performed). Figure 4 shows representative TEM images of different skin areas acquired by an investigator selecting manually the areas of interest. The examination of the samples at the TEM is normally performed by a trained person searching for the structures of interest. Modern microscopes are capable of automatically acquire and stitch images (Thermo Fisher Talos 120 with the integrated Maps software) allowing to obtain an overview of the grid. The investigator can draw areas of interest in overview images and acquire the selected area of interests at higher resolution subsequently (*see* Fig. 5). This multiscale and panoramic approach will help to better understand the high-resolution images, to locate them in the context of the tissue and will reduce artifacts by otherwise looking at very small parts of the sample. The interpretations of the results are easier, quicker, and less biased because of the larger context information contained in the overview.

Acknowledgments

Many thanks to Drs. José María Mateos, Andres Kaech for critically reading of the manuscript, for thoughtful comments and many advices. Special thanks to my former colleague Dr. Bruno Guhl,

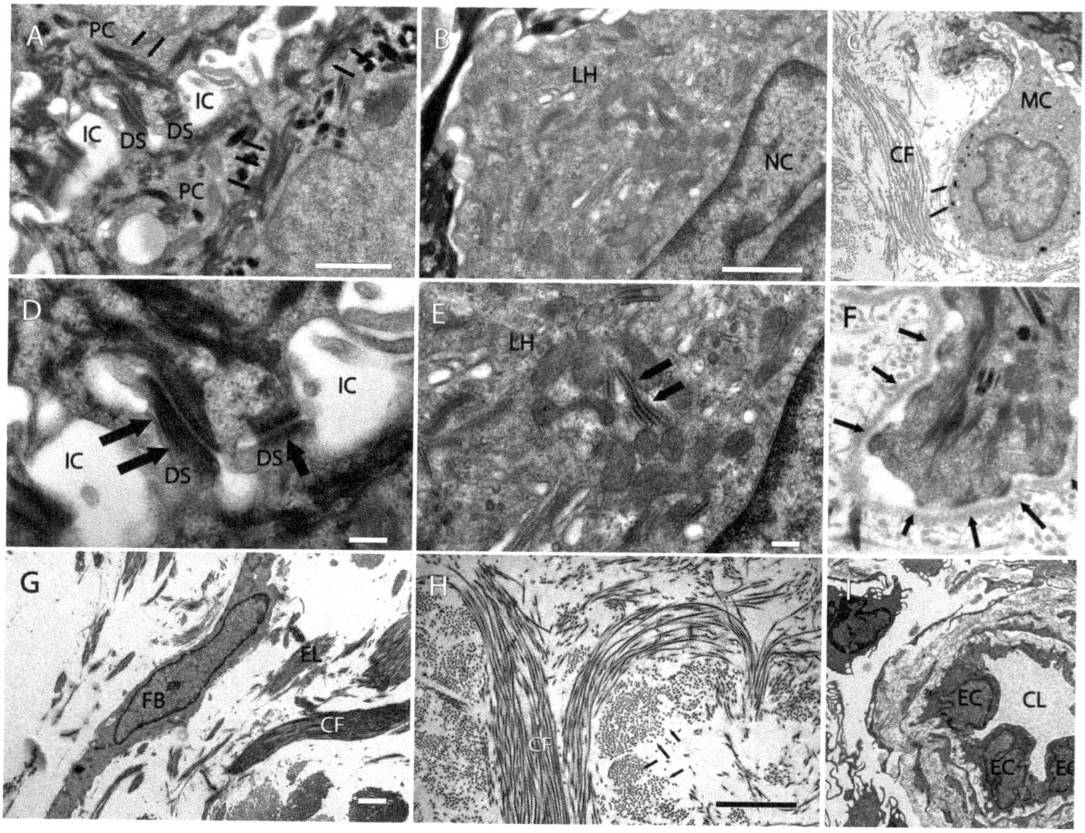

Fig. 4 Transmission Electron Microscopy images of human skin sample: **a** In the layer of the stratum spinosum: prickle cells (PC) with tonofilaments (arrows) and desmosomes (DS), in between of the two prickle cells (PC) intercellular space (IC) **b** Dendritic cell (Langerhans cell cytoplasm (LH), nucleus (NC) **c** In the layer of the stratum basale: melanocyte (MC) with melanin granules (arrows), collagen fibers (CF) **d** Same region as A at higher magnification desmosomes are well observed (arrows) **e** On higher magnification, same region as B, Langerhans cell (LH) with Birrbeck-Bodies (arrows) **f** Same region as C, hemidesmosomes (arrows) anchored at the basal lamina **g–h** Dermis **g** Fibroblast (FB), collagen fibers (CG), elastin (EL) **h** Collagen fibers longitudinal sectioned (CG), collagen fibers cross sectioned (arrows) **i** Capillary with endothelial cells (EC), capillary lumen (CL). Scales: A-C 1 μm, D-F 200 nm, G-I 2 μm

who trained me with his outstanding craftsmanship about the techniques of electron microscopy. I thank the whole team of the Center for Microscopy and Image Analysis for the excellent collaboration over the years. Illustrations: University of Zurich, Information Technology, MELS/SIVIC, Nadja Baltensweiler.

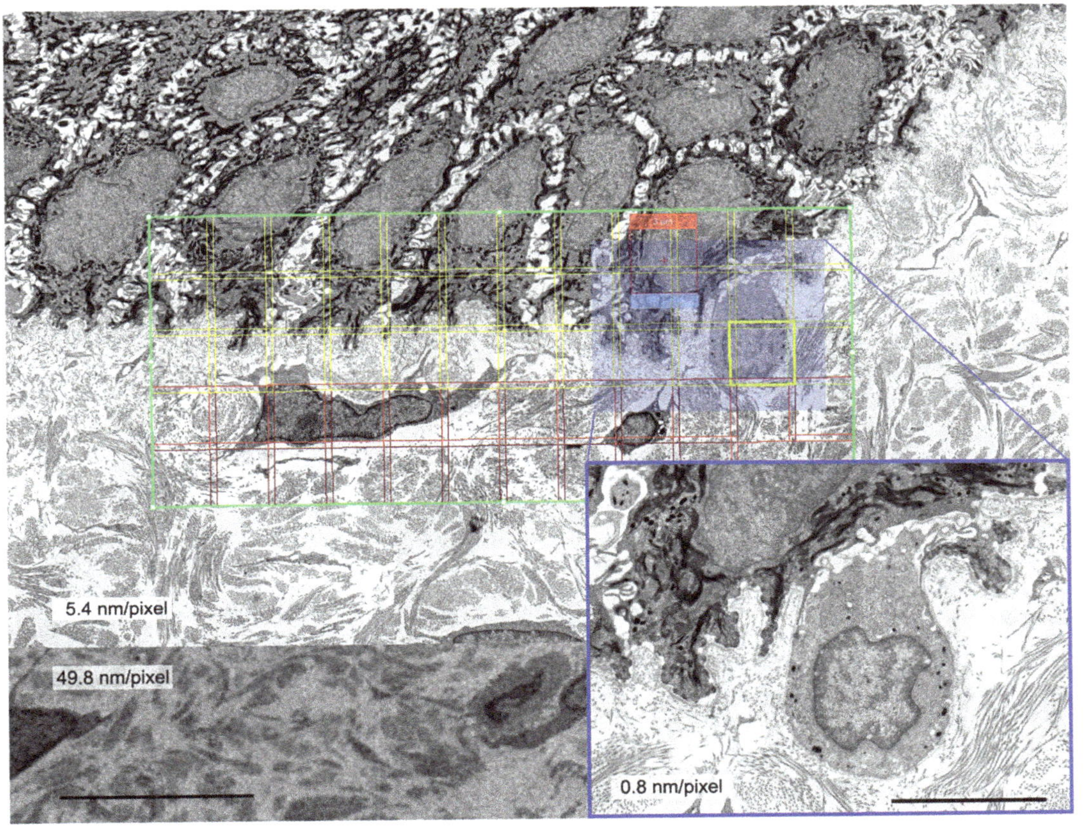

Fig. 5 TEM imaging using an overview approach with selected higher resolution areas of interest of the same human skin tissue: an ultra-thin cross-section, 70 nm, of a normal human skin is seen. The epidermis and dermis are shown. The overview of the complete section is created first with 49.8 nm/pixel (bottom left). Areas of interest are selected and acquired at intermediate resolution of 5.4 nm/pixel. High resolution images are acquired with 0.8 nm/pixel, see green rectangular area with smaller squares, red and yellow. Bottom right image: stitched images of blue area, scale bar 4 μm. Overview and intermediate image scale 10 μm

References

1. Hayat MA (1970) Principles and techniques of electron microscopy. Van Nostrand Reinhold Company, New York
2. Griffiths G (1993) Fixation for fine structure preservation and immunocytochemistry. In: Griffiths G (ed) Fine structure immunocytochemistry. Springer, Berlin Heidelberg. https://doi.org/10.1007/978-3-642-77095-1_3
3. Kaech A (2008) Bal-Tec Hpm 010 high-pressure freezing machine. In: Cavalier A, Spehner DS, Humbel BM (eds) Handbook of cryo-preparation methods for electron microscopy. CRC Press, Boca Raton
4. Kaech A, Ziegler U (2014) High-pressure freezing: current state and future prospects. Methods Mol Biol 1117:151–171
5. Pontiggia L, Biedermann T, Böttcher-Haberzeth S et al (2014) De novo epidermal regeneration using human eccrine sweat gland cells: higher competence of secretory over absorptive cells. J Invest Dermatol 134(6):1735–1742
6. Biedermann T, Klar AS, Böttcher-Haberzeth S et al (2014) Tissue-engineered dermo-epidermal skin analogs exhibit de novo formation of a near natural neurovascular link 10 weeks after transplantation. Pediatr Surg Int 30(2): 165–172
7. Biedermann T, Böttcher-Haberzeth S, Klar AS et al (2015) The influence of stromal cells on the pigmentation of tissue-engineered dermo-epidermal skin grafts. Tissue Eng Part A 21(5–6):960–969
8. Böttcher-Haberzeth S, Klar AS, Biedermann T et al (2013) 'Trooping the color': restoring the original donor skin color by addition of melanocytes to bioengineered skin analogs. Pediatr Surg Int 29(3):239–247
9. Böttcher-Haberzeth S, Biedermann T, Pontiggia L et al (2013) Human eccrine sweat gland cells turn into melanin-uptaking keratinocytes in dermo-epidermal skin substitutes. J Invest Dermatol 133(2):316–324
10. Magne B, Demers A, Savard É et al (2023) Speeding up the production of clinical-grade skin substitutes using off-the-shelf decellularized self-assembled dermal matrices. Acta Biomater 167:249–259
11. Velamoor S, Mitchell A, Bostina M et al (2022) Processing hair follicles for transmission electron microscopy. Exp Dermatol 31(2): 110–121. https://doi.org/10.1111/exd.14439
12. Sierra-Sánchez Á, Magne B, Savard E et al (2023) In vitro comparison of human plasma-based and self-assembled tissue-engineered skin substitutes: two different manufacturing processes for the treatment of deep and difficult to heal injuries. Burns Trauma 11:tkad043
13. Laws R, Steel DH, Rajan N (2022) Research techniques made simple: volume scanning electron microscopy. J Invest Dermatol 142(2): 265–271.e1
14. Reynolds ES (1963) The use of lead citrate at high pH as an electron-opaque stain in electron microscopy. J Cell Biol 17:208–212
15. Maunsbach AB, Afzelius BA (1999) Biomedical electron microscopy. Academic Press, Cambridge
16. Peachey LD (1960) Section thickness and compression. In: Bargmann W, Möllenstedt G, Niehrs H, Peters D, Ruska E, Wolpers C (eds) Vierter Internationaler Kongress für Elektronenmikroskopie/Fourth International Conference on Electron Microscopy/Quatrième Congrès International de Microscopie Électronique. Springer, Berlin, Heidelberg

Chapter 18

Methods for Assessing Scaffold Vascularization with Human Endothelial Cells

Anne M. Kong, Guei-Sheung Liu, Geraldine M. Mitchell, and Shiang Y. Lim

Abstract

The success of tissue engineering relies heavily on the efficient and rapid formation of blood vessels in the newly engineered tissues. To achieve this, it is crucial to use scaffolds that not only support the survival of vascular cells but also facilitate the development and maturation of a capillary network in vivo. Utilizing advanced biomaterials and scaffold designs is essential to ensure that the engineered tissues can integrate effectively with the host tissues and maintain their functionality over time. Here, we present a method for generating endothelial cells from induced pluripotent stem cells and pre-vascularizing scaffolds by seeding them with endothelial cells in a 3D porous structure. These pre-vascularized scaffolds can then be implanted in a rat subcutaneous model to create vascularized tissue constructs. This approach holds great promise for generating clinically viable constructs that can be used in regenerative medicine and drug development.

Key words Endothelial cells, Extrinsic vascularization, Intrinsic vascularization, Tissue engineering, Scaffold vascularization

1 Introduction

Vascularization is crucial in tissue engineering as it ensures the survival, integration, and functionality of the engineered tissues in vivo [1, 2]. The choice of scaffold is critical for successful vascularized tissue engineering, involving a balance of multiple factors, including biocompatibility, biodegradability, mechanical properties, and porosity. Both natural and synthetic materials offer distinct advantages, and advanced fabrication techniques allow for the creation of scaffolds that closely mimic the native tissue environment [3].

The scaffold serves as a framework supporting cell attachment, proliferation, and differentiation, and it must also facilitate

Geraldine M. Mitchell and Shiang Y. Lim contributed equally with all other contributors.

Thomas Biedermann and Sophie Böttcher-Haberzeth (eds.), *Skin Tissue Engineering: Methods and Protocols*, Methods in Molecular Biology, vol. 2922, https://doi.org/10.1007/978-1-0716-4510-9_18,

vascularization to ensure the survival and integration of the engineered tissue [2, 3]. Scaffold vascularization can be achieved in vitro by seeding with endothelial cells (pre-vascularization), and in vivo via extrinsic vascularization, where blood vessels grow from outside the implanted scaffold into it [2, 4, 5]. Both scaffold pre-vascularization in vitro and extrinsic vascularization in vivo can be leveraged to enhance vascularization for developing more complex, thicker tissue constructs.

In this chapter, we outline the steps required for generating human endothelial cells from induced pluripotent stem cells [6], preparing porous scaffolds containing human endothelial cells [6, 7], and an in vivo rat model of subcutaneous scaffold implantation [7]. This methodology can be readily adapted as an assay model for studies on angiogenesis and complex interactions between different cell types and the extracellular matrix in a 3D context. Additionally, this method can also be employed to engineer vascularized tissue constructs for regenerative medicine, and drug development, and testing.

2 Materials

2.1 Porous Scaffolds

1. NovoSorb® polyurethane porous scaffold, circular shape with 6 mm diameter and 1.35 mm thickness (*see* **Note 1**).

2.2 Primary Human Endothelial Cells

1. Primary human dermal microvascular endothelial cells.
2. EGM-2 MV Microvascular Endothelial Cell Growth Medium-2 BulletKits (EGM-2 MV).

2.3 Human Induced Pluripotent Stem Cell-Derived Endothelial Cells

1. Human induced pluripotent stem cells (iPSCs).
2. TeSR-E8 medium (*see* **Note 2**).
3. DMEM/F12 GlutaMAX medium.
4. StemPro-34 SFM complete medium.
5. N-2 supplement.
6. B-27 supplement.
7. VEGF-165.
8. Matrigel Growth Factor Reduced Basement Membrane Matrix.
9. Y-27632 dihydrochloride.
10. CHIR99021.
11. BMP4.
12. Forskolin.
13. Human fibronectin.

14. FITC mouse anti-human CD31 antibody (clone WM5).
15. EGM2-MV Microvascular Endothelial Cell Growth Medium-2 BulletKit.

2.4 Cell Seeding in Scaffold

1. Fibrinogen from human plasma.
2. Aprotinin.
3. Thrombin from human plasma.

2.5 Rat Subcutaneous Scaffold Implantation

1. Nude rats (*see* **Note 3**).
2. Chlorhexidine: 0.5% Chlorhexidine in 70% ethanol.
3. 4–0 silk suture.
4. Carprofen (Rimadyl®).
5. Lethobarb (Virbac).

2.6 Histological Assessment of Tissue Vascularization

1. Paraformaldehyde (PFA): 4% PFA in dH_2O.
2. ≥99.8% Ethanol.
3. Histolene.
4. Paraffin wax.
5. Optimal Cutting Temperature (OCT) compound.
6. 30% hydrogen peroxide.
7. Citrate buffer (10 mM citric acid, pH 6.0).
8. Proteinase K.
9. Serum-free Protein block solution.
10. Antibody diluent.
11. Mouse anti-human CD31 (clone JC70A).
12. Biotinylated *Griffonia Simplicifolia* lectin.
13. Biotinylated rabbit anti-mouse IgG.
14. Vectastain Elite ABC kit.
15. Horseradish peroxidase-Streptavidin.
16. Diaminobenzidine chromogen.
17. Vector VIP substrate kit.
18. Entellan mounting medium.

3 Methods

3.1 Generation of Human Endothelial Cells from Induced Pluripotent Stem Cells

1. Dissociate human iPSCs into single cells and seed onto Matrigel-coated plates at a density of 1×10^5 cells/cm^2 in TeSR-E8 medium supplemented with 10 μM of Y-27632 for 24 h.

2. Replace medium with DMEM/F12 GlutaMAX medium containing 1× N-2 supplement, 1× B-27 supplement, 8 μM of CHIR99021 and 25 ng/mL of BMP4 for 3 days.
3. Replace medium with StemPro-34 SFM complete medium supplemented with 200 ng/mL of VEGF-165 and 2 μM of forskolin for 3 days.
4. Dissociate into single cells and label with FITC mouse anti-human CD31 antibody.
5. Sort CD31 positive cells by Fluorescence Activated Cell Sorting.
6. Seed and expand CD31 positive cells on human fibronectin-coated plates in EGM2-MV medium supplemented with 50 ng/mL of VEGF-165.

3.2 Preparation of Endothelial Cell-Seeded Scaffolds

1. Pre-soak sterile NovoSorb® scaffolds (Fig. 1) in sterile phosphate-buffered saline (PBS) at room temperature for a minimum of 30 min.
2. Prior to cell seeding, squeeze out excess PBS using sterile forceps.
3. Detach human endothelial cells (primary or human iPSC-derived) and resuspend 4 × 10^5 cells (*see* **Note 4**) in 30 μL of human fibrinogen (30 mg/mL in PBS) containing 10 μg/mL of aprotinin (*see* **Note 5**).

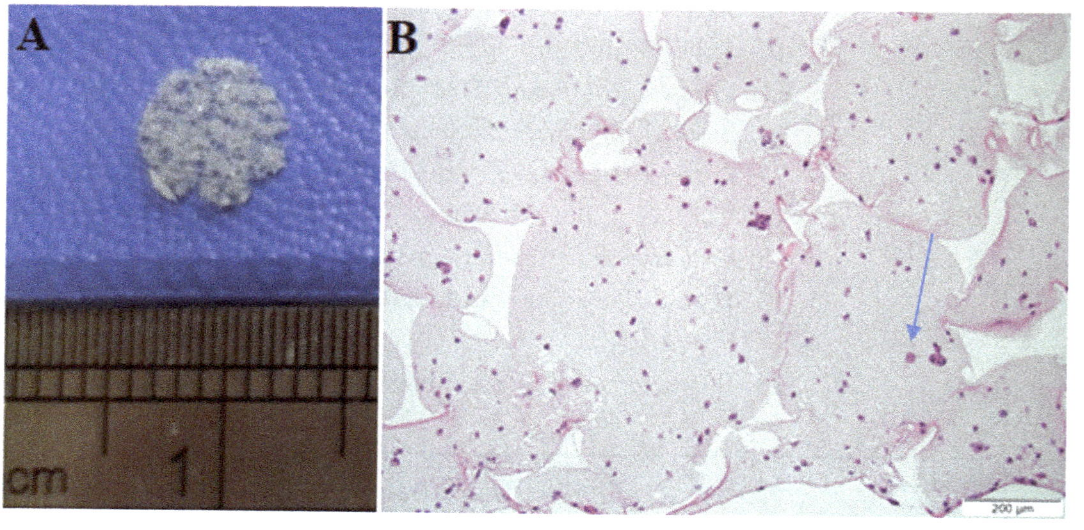

Fig. 1 The microstructures of porous NovoSorb® scaffolds. **a** A representative bright-field image of a NovoSorb® scaffold prior to implantation. **b** A representative cross-sectional image of a NovoSorb® scaffold seeded with endothelial cells before implantation and stained with haematoxylin and eosin. Arrow indicates small groupings of endothelial cells supported in the scaffold pores by human fibrin (light pink areas). Scale bar: 200 μm

4. Transfer cells as a single droplet to a 10 cm cell culture dish.
5. Add 6 μL of human thrombin (25 U/mL) to the cell and fibrinogen mixture. Immediately compress the scaffold from the sides with forceps, lower it onto the top of the droplet, and release it to allow uptake of the cell and fibrin mixture.
6. Allow the fibrin to solidify at room temperature for 5 min and then transfer the scaffold to a 24-well plate containing 1 mL of EGM2-MV media supplemented with 50 ng/mL of VEGF-165 and 10 μg/mL of aprotinin (*see* **Note 6**).

3.3 Subcutaneous Implantation of Endothelial Cell-Seeded Scaffolds

1. Obtain ethical approval for all experimental work involving animals.
2. Prepare the surgical sites with chlorhexidine solution.
3. Anesthetize rats with 4% isoflurane inhalation and keep the rats adequately anesthetized throughout the procedure with 2% isoflurane.
4. Make a 0.5 cm mid-line incision on the dorsal skin with a sterile blade to open subcutaneous pockets to either side of the incision using sterile forceps with blunt ends.
5. Insert a scaffold into each subcutaneous pocket (*see* **Note 7**).
6. Close the incision with sterile stainless steel wound clips.
7. Administer a single dose of Carprofen (5 mg/kg) subcutaneously as analgesic.
8. Euthanize rats at the desired time point (1 or 2 weeks) of the experiment using a lethal dose of intraperitoneal pentobarbital (300 mg/kg) injection.
9. Harvest the implanted scaffolds for histological analysis.

3.4 Histological Analysis of Vascularization

3.4.1 Tissue Processing

1. Fix tissue in 4% paraformaldehyde overnight at 4 °C.
2. Rinse the tissue with PBS for 5 min and dehydrate the tissue through serial changes of ethanol (70%, 80%, 95%, and 100%), clear in histolene and infiltrate with paraffin wax in a tissue processor.
3. Section the paraffin-embedded tissue blocks at 5 μm thickness and transfer the sections onto Polysine glass slides for immunohistochemistry (*see* **Note 8**).

3.4.2 CD31 and Lectin Immunohistochemical Staining of Blood Vessels

1. Deparaffinize tissue sections in histolene for two 5-min periods and rehydrate tissue sections through graded ethanol (70%, 100%, and 100%) to tap water for 5 min each.
2. Retrieve antigen by incubating tissue sections with Proteinase K for 6 min at room temperature.
3. Incubate tissue sections in 3% hydrogen peroxide solution for 10 min to quench endogenous peroxidase activity.

4. After washing twice with PBS for 5 min each time, add protein block solution onto the tissue sections for 30 min to mask non-specific antigens. Remove blocking solution by tapping slides on absorbent paper towel.
5. To detect human endothelial cells, incubate tissue sections with mouse anti-human CD31 (0.2 mg/mL in antibody diluent, for human endothelial cells) at 4 °C overnight.
6. After washing three times with PBS for 5 min each time, incubate tissue with 100 μL of biotinylated rabbit anti-mouse IgG (0.01 mg/mL in antibody diluent, for detection of human endothelial cells) at room temperature for 30 min.
7. After washing three times with PBS for 5 min each time, incubate the sections with avidin-biotinylated-peroxidase complex (Vectastain Elite ABC kit; for detection of human endothelial cells) at room temperature for 30 min.
8. After washing three times with PBS for 5 min each time, add diaminobenzidine chromogen solution to the tissue sections on the slides to visualize peroxidase activity (*see* **Note 8**).
9. To detect rat blood vessels, incubate tissue sections with biotinylated *Griffonia Simplicifolia* lectin (6.67 μg/mL in antibody diluent, for mouse blood vessels) for 30 min (*see* **Note 9**).
10. After washing three times with PBS for 5 min each time, incubate tissue sections with horseradish peroxidase-streptavidin conjugate (1:400) for 30 min at room temperature.
11. After washing three times with PBS for 5 min each time, incubate tissue sections with VIP substrate solution for 5 min at room temperature.
12. Counterstain the tissue sections with hematoxylin and dehydrate the tissue sections through graded ethanol (70%, 100% and 100%) to histolene for 5 min each.
13. Coverslip the tissue section in Entellan mounting medium and dry overnight before visualization of stained blood vessels (Fig. 2).

4 Notes

1. Alternative scaffold types include collagen porous scaffolds, which can be fabricated as described previously [8, 9], or purchased commercially, such as the SpongeCol® collagen porous scaffold and Cellusponge-Collagen porous scaffold.
2. Alternative media for maintaining pluripotent stem cells include mTeSR™ medium and StemFlex medium.

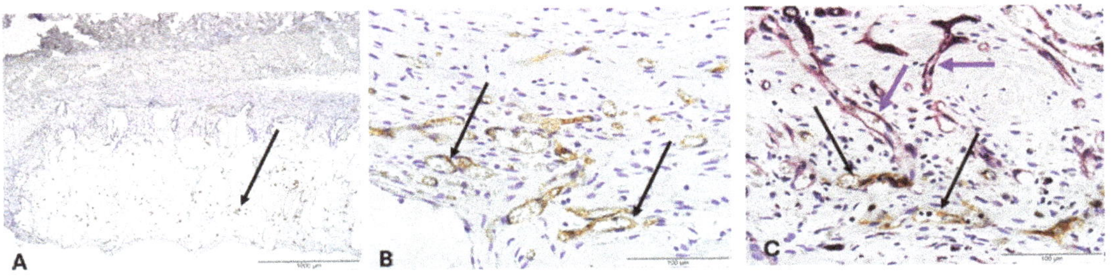

Fig. 2 The vascularized scaffolds harvested from rat post-implantation. **a** A representative low-magnification image displaying the scaffold (lower half of image), where the scaffold pores are filled with fibrin (light grey) and human CD31-positive developing blood vessels (arrow). The top half of the image is sub-dermal rat skin. Scale bar: 1000 μm. **b** A representative high-magnification image showing human CD31-positive blood vessels (brown, arrows) within the sub-dermal scaffold pores. Scale bar: 100 μm. **c** Host rat blood vessels stained with *Griffonia Simplicifolia* (purple arrows indicate purple-stained blood vessels) within the scaffold pores. Human blood vessels, located near the rat blood vessels, are stained with human CD31 antibody and appear brown (black arrows). Scale bar: 100 μm

3. The method can be adapted for scaffold implantations in SCID mice.
4. Optimal cell concentration depends on the size of the scaffold.
5. Aprotinin is a serine protease inhibitor used to delay fibrinolysis.
6. Scaffolds can be cultured in vitro for 2–3 days before implantation.
7. Scaffolds can be securely sutured to the muscle layer of the deep surface of the skin to prevent displacement after implantation.
8. Allow the slides to dry overnight and store slides at room temperature until ready for use. Diaminobenzidine is carcinogenic. Care should be taken when handling and disposing of diaminobenzidine.
9. *Griffonia Simplicifolia* lectin specifically binds to rodent vessels and does not interact with human endothelial cells.

Acknowledgments

The authors declare no conflict of interest. This work was supported by grants from the St Vincent's Hospital (Melbourne) Research Endowment Fund and Stafford Fox Medical Research Foundation, Diabetes Australia Research Program, and the National Stem Cell Foundation Australia. The O'Brien Institute Department of St Vincent's Institute of Medical Research and the Centre for Eye Research Australia receives Operational Infrastructure Support from the Victorian State Government's Department of Innovation, Industry and Regional Development.

References

1. Lopes SV, Collins MN, Reis RL et al (2021) Vascularization approaches in tissue engineering: recent developments on evaluation tests and modulation. ACS Appl Bio Mater 4(4): 2941–2956
2. Yang G, Mahadik B, Choi JY et al (2020) Vascularization in tissue engineering: fundamentals and state-of-art. Prog Biomed Eng (Bristol) 2(1):012002
3. Eldeeb AE, Salah S, Elkasabgy NA (2022) Biomaterials for tissue engineering applications and current updates in the field: a comprehensive review. AAPS PharmSciTech 23(7):267
4. Laschke MW, Menger MD (2016) Prevascularization in tissue engineering: current concepts and future directions. Biotechnol Adv 34(2): 112–121
5. Lim SY, Hsiao ST, Lokmic Z et al (2012) Ischemic preconditioning promotes intrinsic vascularization and enhances survival of implanted cells in an in vivo tissue engineering model. Tissue Eng Part A 18(21–22):2210–2219
6. Kong AM, Yap KK, Lim SY et al (2019) Bio-engineering a tissue flap utilizing a porous scaffold incorporating a human induced pluripotent stem cell-derived endothelial cell capillary network connected to a vascular pedicle. Acta Biomater 94:281–294
7. Kong AM, Lim SY, Palmer JA et al (2022) Engineering transplantable human lymphatic and blood capillary networks in a porous scaffold. J Tissue Eng 13:20417314221140979
8. Wang JH, Chen J, Kuo SM et al (2019) Methods for assessing scaffold vascularization in vivo. Methods Mol Biol 1993:217–226
9. Chan EC, Kuo SM, Kong AM et al (2016) Three dimensional collagen scaffold promotes intrinsic vascularisation for tissue engineering applications. PLoS One 11(2):e0149799

Chapter 19

Histological and Histochemical Methods for the Evaluation of Tissue Engineered Skin Substitutes

Fernando Campos, Víctor Carriel, Fabiola Bermejo-Casares, Jesús Chato-Astrain, and Miguel Alaminos

Abstract

A thorough evaluation and characterization of advanced therapy medicinal products (ATMPs), including tissue-engineered skin substitutes (TESS), is an essential requirement of the European and National Medicines Agencies. Although clear regulatory criteria have been established for different parameters such as stability, sterility or viability, the histological characterization of bioengineered tissues remains inadequately defined. Histological analyses are crucial for assessing both the in vitro features and in vivo behavior of TESS. This chapter describes the essential histological methods used for assessing TESS, including hematoxylin and eosin staining, and several methods focused on the analysis of specific components of the skin cells and extracellular matrix, such as Picrosirius Red for collagen fibers, Fontana Masson Picrosirius for melanin and collagen, Verhoeff for elastic fibers, Alcian Blue for proteoglycans, and Periodic acid–Schiff for glycoproteins. Relevant immunohistochemical methods for the identification of some specific markers are also included in this chapter. These methods provide a comprehensive practical approach to assessing both in vitro and in vivo histological features and evolution of TESS.

Key words Skin histological analysis, Skin tissue engineering, Histochemistry, Immunohistochemistry, Advanced therapy medicinal products

1 Introduction

Evaluation of tissues and organs generated as advanced therapies medicinal products is a crucial requirement of the European Medicines Agency and all National medicines agencies [1]. Although regulation has established clear criteria regarding product stability, sterility, cell viability, immunophenotype and functionality [2], criteria related to the histological characterization of bioengineered tissues, including human skin substitutes, are still poorly defined.

Fernando Campos and Víctor Carriel contributed equally

Thomas Biedermann and Sophie Böttcher-Haberzeth (eds.), *Skin Tissue Engineering: Methods and Protocols*, Methods in Molecular Biology, vol. 2922, https://doi.org/10.1007/978-1-0716-4510-9_19,

There is a close relationship between histological features and functional properties of native skin and tissue-engineered skin substitutes (TESS). As in diagnostic pathology, the histological analysis of these substitutes is essential, as it allows us to confirm their molecular, cellular, and histoarchitectural features in vitro, as well as their evolution, behavior and therapeutic efficacy once implanted in vivo [3–5].

In general, in vitro histological analyses of TESS should determine the correct development of the three essential components of the human skin: the epithelial layer or epidermis, the stromal layer or dermis, and the basement membrane which keep both structures connected [6]. Furthermore, in vivo histological analyses should elucidate TESS integration, epithelial and stromal regeneration, and host response in order to confirm their therapeutic potential [3] and clinical efficacy [4].

For this purpose, a wide range of histological, histochemical, and immunohistochemical methods are available [6]. These techniques can be applied on either cryosections or paraffin-embedded samples. Cryosections are valuable option for rapid histological diagnosis, immunofluorescence techniques, and enzyme histochemistry. However, the morphology in this technique is compromised due to the thickness of the sections and the formation of ice crystals [7]. Nevertheless, cryosection morphology can be significantly improved if the samples are first fixed with formalin, cryoprotected, and then rapidly frozen [8]. Consequently, formalin fixed and paraffin-embedded samples are the most commonly used [6, 9].

General histological assessment normally begins with hematoxylin and eosin (HE) staining [10]. This method allows for the evaluation of the main histological features of native tissues or tissue-engineered tissues (TESS). Moreover, most histological findings associated to the implantation of engineered substitutes, host immunological response, or histopathological findings are normally assessed by HE. However, for the identification of specific cellular, tissue, or molecular components, histochemical and immunohistochemical techniques must be used [11]. Numerous reports highlight the usefulness of immunohistochemistry for conducting a molecular characterization of TESS [12]. Nevertheless, the potential of histochemical techniques has not been fully exploited, despite their significant value.

The aim of this chapter is to describe seven relatively simple and useful histological methods that could be used for the in vitro and in vivo evaluation of TESS generated as advanced therapies medicinal products (ATMP) for clinical use.

The first procedure included here is conventional HE staining, which is the starting point for the histological analysis of TESS. Given the structural and biomechanical importance of the extracellular matrix (ECM) in native skin and TESS, we here describe five histochemical methods to evaluate three main components: collagen fibers, proteoglycans (PGs), and basement membrane glycoproteins (GPs), as previously described [3, 4, 6]. Fibrillar collagen fibers are essential dermal ECM proteins that can be studied using Picrosirius Red staining method (PS). This method uses two anionic dyes, Sirius Red F3B and picric acid, which compete for molecular binding sites, resulting in an intense red staining of these fibers. Notably, Sirius Red also enhances the natural birefringence of collagen fibers, allowing their specific evaluation under polarized light microscopy [13]. Although conventional PS staining is frequently employed as a standalone method, it can be effectively combined with the Fontana Masson (FM) metal reduction technique for melanin. This combined approach, known as Fontana Masson Picrosirius (FMPS) technique, enables the simultaneous assessment of histological features, melanin content (black silver precipitate), and collagens, providing a comprehensive evaluation of tissue characteristics [14]. To analyze the elastic fibers, responsible of the dermal elasticity, there are some histochemical alternatives, such as orcein, aldehyde fuchsine or Verhoeff staining methods [15]. Here, we describe a modified Verhoeff technique combined with PS that allows the specific and simultaneous study of elastic and collagen dermal fibers.

Non-fibrillar ECM molecules, such as PGs and GPs, are key for the structural, biomechanical and biological stability of tissues and organs, including the native skin and bioengineered tissues [16]. PGs constitute a complex and extensive family of molecules, distinguished by the presence of negatively charged groups that provides hydrophilic properties. Specific molecules must be studied by immunohistochemistry, but a general evaluation of the PGs content can be conducted by histochemical methods, such as Alcian blue (AB), toluidine blue or Safranine O [13, 16]. Here, we describe the AB histochemical method, which is accurate enough and recommended for the evaluation of PGs in non-cartilaginous tissues [16]. In relation to the GPs, these molecules play key biological and structural roles, providing multiple motifs to interact with cells and other ECM molecules. GPs are well-known for their role in stimulating cell adhesion, proliferation or migration, and for being essential components of most basement membranes [17, 18]. As with PGs, identification of specific GPs can be performed by immunohistochemistry. In this regard, the periodic acid Schiff (PAS) histochemical method, which is described here, is particularly useful for the study of basement membrane GPs.

To evaluate specific epithelial and stromal components, immunohistochemistry must be conducted using specific antibodies. Here, we describe a general immunohistochemical procedure, and some antibodies that are relevant to assess the epithelial and stromal phenotypes are described. Finally, this methodological chapter offer seven easy, fast, and economic histological approaches suitable for in vitro and in vivo complete evaluation of TESS.

2 Materials

This chapter details various methods that require specific reagents and equipment. It is crucial to carefully read the specifications and safety information for each reagent used. Always wear gloves and a lab coat, work under a chemical fume hood, and use a mask and/or safety goggles when necessary. Ensure the use of analytical-grade reagents and distilled water for all aqueous solutions.

2.1 Histochemical Methods

1. Xylene.
2. 99%, 96%, 70% ethanol.
3. Distilled water.
4. Hydrophobic mounting medium.

2.1.1 Hematoxylin-Eosin

1. Ready-to-use Harris' hematoxylin solution: This solution can be bought or prepared by using hematoxylin (C.I. 75,290) powder and following well-described procedures [13].
2. Ready-to-use Eosin Y alcoholic solution: This solution can be bought or prepared by using hematoxylin (C.I. 45,380) powder and following well-described procedures.

2.1.2 Picrosirius Staining

1. Aqueous saturated picric acid solution.
2. 0.2% Picrosirius staining solution: Dilute 2 g of Sirius Red F3B (C.I. 35,780) powder in aqueous saturated picric acid solution and stirring this solution gently. This working solution remains stable and can be stored for 2–3 years.
3. Ready-to-use Harris' hematoxylin solution: This solution can be bought or prepared by using hematoxylin (C.I. 75,290) powder and following well-described procedures [13].

2.1.3 Fontana Masson-Picrosirius Technique

1. Fontana-Masson (FM) stock solution: Mix 95 mL of 10% silver nitrate with ammonium hydroxide, adding the ammonium hydroxide drop by drop until the solution initially precipitates and then clears again. Add an additional 1 mL of 10% silver nitrate to induce slight cloudiness. Allow the mixture to stand undisturbed for 4–24 h.
2. Fontana-Masson (FM) staining solution: Dilute 2.5 mL of FM stock solution in 47.5 mL of distilled water.
3. 2% (w/v) aqueous gold chloride solution.
4. 2% (w/v) aqueous sodium thiosulfate solution.
5. Use reagents described in Subheading 2.1.2.

2.1.4 Verhoeff-Picrosirius Staining

1. 5% Alcoholic Hematoxylin: Dissolve 5 g of hematoxylin (C.I. 75,290) powder in 100 mL of 100% ethanol. Heat gently and filter the solution.
2. 10% Ferric Chloride: Dissolve 10 g of Ferric Chloride in 100 mL of distilled water.
3. Weigert's Iodine Solution: Dissolve 2 g of potassium iodide and 1 g of iodine in 100 mL of distilled water (*see* **Note 1**).
4. Verhoeff Staining Solution: Combine 100 mL of 5% Alcoholic Hematoxylin with 40 mL of ferric chloride and 40 mL of Weigert's Iodine Solution (*see* **Note 2**).
5. 1% Acid Alcohol Solution: Mix 1 mL of hydrochloric acid (HCl) with 99 mL of 70% ethanol. Prepare fresh as needed.
6. 2% (w/v) aqueous sodium thiosulphate.
7. Use reagents described in Subheading 2.1.2.

2.1.5 Alcian Blue Staining

1. 3% acetic acid aqueous solutions: Prepare using 99.5% analytic-degree glacial acetic acid to achieve the desired concentrations.
2. 1% Alcian Blue 8GX (C.I. 74,240) staining solution: Dissolve 1 g of AB powder in 100 mL of 3% acetic acid aqueous solution.
3. 5% aluminum sulfate aqueous solution: Dissolve 5 g of aluminum sulfate in 100 mL of distilled water.
4. 0.1% Nuclear Fast Red (NFR) staining solution (C.I. 60,760): Dissolve 0.1 g of NFR powder in 5% aluminum sulfate aqueous solution. Heat with stirring until nearly boiling, then cool overnight at room temperature (RT).

2.1.6 Periodic Acid–Schiff Staining

1. 1% (w/v) aqueous periodic acid solution.
2. Commercially available ready-to-use Schiff's reagent solution. This solution can be prepared following well-described procedures [13].
3. Ready-to-use Harris' hematoxylin solution: This solution can be bought or prepared by using hematoxylin (C.I. 75,290) powder and following well-described procedures.

2.2 Immunohistochemistry

1. 0.1 M Phosphate-Buffered Saline (PBS) Solution, pH 7.4: Dissolve 0.245 g of KH_2PO_4, 1.463 g of $Na_2HPO_4{\cdot}2H_2O$, and 8 g of NaCl in 90 mL of distilled water. Once the solutes are fully dissolved, adjust the volume to 100 mL with distilled water and store at 4 °C.
2. PBS-Tween 20 Solution: Dilute 0.3 mL of Tween 20 in 100 mL of PBS.
3. Citrate Buffer Solution, pH 6.0: Mix 9 mL of 7.35% (w/v) sodium citrate aqueous solution with 41 mL of 5.25% (w/v)

citric acid aqueous solution. Add distilled water to reach a volume of 400 mL. Adjust the pH to 6.0, then add water to reach a final volume of 500 mL.

4. EDTA Buffer Solution, pH 8.0: Dissolve 185 mg of ethylenediaminetetraacetic acid in 400 mL of distilled water (1 mM EDTA). Adjust the pH to 8.0, then add water to reach a final volume of 500 mL.
5. 3% Hydrogen Peroxide (H_2O_2) Solution in PBS: Prepare a 3% H_2O_2 solution in PBS for blocking endogenous peroxidase activity.
6. Casein Blocking Solution: Use a commercially available 1% casein solution to block non-specific antigen-binding sites (*see* **Note 3**).
7. Primary and Secondary Antibodies: Prepare as per the protocols provided in Table 1 and *see* **Note 4**.
8. Enzymatic-Based Detection System: Use a ready-to-use secondary antibody and 3,3′-diaminobenzidine tetrahydrochloride (DAB) Substrate Peroxidase Kit for detecting antigen-antibody complexes (*see* **Note 5**).
9. Harris (or Mayer) Hematoxylin: Use a ready-to-use hematoxylin solution for counterstaining.

2.3 Equipment

1. Poly-L-lysine-treated slides (*see* **Note 6**).
2. Coverslips.
3. Hydrophobic mounting medium.
4. Forceps or tweezers and needles.
5. Glass Coplin jars with cap.
6. Plastic or glass staining jars set (for dewax, hydration and dehydration procedures).
7. Micropipettes (1–10; 50–200; 200–1000 μL) and tips.
8. Electrothermal drying oven or hybridization oven at 37 °C.
9. Hydrophobic pen for immunohistochemistry.
10. Humidified chamber for histological use.
11. Steamer and plastic Coplin jars with cap (for antigen retrieval).

3 Methods

For the quality control of the TESS generated as advanced therapy medicinal products (ATMPs), it is essential to characterize the histological patterns and both the fibrillar and non-fibrillar components of the extracellular matrix (ECM) of the resulting products. This methods section will describe different histochemical

Table 1
Antibodies and their specifications for pretreatment and dilution in the identification of various markers in grafted tissue-engineered skin substitutes (TESS)

Antigen	Species	Cat N°	Company	Pretreatment	Dilution	Incubation
Primary antibodies						
Anti-Cytokeratin10	Mouse	MAD-000150QD	Master Diagnóstica	Citrate buffer, pH 6, 95 °C for 25 min	Ready-to-use	Overnight 4 °C
Anti-Filaggrin	Mouse	ab17808	Abcam	Citrate buffer, pH 6, 95 °C for 20 min	1:50	1 h, RT
Anti-Human Ki-67	Rabbit	MAD-000310QD	Master Diagnostica	EDTA, pH 8, 95 °C for 25 min	Ready-to-use	1 h, RT
Anti-SMA-ACT	Mouse	MAD-001195QD	Master Diagnostica	Citrate buffer, pH 6, 95 °C for 25 min	Ready-to-use	1 h, RT
Anti-decorin	Goat	AF143SP	R&D System	Citrate buffer, pH 6, 95 °C for 25 min	1:500	Overnight 4 °C
Anti-Melan-A	Rabbit	MAD-000695QD	Master Diagnostica	EDTA, pH 8, 95 °C for 25 min	Ready-to-use	1 h, RT
Anti-CD31	Mouse	MAD-002048QD	Master Diagnostica	EDTA, pH 8, 95 °C for 25 min	Ready-to-use	1 h, RT
Secondary antibodies						
Anti-mouse IgG (peroxidase)	Horse	MP-7402	Vector laboratories		Ready-to-use	30 min, RT
Anti-rabbit IgG (peroxidase)	Horse	MP-7401	Vector laboratories		Ready-to-use	30 min, RT
Anti-goat IgG (peroxidase)	Horse	MP-7405	Vector laboratories		Ready-to-use	30 min, RT

techniques used for this purpose using the most used techniques starting with routine hematoxylin and eosin staining and including techniques for analyzing extracellular matrix components like collagen, elastic fibers, proteoglycans, and glycoproteins alongside immunohistochemical methods for identifying specific markers.

3.1 Hematoxylin-Eosin

1. Dewax tissue sections in xylene or an equivalent solution with two changes of 10 min each.
2. Hydrate sections in decreasing ethanol concentrations (99%, 96%, and 70%) and then place them in distilled water for a few minutes.
3. Immerse sections in filtered Harris hematoxylin solution for 5 min at room temperature.
4. Rinse sections in tap water for 10 min.
5. Immerse sections in eosin solution for 30 s at room temperature.
6. Quick wash in tap water to remove excess eosin.
7. Rapidly dehydrate the sections by immersing them in increasing concentrations of ethanol (70%, 96%, and 99%), then rinse in xylene with two changes of 2 min each, and mount with hydrophobic mounting medium. Representative images are shown in Fig. 1.

3.2 Picrosirius Staining

1. Follows **steps 1–2** in Subheading 3.1.
2. Immerse tissue sections in Picrosirius Red staining solution for 45 min at room temperature.
3. Wash sections several times in distilled water to remove rests of PS solution and unspecific staining.
4. Counterstain sections using Harris hematoxylin for 1 min at room temperature.
5. Rinse sections in tap water for at least 5 min.
6. Dehydrate sections by immersing them in increasing concentrations of ethanol (96% and 99%) for 3–5 min each, then clear in xylene with two changes of 2 min each, and mount with a hydrophobic mounting medium. Representative images are shown in Figs. 2 and 3.

3.3 Fontana Masson-Picrosirius Technique

1. Follows **steps 1–2** in Subheading 3.1.
2. Incubate the sections in the FM working solution in a microwave for 3 cycles of 50 s at maximum power (*see* **Note 7**).
3. Rinse the sections at least three times in distilled water.
4. Differentiate the sections in 2% gold chloride solution for 5 min.

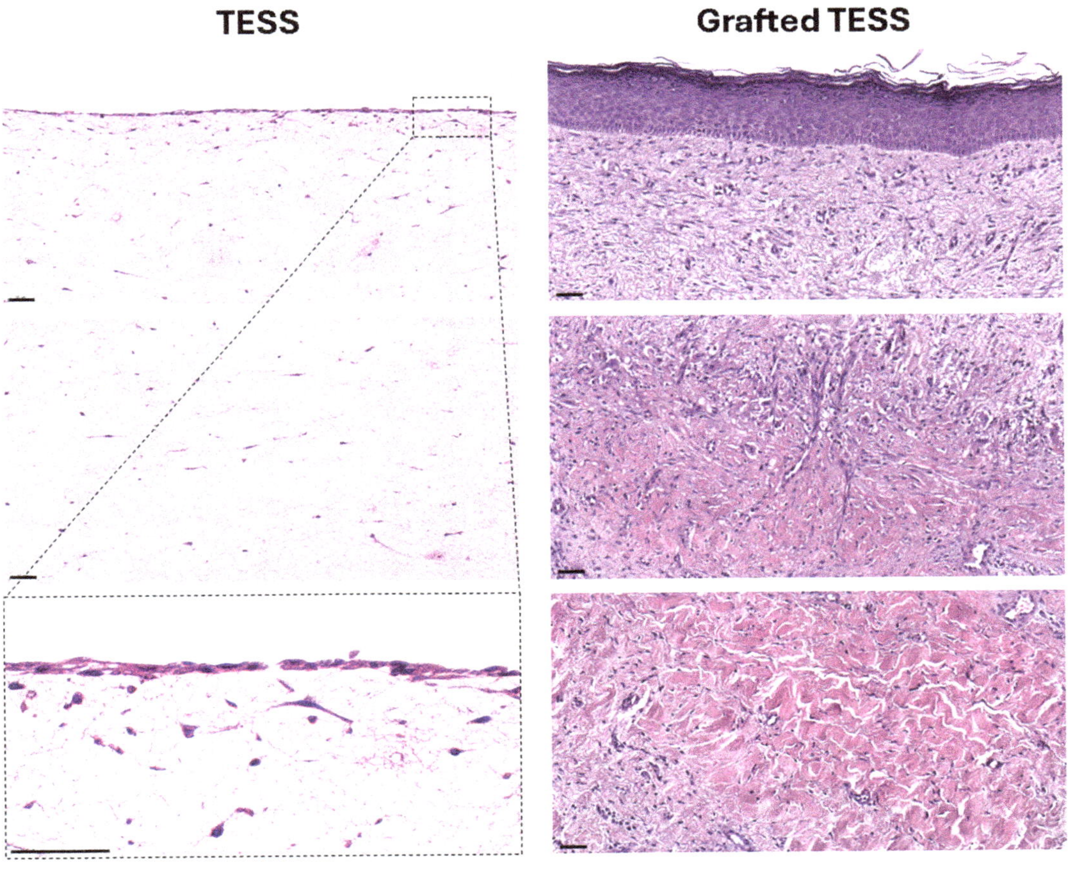

Fig. 1 Hematoxylin-eosin staining of tissue engineered skin substitute (TESS) and grafted TESS in a burn patient. Representative images showing hematoxylin-eosin staining of a Tissue Engineered Skin Substitute (TESS) prior to grafting and after 30 days of TESS treatment of a burn patient. The images illustrate the structural integrity and cellular distribution within the engineered tissue and its integration into the host tissue. Scale bar: 50 μm

5. Rinse the sections in distilled water.
6. Fix the silver in the sections with 2% sodium thiosulfate solution for 1 min.
7. Rinse the sections three times in distilled water.
8. Follows **steps 2–6** in Subheading 3.2. Representative images are shown in Fig. 2.

3.4 Modified Verhoeff-Picrosirius Staining

1. Follows **steps 1–2** in Subheading 3.1.
2. Immerse the sections in Verhoeff solution in darkness for 30 min to allow comprehensive staining.
3. Rinse the sections thoroughly three times with tap water for 1 min each rinse to remove excess stain.

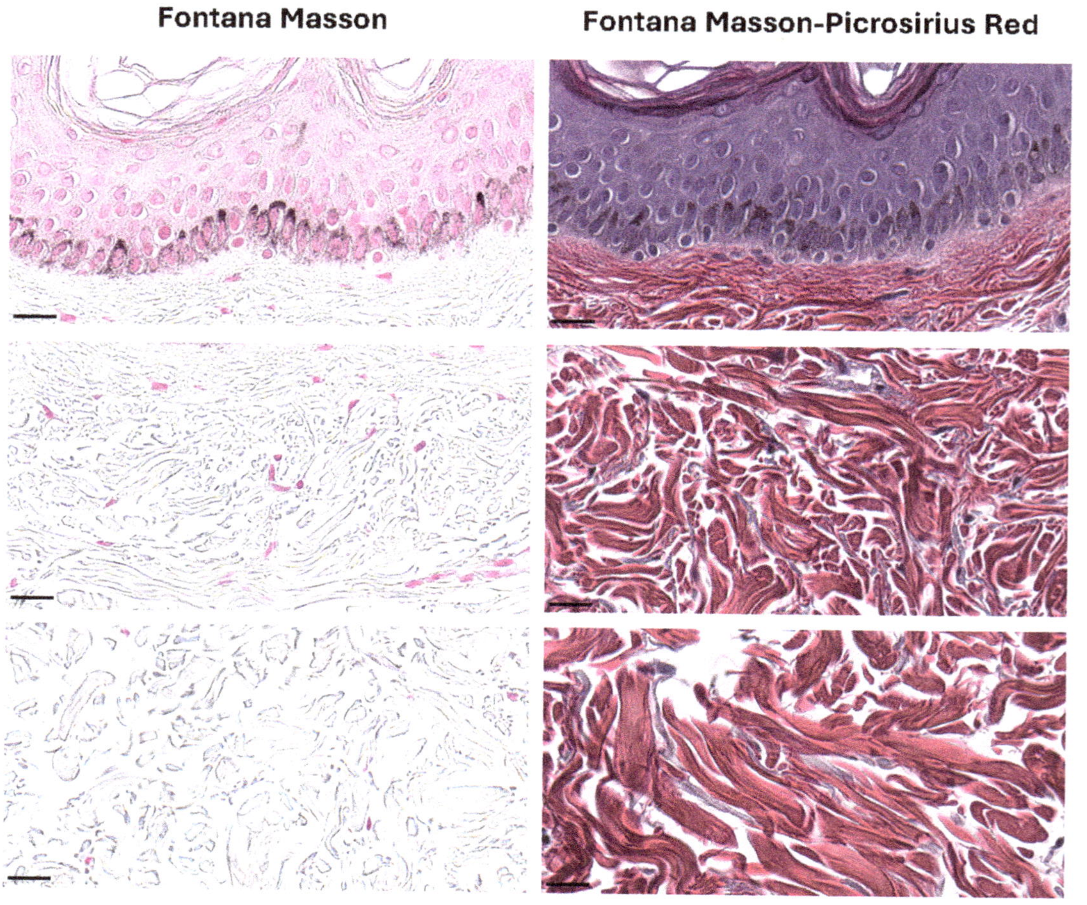

Fig. 2 Fontana-Masson staining (FMS) and combined with Picrosirius Red (FMS-PS) of skin sections. Images showing skin sections stained with Fontana-Masson staining (FMS) to detect melanin and combined with Picrosirius Red (FMS-PS) to visualize collagen fibers. Scale bar: 20 μm

4. Differentiate the sections in alcohol acid for 20 s to enhance contrast.
5. Immediately rinse thoroughly with tap water and then with distilled water to halt the staining process effectively.
6. Immerse the sections in 2% sodium thiosulfate for 1 min to fix the stained sections.
7. Wash the sections in running water and then twice in distilled water for 1 min each to remove residual chemicals.
8. Follows **steps 2–6** in Subheading 3.2. Representative images are shown in Fig. 3.

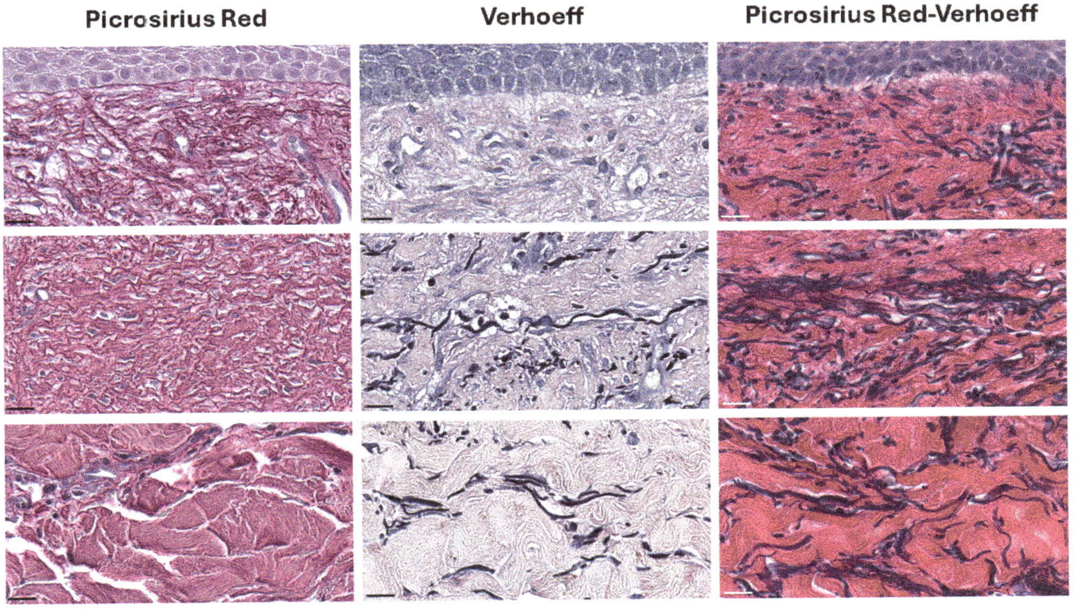

Fig. 3 Picrosirius and Verhoeff staining of grafted tissue engineered skin substitute (TESS). Representative images demonstrating staining of grafted Tissue Engineered Skin Substitute (TESS) with Picrosirius staining showing collagen fibers in red, Verhoeff staining highlighting elastic fibers in black (without contrast), and the combined Picrosirius red-Verhoeff staining displaying collagen fibers in red and elastic fibers in black. Scale bar: 20 μm

3.5 Alcian Blue pH 2.5

1. Follows **steps 1–2** in Subheading 3.1.
2. Rinse tissue sections in a 3% aqueous acetic acid solution for 3 min.
3. Stain sections with Alcian Blue staining solution for at least 30 min at room temperature.
4. Wash sections in tap water for 10 min and then immerse them in distilled water.
5. Contrast sections using a filtered Nuclear Fast Red staining solution for 5 min at room temperature.
6. Wash sections in tap water for 1 min.
7. Dehydrate sections by immersing them in increasing concentrations of ethanol (96% and 99%) for 3–5 min each, then clear in xylene with two changes of 2 min each, and mount with a hydrophobic mounting medium. Representative images are shown in Fig. 4.

3.6 Periodic Acid–Schiff

1. Follows **steps 1–2** in Subheading 3.1.
2. Immerse the sections in 1% periodic acid solution for 10 min.
3. Rinse the sections thoroughly in tap water for 10 min, followed by a rinse in distilled water.

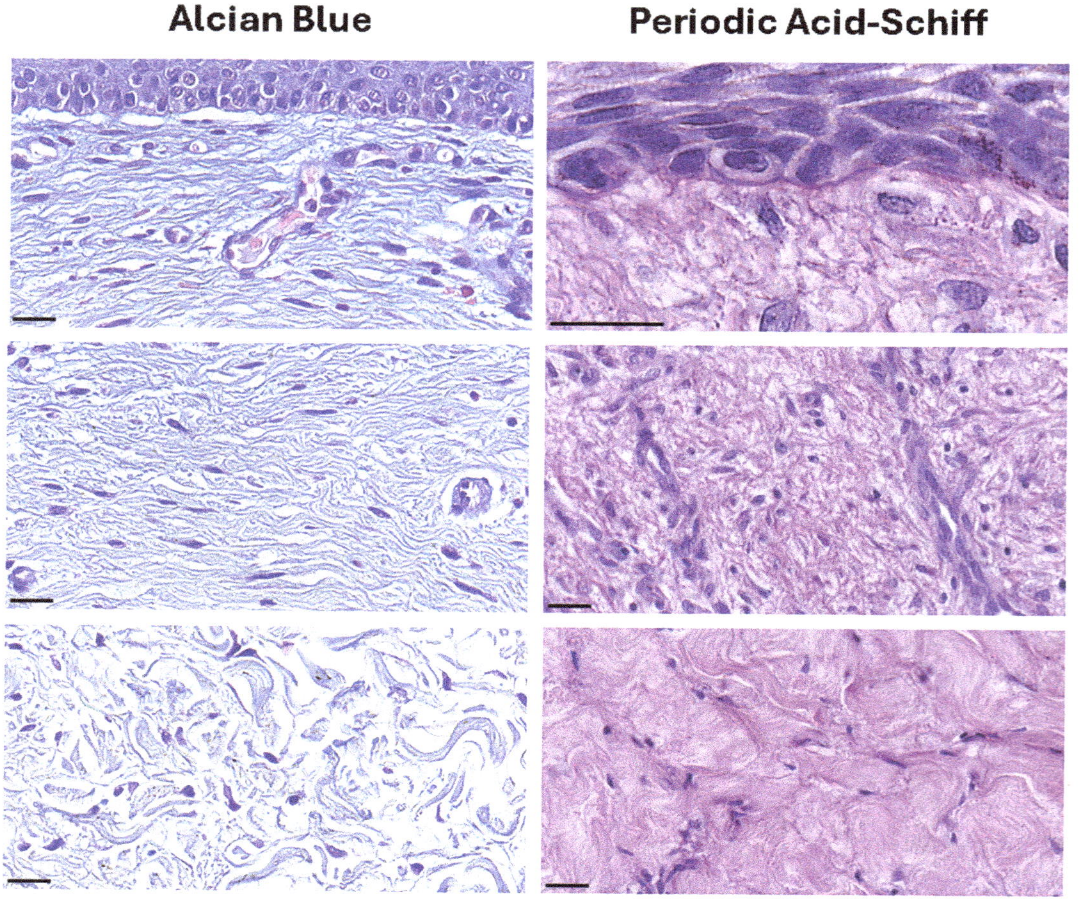

Fig. 4 Alcian Blue and periodic acid-Schiff Staining of grafted tissue engineered skin substitute (TESS). Representative images showing Alcian Blue staining and Periodic Acid-Schiff staining of grafted Tissue Engineered Skin Substitute (TESS). Alcian Blue staining highlights proteoglycans in blue, while PAS staining indicates carbohydrates in purple. Scale bar: 20 μm

4. Stain the sections in Schiff reagent for 20–30 min at room temperature.
5. Rinse the sections several times in distilled water until the water runs clear.
6. Counterstain the sections with Harris hematoxylin for 1 min at room temperature.
7. Rinse the sections in tap water for at least 2 min.
8. Dehydrate the sections through a series of increasing ethanol concentrations (96% and 99%) for 3–5 min each, clear in xylene with two changes of 2 min each, and mount with a hydrophobic mounting medium. Representative images are shown in Fig. 4.

3.7 Immunohistochemistry

1. Follow **steps 1** and **2** as outlined in Subheading 3.1.
2. Gently dry the slide with absorbent paper, outline the section (s) with a hydrophobic pen, and wash the slide three times with PBS-Tween 20.
3. Antigen Retrieval: For antigen retrieval, perform heat-induced epitope retrieval by incubating the sections in citrate buffer (pH 6.0) or EDTA buffer (pH 9) at 90–95 °C for 25–30 min using a steamer. Allow the sections to cool to room temperature in the same buffer for 20 min (refer to Table 1 for detailed protocol).
4. Wash the sections three times in PBS-Tween 20, with each wash lasting 5 min.
5. Incubate the sections in a 3% H_2O_2 solution in PBS for 10 min with gentle agitation and protection from light to block endogenous peroxidase activity (refer to *see* **Note 8**).
6. Wash the sections three times in PBS-Tween 20, with each wash lasting 5 min.
7. Treat the sections with 1% casein blocking solution for at least 15 min at room temperature (*see* **Note 3**).
8. Remove the blocking solution and apply the primary antibody solution to the sections.
9. Incubate primary antibody in a humidified chamber.
10. Wash the sections three times in PBS-Tween 20, with each wash lasting at least 5 min at room temperature.
11. Apply the secondary antibody solution to the sections and incubate for 30 min at room temperature.
12. Wash the sections three times in PBS, with each wash lasting 5 min.
13. Incubate the sections with DAB working solution and control the enzymatic histochemical reaction under a light microscope (refer to *see* **Note 9**).
14. Rinse the sections in distilled water.
15. Perform a slight counterstaining with ready-to-use Harris hematoxylin working solution.
16. Dehydrate the sections through increasing ethanol concentrations (96% and 99%, for 3–5 min each), clear in xylene (two changes, 2 min each), and mount using a hydrophobic mounting medium. Representative images are shown in Fig. 5.

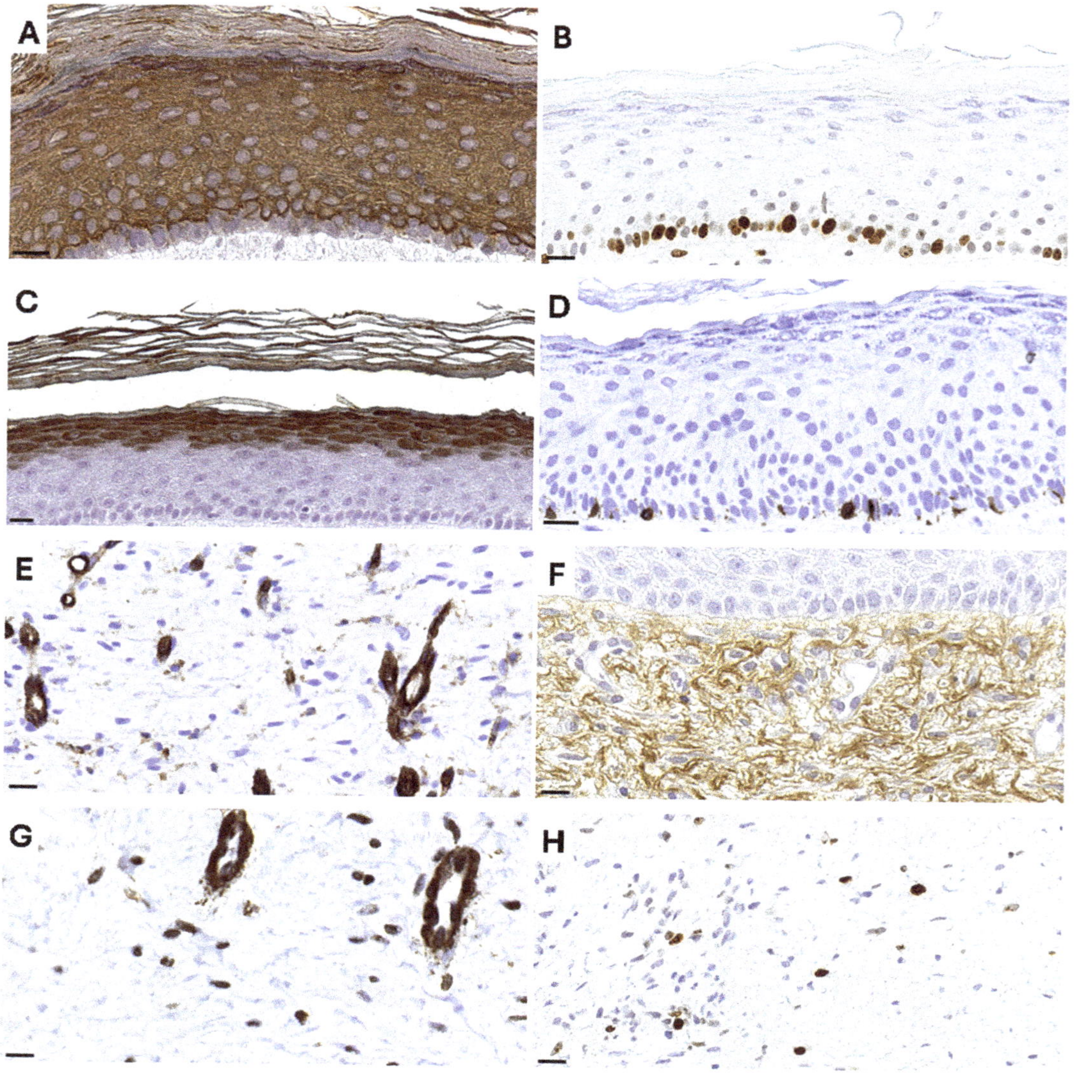

Fig. 5 Immunohistochemical detection of various markers in Grafted Tissue Engineered Skin Substitute (TESS). This figure provides examples of different immunohistochemical markers used to analyze TESS. **a** Shows cytokeratin 10 expression, while **b** and **h** display Ki67, a marker of proliferation in the epithelial and stromal layer respectively. Image E illustrates CD31, a marker for endothelial cells, and panel G shows smooth muscle actin. Finally, image F depicts the distribution of the proteoglycan decorin. Scale bar represents 20 μm

4 Notes

1. For optimal results, dissolve potassium iodide in 4 mL of distilled water first, followed by adding iodine. Once iodine is dissolved, add the remaining 96 mL of distilled water. This solution can be freshly prepared as needed or stored at a higher concentration in a light-protected bottle.

2. This solution should be freshly prepared for optimal staining results. It will not achieve satisfactory staining if stored for more than one working day. The solution should appear jet black. Use immediately and discard after use.
3. Various alternative solutions exist to block the nonspecific binding of primary antibodies. In this chapter, a commercially available casein solution was utilized.
4. When handling concentrated antibodies, it is essential to reconstitute and aliquot them according to the manufacturer's guidelines. Each laboratory must determine the working concentration of the antibodies. Antibodies can be diluted in blocking solutions, serum solutions (from commercially available secondary antibodies), or in PBS, as employed in this chapter.
5. Various detection systems are available for immunohistochemistry. In this chapter, we utilized a ready-to-use, one-step horse secondary antibody directly conjugated with peroxidase. Antigen-epitope interactions were visualized using a conventional enzymatic histochemical reaction with DAB as the chromogen.
6. Conventional glass slides are appropriate for routine use and some histochemical reactions. However, certain tissue samples tend to detach from the slides, especially when exposed to aggressive (acidic and alkaline) reagents or during antigen retrieval procedures. To address this issue, it is recommended to utilized poly-L-lysine-coated slides.
7. This step must be conducted as a water bath to prevent overheating of the solution. Place the slides in a glass coupling jar containing the Fontana working solution and then place the jar in a beaker filled with water, ensuring that the water level matches the level of the Fontana solution. For optimal results, monitor the reaction after each microwave cycle. If a positive brownish reaction is observed with the naked eye, additional cycles are unnecessary.
8. Endogenous peroxidase activity can alternatively be blocked using 5% H_2O_2 in methanol. However, methanol is not suitable for identifying cell-surface antigens.
9. The optimal staining reaction should be assessed by the user by examining the positive reaction under a microscope, using external or internal positive controls. It is crucial to use positive controls from tissues where the target molecules are present in low abundance but are highly specific to ensure accurate results.

Acknowledgements

This work was supported by grant C-CTS-032-UGR23 (Plan Propio de Investigación y Transferencia de la Universidad de Granada 2023, programa operativo FEDER Andalucía 2021–2027, University of Granada and Consejería de Universidad, Investigación e Innovación, Junta de Andalucía); grant PE-0395-2019 from Consejería de Salud y Familias, Junta de Andalucía, Spain; cofinanced by the European Regional Development Fund (ERDF) through the "Una manera de hacer Europa" program.

References

1. Rico-Sánchez L, Garzón I, González-Andrades M et al (2019) Successful development and clinical translation of a novel anterior lamellar artificial cornea. J Tissue Eng Regen Med 13: 2142–2154. https://doi.org/10.1002/term.2951
2. Capelli C, Frigerio S, Lisini D et al (2022) A comprehensive report of long-term stability data for a range ATMPs: a need to develop guidelines for safe and harmonized stability studies. Cytotherapy 24:544–556. https://doi.org/10.1016/j.jcyt.2021.12.004
3. Carriel V, Garzón I, Jiménez JM et al (2012) Epithelial and stromal developmental patterns in a novel substitute of the human skin generated with fibrin-agarose biomaterials. Cells Tissues Organs 196:1–12. https://doi.org/10.1159/000330682
4. Martin-Piedra MA, Carmona G, Campos F et al (2023) Histological assessment of nanostructured fibrin-agarose skin substitutes grafted in burnt patients. A time-course study. Bioeng Transl Med 8:e10572. https://doi.org/10.1002/btm2.10572
5. Mazzarini M, Falchi M, Bani D et al (2021) Evolution and new frontiers of histology in bio-medical research. Microsc Res Tech 84: 217–237. https://doi.org/10.1002/jemt.23579
6. Linares-Gonzalez L, Rodenas-Herranz T, Campos F et al (2021) Basic quality controls used in skin tissue engineering. Life Basel 11: 1033. https://doi.org/10.3390/life11101033
7. Desciak EB, Maloney ME (2000) Artifacts in frozen section preparation. Dermatol Surg 26: 500–504. https://doi.org/10.1046/j.1524-4725.2000.99246.x
8. Sánchez-Porras D, Bermejo-Casares F, Carmona R et al (2023) Tissue fixation and processing for the histological identification of lipids. Methods Mol Biol 2566:175–186. https://doi.org/10.1007/978-1-0716-2675-7_14
9. Musiał A, Gryglewski RW, Kielczewski S et al (2016) Formalin use in anatomical and histological science in the 19th and 20th centuries. Folia Med Cracov 56:31–40
10. Hagenaars SC, Vangangelt KMH, Van Pelt GW et al (2022) Standardization of the tumor-stroma ratio scoring method for breast cancer research. Breast Cancer Res Treat 193:545–553. https://doi.org/10.1007/s10549-022-06587-3
11. Ramos-Vara JA (2017) Principles and methods of immunohistochemistry. Methods Mol Biol 1641:115–128. https://doi.org/10.1007/978-1-4939-7172-5_5
12. Dixon AR, Bathany C, Tsuei M et al (2015) Recent developments in multiplexing techniques for immunohistochemistry. Expert Rev Mol Diagn 15:1171–1186. https://doi.org/10.1586/14737159.2015.1069182
13. Kiernan J (2015) Histological and histochemical methods, 5th edn. Scion Publishing Ltd, Banbury
14. Carriel VS, Aneiros-Fernandez J, Arias-Santiago S et al (2011) A novel histochemical method for a simultaneous staining of melanin and collagen fibers. J Histochem Cytochem Off J Histochem Soc 59:270–277. https://doi.org/10.1369/0022155410398001
15. Chato-Astrain J, Garcia Garcia O, Campos F et al (2021) Basic nerve histology and histological analyses following peripheral nerve repair and regeneration. In: Phillips JB, Hercher D, Hausner T (eds) Peripheral nerve tissue engineering and regeneration. Reference series in biomedical engineering. Springer, Cham. https://doi.org/10.1007/978-3-030-21052-6_14
16. Sánchez-Porras D, Varas J, Godoy-Guzmán C et al (2023) Histochemical and

immunohistochemical methods for the identification of proteoglycans. Methods Mol Biol 2566:85–98. https://doi.org/10.1007/978-1-0716-2675-7_7

17. García-García ÓD, Chato-Astrain J, AHD H et al (2023) Decellularized tissue-derived materials for grafts development. In: Maia FRA, Oliveira JM, Reis RL (eds) Handbook of the extracellular matrix: biologically-derived materials. Springer, Cham, pp 1–35. https://doi.org/10.1007/978-3-030-92090-6_49-1

18. Sánchez-Porras D, Durand-Herrera D, Carmona R et al (2023) Expression of basement membrane molecules by Wharton Jelly Stem Cells (WJSC) in full-term human umbilical cords, cell cultures and microtissues. Cells 12:629. https://doi.org/10.3390/cells12040629

Chapter 20

Pig Model to Test Tissue Engineered Skin

Christian Tapking and Ludwik K. Branski

Abstract

Tissue engineering of skin is a field with high research activities and major importance for wound healing, especially following burn injuries. Animal models enable to test tissue engineered skin as well as different types of cells in a realistic setting. Although there are several challenges in working with pigs, it is a good model because of its similarity to the human skin and a comparable wound regeneration. Here, we explain our routinely used methods for using pig models to test tissue engineered skin in burn injuries.

Key words Tissue engineering, Skin graft, Pig model, Burn injury, Wound healing

1 Introduction

There is a high need for effective and safe wound coverage in different clinical settings such as genetic disorders (e.g., bullous conditions), acute trauma, and chronic wounds. One of the major reasons for the need of tissue repair is thermal trauma since the possibility of skin regeneration in this type of injury is often very limited. This may lead to deep wounds and massive (hypertrophic) scarring, which can affect the patients in terms of a decreased quality of life or delayed reintegration into society during their lifetime [1–3]. Wound healing is of major importance in the clinical outcome and survival of burn patients. Early excision and grafting reduces metabolic responses and increased survival after burn injury [4–7]. However, early excision leaves large uncovered wounds that need to be covered with either the patient's own non-burned skin, cadaver skin or other synthetic/biologic material. Due to the limited availability of autografts and allografts, bioengineered skin has become an active area of research and has shown great promise, especially in the treatment of burns.

Clearly, human studies are the most accurate way of investigating the effectiveness of wound-treatment, but are often impractical because of the limited histological assessment, which requires

Thomas Biedermann and Sophie Böttcher-Haberzeth (eds.), *Skin Tissue Engineering: Methods and Protocols*, Methods in Molecular Biology, vol. 2922, https://doi.org/10.1007/978-1-0716-4510-9_20,

frequent biopsies. Furthermore, ethical concerns prevent the use in a regular basis [8]. There are numerous in vitro and in vivo wound-healing models. Many models can only isolate specific mechanisms or events and not the impact on the overall human organism. Studies suggest that cultured skin cells returned to patients may keep the competence to self-renew for a lifetime [9]. Currently available bioengineered skin grafts still deal with problems such as patients safety, clinical efficacy and sheets with keratinocytes are often complicated to handle [2]. The selection of an animal model depends on factors such as availability, anatomical, and physiological similarity and costs. Small animals such as rat, rabbit, or mouse are frequently used, inexpensive and good to handle. However, the small animals differ significantly from humans regarding the anatomical and physiological structure of their skin [10, 11]. Using pigs as a model is associated with high costs and the need of a good infrastructure for the handling of large animals. Nevertheless, pig is a suitable model because of its anatomic and physiological similarity to human skin in many terms [10, 12]. Epidermis thickness in pigs range from 30 to 140 μm in pigs and from 50 to 120 μm in humans [13]. In addition, both show similar dermal collagen, papillary bodies, and abundant subdermal adipose tissue, allowing accurate evaluation of hypertrophic scars in pigs [14–16]. Furthermore, the skin of other animals such as rats primarily heals by contraction, whereas the skin of Yorkshire and Red Duroc pigs heals by re-epithelialization, just as seen in the human wound. Furthermore, pigs are better for studies on wound healing due to their size and the instruments, which are used in the daily clinical setting.

In this chapter, we describe our routinely used methodology for pig models to test tissue engineered skin using the example of burn injuries.

2 Materials

2.1 Animals

All preparations and procedures are done at room temperature if not indicated otherwise.

For the experiments, we usually use Yorkshire or Red Duroc pigs. Housing and animal care are supplied in line with the National Research Council guidelines [17].

1. The healthy 10–40 kg (1–4 months old) pigs are housed individually for at least 1 week for acclimatization at the large animal resource center before starting the surgical procedures.
2. For the duration of the experiment, the animals are held in individual cages with about 3 m^2 size. Cages are cleaned at least once daily.

3. A standard ration of porcine diet is given twice daily and water ad libitum.
4. Before undergoing any anesthesiological or surgical procedure (e.g., burning, dressing changes, biopsies), the animals are fasted overnight to achieve safe airway conditions during the anesthesia.

2.2 Surgical Materials and Veterinary Drugs

1. TKX Solution: 1 mL TKX contains 0.5 mg Ketamine, 0.5 mg Xylazine, and 1 mg Telamine/Zolazepam.
2. Isoflurane.
3. Oxygen-nitrous oxide mixture (2:1).
4. Laryngoscope blade.
5. Endotracheal tubes (size 7 or 8).
6. Ear oximeter.
7. Rectal thermometer.
8. Bear Hugger® blanket.
9. Cefazolin.
10. 500 mg Levofloxacin tablets.
11. Transdermal fentanyl patches.
12. Aluminum bar.
13. Meeker gas burner.
14. 50 mL syringe.
15. Ringer's Lactate solution.
16. Oil emulsion dressing.
17. 1% Nystin.
18. 2% Polymyxin B/Bacitracin.
19. Opsite®.
20. Nylon dressing.
21. Planimetrical software (Lucia G®, Version 4.8).

2.3 Materials and Solutions for Tissue Engineered Skin

1. 0.1% w/v trypsin.
2. 1% trypsin.
3. 5 mM EDTA.
4. 0.5% collagenase A.
5. Keratinocyte cultured medium.
6. Fibroblast culture medium again overnight.
7. Phosphate-buffered saline.

2.4 Histopatholgy and Hematology

1. Hematoxylin-eosin.
2. Hematology analyser (Drew-Scientific Hema Vet 950).
3. ELISA using porcine kits (P6000 Porcine IL-6 Quantikine Elisa Kit and PTA00 Porcine TNF- α Quantikine Elisa Kit, R&D Systems, Minneapolis, MN, USA).
4. Masson's trichrome.

3 Methods

In our institution, all animal experiments are conducted in a central animal research facility that provides all required and needed materials. All animals are monitored continuously by professional staff in order to correct postoperative problems such as dressing failures, pain, or bleedings early and effectively.

3.1 Anesthesia and Perioperative Care

Appropriate methods for sedation, inducing anesthesia and pain control are of major relevance throughout the whole experiments and require professional and experienced staff and guidelines. Animals were fasted overnight to avoid vomiting and aspiration.

1. Anesthesia is inducted by an intramuscular injection of 0.04 mL/kg TKX (*see* **Note 1**).
2. The animals are weighted and clipped before beginning any surgical procedure (*see* **Note 2**).
3. After transfer to the operating room, the pigs are placed on the operation table in prone position, pre-oxygenated with 100% oxygen via inhalation mask, and intubated with a straight laryngoscope blade using endotracheal tube of size 7 or 8 depending on the pigs' size.
4. Isoflurane is delivered after successful intubation as oxygen-nitrous oxide mixture (2:1) to maintain anesthesia (Aestive® 5, *see* **Note 3**).
5. Before beginning with any surgical procedure, animals are depilated using a disposable razor blade.
6. A venous catheter in the right jugular vein, tunnelled underneath the skin of the lateral neck and released through the dorsal neck region was placed before the initial burn (*see* **Note 4**).
7. During the surgical procedure, heart rate and oxygen saturation is monitored via ear oximeter whereas temperature is controlled via rectal thermometer. A disposable Bear Hugger® blanket is used to maintain a body core temperature of 40 °C.
8. Pigs are allowed to recover on the operating table after each procedure. Once spontaneous breathing is stabilized and gag

reflex is apparent, they need to be extubated and transferred to their individual cage. Temperature control is continued in the cage using a heat lamp or blankets. Continuous clinical controls are required by the staff.

9. Perioperative antimicrobial prophylaxis is provided with 1 g Cefazolin i.v. before each operation and dressing and 500 mg Levofloxacin tablets p.o. for 7 days after burn.
10. Postoperative analgesia is provided with transdermal fentanyl patches applied on the dorsal neck behind the ear and kept in place using surgical staples and an additional protective layer of adhesive transparent dressing (*see* **Note 5**).
11. After surgical procedures, the food intake remains stable without aspiring weight loss.

3.2 Burn Injury

1. Full-thickness contact burns are applied paravertebrally under aseptic conditions by using a heated aluminium bar on the dorsum of the animal. The pigs receive up to 16 bilateral contact burn on the dorsum of 3 × 3 cm^2. A temperature of 200 °C of the aluminium bar is achieved with a Meker gas burner and monitored with a digital thermometer (*see* **Note 6**).
2. The heated bar has contact to the animal for 30 s and application pressure is measured with a 50 mL syringe attached to the aluminium bar (*see* **Note 7**).
3. Depending on the study protocol, biopsies from the burned tissue are obtained several times after burn to assess the depth of the burn under experimental conditions (*see* **Note 8**).
4. Throughout and right after the burn procedure, the pigs are provided with Ringer's Lactate solution a rate of 10 mL/kg/h, usually at a total intravenous resuscitation volume of 20 mL/kg. After that, they have free access to water.
5. To mimic the clinical setting, the animals undergo full-thickness excision and grafting 24–48 h after burn. Excision is done in a size of 7 × 7 cm^2 with the burn area being in the center of the wound.
6. As a supportive procedure after the excision, the area surrounding the wounds can be marked by tattoos to track the rate of wound coverage and re-epithelialization. This can simplify the assessments of scar in the further course of the experiment.

3.3 Producing and Applying the Tissue Engineered Skin

Producing tissue engineered skin is currently under intense research and a high amount of different methods are described in the literature. Here, we describe a method that was used at our institution, but can differ from the usually used method at other institutions.

1. Small samples from porcine skin (2 × 2 cm) are taken, cut into pieces of <0.5 cm^2 and then incubated at 4 °C overnight in 0.1% w/v trypsin.
2. After incubating the epidermal and dermal layers are separated carefully using an anatomic forceps (*see* **Note 9**).
3. The epidermal cells are isolated by incubating at 37 °C in 1% trypsin and 5 mM EDTA for 3 min and centrifuged at 300 g for 5 min.
4. Then the cells are resuspended into a keratinocyte cultured medium and cultured at 37 °C (*see* **Note 10**).
5. The dermal cells are incubated at 37 °C in a 0.5% collagenase A solution overnight and centrifuged at 300 g for 10 min the next day. Then the cells are resuspended in fibroblast culture medium again overnight.
6. Dead cells are then washed out using phosphate-buffered saline.
7. Fibroblasts are seeded on a reticular surface of de-epidermized dermis at a concentration of 1×10^5 cells (*see* **Note 11**).
8. The gels with fibroblasts are growing in a fibroblast growth medium for 5–7 days and then 1×10^6 keratinocytes are seeded on the surface.
9. After that, the tissue engineered skin is exposed to air-liquid interface for another 10–14 days with medium changes every 2–3 days.
10. The developed and stable tissue engineered skin plates are applied on the burned area using surgical staples for fixation.
11. Alternatively to tissue engineered skin, 1:4 autograft split-thickness mesh grafts can be used as transfer-medium for adipose derived stem cells and other cells.

3.4 Treatment of Donor Site and Wound Coverage

1. Prior to wound coverage, donor sites are labeled in chronologic order from cranial to caudal with ink.
2. Skin biopsies are taken (two central and two peripheral) from each wound site and donor site. If autografts are used to cover the wounds, skin graft donor sites (considered as partial thickness dermal wounds) are covered with wound dressings.
3. The first dressing layer, directly placed on the autograft, consists of non-sticking (silicone) oil emulsion dressing. This layer is covered with a thin protective layer of antimicrobial ointment consisting of 1% Nystin and 2% Polymyxin B/Bacitracin in a Vaseline formulation (*see* **Note 12**).
4. The second layer of double polyurethane film, such as Opsite®, is placed over the first dressing layer to seal off the wound fields while providing vapor permeability (*see* **Note 13**).

5. A nylon dressing is used as an outer layer for protection of the inner dressing layers.
6. After treatment, the animals should be seen twice daily within the first 2 weeks and daily after 2 weeks. Dressings are checked for being clean and intact to prevent unintended damage to wound dressings caused by normal animal movements (*see* **Note 14**). An adequate and systematic pain assessment as well as frequent check of wound dressing and wounds are essential the animals' well-being for the outcome of the study.
7. The frequency of dressing changes depends on the specific study protocol (*see* **Note 15**) and is a complex balance between giving the grafts the time to heal and avoiding infections under the dressings. At days 2 through 14 (around days 2, 5 and 14), dressings are changed and skin biopsies (two central and two peripheral for each donor site). Before every dressing change and/or biopsy, the animals' weight should be assessed.
8. When donor sited are healed as well as at time of scarification (anticipated between 3 and 6 months post injury), biopsies are taken again (*see* **Note 16**).

3.5 Outcome Measurements

There are several ways to measure the results of dressings, skin substitutes, or tissue engineered skin in animal models. Some of them are similar to the methods used in humans due to their non-invasive nature. For example, other studies that require routine biopsies can only be performed experimentally using animal models.

3.5.1 Clinical Evaluation

Most of the studies concerning tissue engineered skin use at least one clinical evaluation tool. A typical score as described by Branski et al. is shown in Table 1 [18]. These scores should be assessed by at least two investigators during the dressing changes.

3.5.2 Planimetry

After initial grafting and during each dressing changes, we usually take standardized photographs. We use a planimetrical software. Total wound size and the size of all open mesh interstices are measured for each wound at each time point (*see* **Note 17**). To assess the healing, total wound size reduction and the open interstices have to be compared to the baseline values (date of injury/graft).

3.5.3 Wound Histopathology

After embedded in paraffin and sectioned, histochemical staining techniques (for example hematoxylin-eosin (H&E) staining can be used to evaluate cellularity, cell morphology, and architectural aspects such as thickness and orientation of the collagen [19]. Differences of the architecture of the collagen, which are very specific for the differentiation between normal skin an scar tissue, are reliably apparent on H&E [20].

Table 1
Clinical evaluation of wounds (Modified from Branski et al. [18])

Grade	Graft dislocation	Graft adherence	Granulation tissue	Hypergranulation	Hematoma	Fibrin deposition
0	None	Full graft loss	None, full depth defect	None	None	None
1	5–10%, tissue still fully viable	Graft 20% adherent, 50% graft loss	Low level of granulation tissue	10%	Up to 10% wound size	Small patches, up to 10% wound size
2	10–25%, one or two spots, q0% tissue not viable	Graft 50% adherent, 30% graft loss	Granulation tissue level at half of initial wound depth	20%	Up to 20% wound size	Thicker, confluent patches, up to 20% wound size
3	25–50%, more than 20% tissue not viable	Graft 75% adherent, 20% graft loss	Granulation tissue level just below surrounding tissue	40%	Up to 40% wound size	Up to 40% wound size
4	50–80%, more than 50% tissue not viable	Graft 90% adherent	Granulation tissue at level with surrounding tissue	65%	Up to 60% wound size, partial graft loss	Up to 60% wound size
5	Graft completely dislocated, full tissue loss	Full graft adherence	Fully granulated wound, exceeds level of surrounding tissue	100%	Up to 100%, total graft loss	Wound fully covered with fibrin and graft skin totally mascerated

Table 2
Normal ranges of blood parameters in pigs

Measurement	Mean	Range
Blood characteristics		
Hematocrit (%)	27.0	23.5–33.0
Hemoglobin (g/dl)	8.5	7.3–10.2
WBC/μL	4500	3800–10,000
Platelet count (10^9/L)	300	250–600
Plasma albumin (g/L)	25.4	21.6–28.0
Sodium (mEq/L)	138	129–143
Potassium (mEq/L)	4.4	3.9–4.1
Blood gas and acid-base status		
Arterial PO2 (mmHg)	82	73–92
Arterial HbO2 (%)	94	92–95
Arterial carboxyhemoglobin (%)	4	4–5
Arterial pH	7.48	7.40–7.53
Arterial PCO2 (mmHg)	40	35–44
Arterial plasma HCO3 (mEq/L)	29	22–33
Arterial buffer base (mEq/L)	45	40–52

3.5.4 Hematology

Selected serum cytokines and hematologic parameters can be used to evaluate the general condition of the animal after surgical procedures and to evaluate systematic effects of the burn or other injury. To later compare the findings with baseline data, blood samples should be obtained prior to the procedure and before grafting. We usually obtain blood samples on days 2, 5, 9, and 14 after the procedure. This can differ in the specific study protocol and should always be discussed before beginning the study. We use a veterinary hematology analyzer to measure parameters such as WBC, hemoglobin and platelets. IL-6 and TNF-α are measured by ELISA using porcine kits. The normal ranges for 20–25 kg Yorkshire/Red Duroc pigs are described in "Swine in the Laboratory" by Swindle and Smith [21]. Some parameters of major interest are shown in Table 2.

4 Notes

1. Other components and methods for sedation and anesthesia can be used. We use this mixture for its quick onset of

anesthesia and the duration of sedation which is of high importance for a safe transportation to the operating facilities. A sufficient sedation is usually achieved after 10–15 min.

2. Weighting and clipping are performed in a separated room with the additional use of 2–3% of isoflurane in room air via face mask.
3. Successful Intubation needs to be verified via auscultation of both lungs.
4. The catheter is used for repeated blood draws and infusion of fluids (Ringer's lactate solution, 5–10 mL/kg/h) during the entire experiment.
5. In our opinion, fentanyl patches are the most effective in both control and reduction of postoperative activity. This is in line with previous literature and of high relevance in order to achieve successful long-term observations of wounds which is only possible with intact dressings [18].
6. Other methods to cause wounds in animals can be used. Another method was described by Middelkoop et al. They used an electrical dermatome with a depth of 2.7 mm [16].
7. The percentage of total surface body area (%TSBA) burned depends on the study protocol and the aims of the study. Most studies have not used a burn wound not greater than 40% TBSA [22]. TBSA for pigs is calculated using the validated Wachtel formula with known weight (TBSA ($m^2 = 0.121 * wt (kg)^{0.575}$) [23].
8. We stain the taken biopsies with Hematoxylin and Eosin and with Masson's trichrome to determine the extent of burn as determined by the depth of collagen degeneration. The described burn procedure leads to the complete coagulation of the superficial and most of the deep collagenous structures.
9. FCS, which contains protease inhibitors, can be added to better separate the two layers.
10. A cell count can be carried out with a haemocytometer. Cell viability may be assessed via trypan blue exclusion.
11. Instead of de-epidermized skin, transwell systems may be used for culturing the engineered skin construct [24].
12. Since pigs and other animals tend to repeatedly scratch their back, a multilayer dressing is necessary to protect the wounds especially in the first days after burn.
13. Since the use of second layer after day 5 frequently leads to hypergranulation and sporadic graft lost, we only use the second layer for the initial dressing application and the first dressing change.

14. Special swine clothing, goat tubes (intended to expand and provide a comfortable fit) may be positioned around the torso of the animal.
15. All dressing changes must be performed under general anesthesia and preferably intubated. Mask anesthesia should only be used for rapid dressing changes.
16. Note that donor sited are larger in area than individual wound sited; therefore, four punch biopsies may be obtained for each donor site, whereas three punch biopsies may be obtained for each burn wound site.
17. The angle to the wound should be as stable as possible using a 90° angle and the same distance at every time point.

References

1. Shevchenko RV, James SL, James SE (2010) A review of tissue-engineered skin bioconstructs available for skin reconstruction. J R Soc Interface 7:229–258
2. Huang S, Fu X (2011) Tissue-engineered skin: bottleneck or breakthrough. Int J Burn Trauma 1(1):1–10
3. Finnerty CC, Jeschke MG, Branski LK et al (2016) Hypertrophic scarring: the greatest unmet challenge following burn injury. Lancet 388(10052):1427–1436
4. Saaiq M, Zaib S, Ahmad S (2012) Early excision and grafting versus delayed excision and gratinf of deep thermal burns up to 40% total boday surface area: a comparison of outcome. Ann Burns Fire Disasters 25(3):143–147
5. Ayaz M, Bahadoran H, Arasteh P et al (2014) Early excision and grafting versus delayed skin grafting in burns covering less than 15% of total body surface area; A non- randomized clinical trial. Bull Emerg Trauma 2(4):141–145
6. Herndon DN, Barrow RE, Rutan RL et al (1989) A comparison of conservative versus early excision. Therapies in severely burned patients. Ann Surg 209(5):547–553
7. De La Tejera G, Corona K, Efejuku T et al (2023) Early wound excision within three days decreases risks of wound infection and death in burned patients. Burns 49(8): 1816–1822
8. Kapp MB (2006) Ethical and legal issues in research involving human subjects: do you want a piece of me? J Clin Pathol 59(4): 335–339
9. Rheinwald JG, Green H (1977) Epidermal growth factor and the multiplication of cultured human epidermal keratinocytes. Nature 265(5593):421–424
10. Sullivan TP, Eaglstein WH, Davis SC et al (2001) The pig as a model for human wound healing. Wound Repair Regen 9(2):66–76
11. Ignacio G, El-Amin I, Mendenhall V (2016) Animal models for wound healing. In: Skin tissue engineering and regenerative medicine, pp 387–400
12. Lunney JK, van Goor A, Walker KE et al (2021) Importance of the pig as a human biomedical model. Sci Transl Med 13(621)
13. Meyer W, Schwarz R, Neurand K (1978) The skin of domestic mammals as a model for the human skin, with special reference to the domestic pig. Curr Probl Dermatol 7:39–52
14. Heinrich W, Lange PM, Stirtz T et al (1971) Isolation and characterization of the large cyanogen bromide peptides from the alpha1- and alpha2-chains of pig skin collagen. FEBS Lett 16(1):63–67
15. Morris GM, Hopewell JW (1990) Epidermal cell kinetics of the pig: a review. Cell Tissue Kinet 23(4):271–282
16. Middelkoop E, van den Boghaerdt AJ, Lamme EN et al (2004) Porcine wound model for skin substitution and burn treatment. Biomaterials 25(9):1559–1567
17. National Research Council (1996) Guide for the care and use of laboratory animals. National Acadamy Press, Washington, DC
18. Branski LK, Mittermayr R, Herndon DN et al (2008) A porcine model of full-thickness burn, excision and skin autografting. Burns 34(8): 1119–1127
19. Lammers G, Verhaegen PDHM, Ulrich MMW et al (2011) An overview of methods for the in vivo evaluation of tissue-engineered skin constructs. Tissue Eng 17(1):33–55

20. Verhaegen PD, van Zuijlen PPM, Pennings NM et al (2009) Differences in collagen architecture between keloid, hypertrophic scar, normotrophic scar, and normal skin: an objective histopathological analysis. Wound Repair Regen 17(5):649–656
21. Swindle MM (ed) (2007) Swine in the laboratory: surgery, anesthesia, imaging, and experimental techniques, 2nd edn. CRC Press. https://doi.org/10.1201/9781420009156
22. Abdullahi A, Amini-Nik S, Jeschke MG (2014) Animal models in burn research. Cell Mol Life Sci 71(17):3241–3255
23. Wachtel TL, McCahan GR, Watson WI et al (1972) Determininh the surface areas of miniature swine and domestic swine by geometric design – a comparative study. USAARL Report, 73–5. Fort Rucker, Alabama
24. Braziulis E, Biedermann T, Hartmann-Fritsch F et al (2011) Skingineering I: engineering porcine dermo-epidermal skin analogues for autologous transplantation in a large animal model. Pediatr Surg Int 7:241–247

Chapter 21

Transplantation of Autologous Dermo-Epidermal Skin Substitutes in a Pig Model

Thea Fleischmann, Flora Nicholls, Miriam Lipiski, Margarete Arras, and Nikola Cesarovic

Abstract

Due to its similarity of skin anatomy and physiology, the pig appears to be a well-suited animal model for pre-clinical studies of skin analogue transplantations. The choice of the location of the skin defect and appropriate post-operative measures are essential for the protection of the transplanted graft. This protocol describes in detail a porcine skin transplantation model including peri- and postoperative measures taken to improve and refine the studies outcome.

Key words Pig, Skin transplantation, Dermo-epidermal skin substitute, Peri-operative refinement, Animal model

1 Introduction

Pigs appear to be the most concordant animal model for pre-clinical studies of skin analogue transplantations for two main reasons [1]. First, pig skin is very similar to human skin in terms of its anatomical structure (e.g., similar dermal-epidermal thickness ratio, fixed subcutaneous layer and hair follicles like humans) [2], its physiology as well as its wound healing characteristics (e.g., in contrary to rodents, primarily by re-epithelialization rather than contraction) [3]. Secondly, the size of the pig allows transplantation of sizeable grafts equal to commercially available cultured epidermal autografts currently used in clinics [4].

The challenges of the model lay in the maximum protection of the skin graft during the first post-operative weeks. Choosing the optimal location for the skin defect is therefore crucial to prevent the animal from scratching the wound site with the mouth or the extremities. As pigs should be, whenever possible, housed in groups, protection of the skin graft from pen mates presents another challenge.

Thomas Biedermann and Sophie Böttcher-Haberzeth (eds.), *Skin Tissue Engineering: Methods and Protocols*, Methods in Molecular Biology, vol. 2922, https://doi.org/10.1007/978-1-0716-4510-9_21,

We here describe a porcine model, which permits the transplantation of skin analogous grafts measuring 7.5 × 7.5 cm [5]. An adequate pain management protocol, single housing of the animals and a custom-made protective garment ensured the survival of the graft in the decisive initial phase [6]. Additionally, to enhance the scientific read out, the use of a silicon chamber to compartmentalize the operation field from the surrounding skin proved to be of great value as it prevented healing from the wound edges themselves.

The protocol described here in detail should be of interests to veterinarians, veterinary technicians, animal caretakers, or other personnel responsible for the peri- and postoperative care of pigs involved in skin transplantation studies.

2 Materials

2.1 Animals

For the development of this protocol, domestic pigs, breed "Large White," aged between 2 and 3 months, weighing ca. 30–35 kg, were purchased from a local breeder. The animals were brought to the facility two weeks prior to the start of the experiment to allow appropriate acclimatization (*see* **Note 1**). Each pig was thoroughly examined within the acclimatization period to ensure that animals are healthy. Pigs were group-housed under standardized conditions: room temperature 19–22 °C, relative humidity 45–55%, light cycle 12 h:12 h. The artificial light cycle was complemented by a natural light source from large skylights. Each animal received pelleted dry food twice daily and had ad libitum access to drinking water. Fruit yoghurt was given as a means of conditioning and reward (*see* **Note 2**). Environmental enrichment like toys, cardboard boxes, wood shavings, etc. should also be used to prevent boredom and keep the pigs occupied. The general condition was checked twice daily throughout the study.

2.2 Veterinary Drugs

1. Anesthesia: Ketamine hydrochloride (Narketan®, 20 mg/kg BW), Azaperone (Stresnil®, 1.25 mg/kg BW), Atropine (0.03 mg/kg BW), Isoflurane (Attane™, 1–2.5 Vol. %), Propofol (Propofol-®Lipuro 1%, 1–2 mg/kg BW).
2. Sedation: Diazepam (Valium®, 0.4 mg/kg BW).
3. Analgesia: Buprenorphine (Temgesic®, 0.01 mg/kg BW), Fentanyl patches (Durogesic® Matrix, 50 μg/h), Carprofen (Rimadyl®, 3 mg/kg BW).
4. Antibiotics: Enrofloxacin (Baytril®, 2.5 mg/kg BW).
5. Antacid: Lansoprazol (Agopton® 30; 30 mg/day per os).
6. Wound treatment: Vita-Hexin®, Bepanthen® Ointment.
7. Euthanasia: Pentobarbital (Eskonarkon®, 50 mg/kg BW).

2.3 Basic Surgical Materials

1. A sterile basic surgical kit including scissors, forceps, scalpel blades, scalpel holder, bowls, needle holder, swabs.
2. Fur clipper and disposable razors.
3. Sterile folio drapes (Foliodrape®).
4. Skin stapler, staple remover.
5. Suture material: 0–0 Ethilon, 1/0 Vicryl.
6. Skin disinfectant (Betadine®).
7. Sodium chloride solution: 0.9% NaCl.
8. Wound spray (Opsite®).

2.4 Materials for Skin Transplantation

1. Wound dressings: Silon- SES®, Acticoat, Mepilex® Ag, Ligasano®, Moltex.
2. Transplantation chamber: custom made silicone consists of a chamber and a cover plate. The elastic chamber has a hardness (shore A) of 80 and the flexible cover has hardness (shore A) of 25. The molded silicone rubber parts are biocompatible, biostable, and heat-resistant (200 °C) and thus amenable to standard sterilization processes, e.g., steam sterilization.
3. Protective suit: It is advisable to purchase a suit to protect the skin graft after transplantation. We obtained a made-to-measure suit with the following special features: Longitudinal and transversal stitching, Neoprene neck element, sufficient length on the ventral side, thick fleece padding around front and hind legs, two lateral zippers divided into a cranial and a caudal part. At least two suits were manufactured for each animal in different sizes taking the growth of the animals into consideration (*see* Fig. 1).

2.5 Medium

1. Transport medium: 500 mL DMEM, 1 g/L D-Glucose, 110 mg/L Pyruvate, 15 mL 10′000 Units/mL Penicillin/ 10′000 μg/mL Streptomycin, 0.75 mL Gentamycin, 0.5 mL 250 μg/mL, Amphotericin B.

3 Methods

3.1 Animal Preparation

1. Prior to the procedure, the pigs are fasted overnight to avoid vomiting and aspiration.
2. Perform a deep sedation in the stable with an intramuscular injection of ketamine hydrochloride (20 mg/kg BW), azaperone (1.25 mg/kg BW) and atropine (0.03 mg/kg BW). To ensure stress-free injection, distract the pig during the injection with the above-mentioned rewarding food.

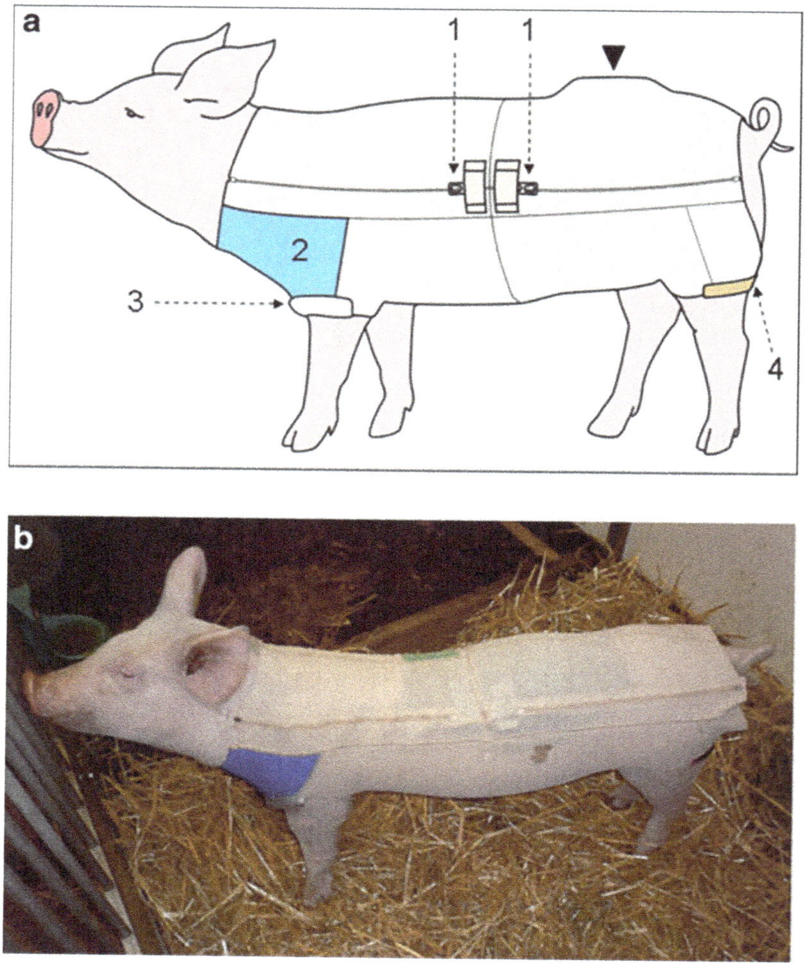

Fig. 1 Customized suit. (**a**) Schematic drawing of the customized garment for pigs, indicating the essential features: *1* two separate zippers, *2* neoprene insertion, *3* border of fleece upholstery, *4* elastic fabric strip. Arrowhead indicates location of shock absorber device under the suit. (**b**) Picture showing the "custom-made" suit on a pig. (Adapted from Ref. [5])

3. After loss of the righting reflex, place a venous catheter in the auricular vein and administer propofol (bolus 1–2 mg/kg BW i.v.) to achieve relaxation and diminishment of the swallowing-reflex for intubation.
4. Place an endotracheal tube (diameter 8 mm) and maintain anesthesia by inhalation anesthesia with isoflurane (1.0–2.5 Vol %) in 50:50 oxygen/air gas mixture (3–5 L/min) under positive pressure ventilation.
5. Administer buprenorphine (0.01 mg/kg BW i.v.) for perioperative pain management and enrofloxacin (2.5 mg/kg BW i.v.) as perioperative antibiosis.

6. Throughout the procedure, the animal should receive fluid supplementation with Ringer Lactate (10 mL/kg/h).
7. Place the pig in dorsal recumbency on the operating table. Install anesthetic monitoring (Pulsoxymetry, temperature measurement, $etCO_2$, O_2, isoflurane, invasive blood pressure (A. coccygeal), ECG) and monitor vital signs throughout the procedure.

3.2 Skin Biopsy Procedure

1. Shave and wash the surgical area with soap, then perform three cycles of disinfection with Betadine®. Cover the surgical area with sterile drapes and neutralize the area with 0.9% NaCl.
2. The skin biopsy should be taken from the ventral abdomen, lateral and parallel to the mammary teats. Use a scalpel blade No. 10 and take a skin biopsy of about 15 cm length, 2 cm widths, total 15 cm^2, depth from epidermis to fascia.
3. Transfer the skin immediately into transport medium.
4. Close the wound in two layers, e.g. subcutaneous closure with resorbable Vicryl 1/0 and cutaneous wound closure with non-resorbable Ethilon 0/0 in single interrupted sutures.
5. Apply Opsite® wound spray and adhesive wound dressing.
6. After the intervention, anesthesia is discontinued and the pig brought back to the stable. For undisturbed recovery, keep the pig individually in a single compartment on soft bedding under a warming lamp and closely monitor until full recovery.
7. The pigs receive 3 mg/kg BW carprofen i.v. post-operatively and a second dose of 0.01 mg/kg buprenorphine i.m. 6 h after the first application. Administration of carprofen can be continued every 12 h if necessary and until the pigs appear free of pain.
8. Check the animals 2×/day for the following 3 weeks to ensure proper wound healing.

3.3 Skin Analogue Transplantation

After the skin biopsy was taken, grant the pigs a recovery time of two to three weeks. Ensure general well-being and appropriate wound healing by checking the animal regularly. During this period, measure each pig (body length, shoulder to elbow, hip to knee) so a made-to-measure protective suit can be manufactured.

3.4 Animal Preparation

1. Perform anesthesia and preparation of the animal as described above for the skin biopsy.
2. Place the pig in sternal recumbency on the operating table. Install anesthetic monitoring (Pulsoxymetry, temperature measurement, $etCO_2$, O_2, isoflurane, invasive blood pressure (A. coccygeal), ECG) and monitor vital signs throughout the procedure.

3.5 Skin Transplantation Procedure

1. Choose a transplantation site medial to the vertebral column over the lumbar area (7.5 × 7.5 cm, square-shaped) to minimize pain or discomfort during normal ambulation after the surgery.
2. Shave and wash the surgical area with soap and mark the implantation site with a waterproof pen. Perform three cycles of disinfection with with Betadine® and cover the surgical area with sterile drapes.
3. Excise the skin in the marked area. The entire subcutaneous fat layer should be excised but be careful to not injure the vascularized layer above the fascia.
4. Mobilize the skin edges to introduce the chamber. Cut the silicon edge of the chamber appropriately, until it fits smoothly into the skin.
5. Fix the chamber with 0–0 Ethilon on all sides.
6. Disinfect the wound inside the chamber with Betadine® and thoroughly neutralize with sterile 0.9% NaCl.
7. Apply the skin transplant inside the chamber and fix with skin clamps around the edges.
8. Apply silicon (Silon-SES®) directly on top of the transplant.
9. Apply Mepilex® Ag (Mölnlycke), 10 × 10 cm, and the silicon lid on top of the chamber.
10. Apply Mepilex® Ag (Mölnlycke), 20 × 20 cm, folio drapes, Ligasano®, 25 × 25 cm, folio drapes® and Moltex (*see* **Note 3** and Fig. 2).
11. Finally, put the made-to-measure suit on while the animal is still anesthetized.
12. Apply a fentanyl patch (Durogesic patch, 50 μg/h) on the skin between the shoulder blades and fixed with adhesive tape. The patch is left in place for three days to provide continuous analgesia.
13. Transport the pig back to the stable to recover from anesthesia. Administer diazepam (0.2 mg/kg BW i.m.) to ensure a smooth recovery and to reduce stress caused by the unfamiliarity of the protective suit. Give all animals diazepam (0.4 mg/kg BW; SID, p.o.) for three days to minimize agitation until the animals are fully accustomed to the suit.
14. Continue antibiosis for two days after the intervention. To reduce the risk of stress-induced gastric ulcers, pigs receive lansoprazol (30 mg/day per os) for 5 days.

3.6 Post-operative Care

1. Check pigs twice daily for general health and well-being with emphasis on mobility, food and water intake, and signs of pain or wound infection and proper fitting of the suit. During the

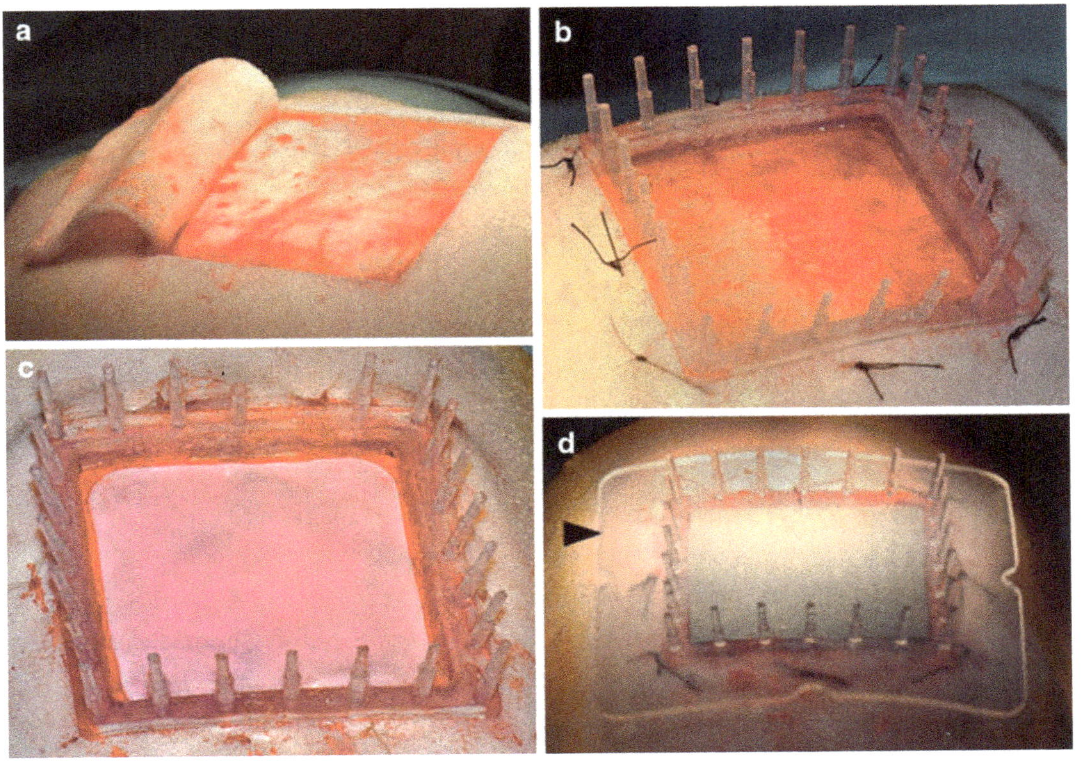

Fig. 2 Surgical procedure of skin substitute transplantation. (**a**) A full-thickness skin defect of 7.5 × 7.5 cm, located over the lumbar area is created. (**b**) The silicone chamber is implanted and secured by fixation sutures. (**c**) The cultured skin substitute is transplanted onto the full-thickness wound. (**d**) The wound is covered with soft silver foam (grey) and a 2-mm-thick silicone cover plate (arrowhead). (Adapted from Ref. [5])

daily inspections, the suit is left in place to avoid disturbance to the healing of the transplanted skin graft.

2. Apply a soothing creme (VitaHexin® or Bepanthen® ointment) under the arm pits and groin on a regular basis to prevent irritation of the skin by the protective suit.
3. House pigs now individually in neighbouring sties to minimize the risk of peer-inflicted damage to the operation site (*see* **Notes 4** and **5**).

3.7 Inspection of Wound and Evaluation of Skin Analogue

Check the transplant site every seven days following transplant surgery.

3.8 Animal Preparation

1. Perform a deep sedation in the stable with an intramuscular injection of ketamine hydrochloride (20 mg/kg BW), azaperone (1.25 mg/kg BW) and atropine (0.03 mg/kg BW).

2. After loss of righting reflex, place a venous catheter in the auricular vein and place the animal in sternal recumbency on the operating table.
3. Maintain anesthesia by inhalation anesthesia with isoflurane (1.0–2.5 Vol %) in 50:50 oxygen/air gas mixture mixture (3–5 L/min) under spontaneous breathing via a face mask.
4. Install anesthetic monitoring (Pulsoxymetry, temperature measurement) and monitor vital signs throughout the procedure.

3.9 Wound Inspection

1. Remove the suit and bandages and clean the wound site.
2. Open the silicon lid and disinfect the skin around the chamber with Betadine®.
3. Exchange Silon-SES® only when needed. Otherwise, it should be left in place without further disinfection.
4. Apply chamber lid and bandage again.
5. Put suit back on.
6. Discontinue anesthesia and let the animal recover.
7. After each inspection of the transplant site, exchange the made-to-measure suit for a clean one while the dirty suit is being washed (*see* **Notes 6** and 7).

3.10 Euthanasia and Excision of Skin Analogue

1. Perform anesthesia and preparation of the animal as described above.
2. Remove the wound drapes and the silicon chamber.
3. Carefully excise the transplant and prepare it for histological examination.
4. Perform euthanasia with intravenous administration of pentobarbital (75 mg/kg BW).

4 Notes

1. It is important to handle the pigs daily during the acclimatization phase to familiarize the animals and to allow unstressed handling. It facilitates daily observations, clinical examination, or repositioning of the suit.
2. We offer our pigs some special treats, e.g., each pig daily receives some yogurt to make administration of anesthesia, clinical examinations (e.g., body temperature measurement) or readjustment of suits easier and stress-free.
3. The several layers of dressing and the silicone chamber act as a "multilayer shock absorbent system," consisting of the silicon chamber and a 2-mm-thick silicone cover plate to shield the area of interest from the wound edges, wound sites are covered

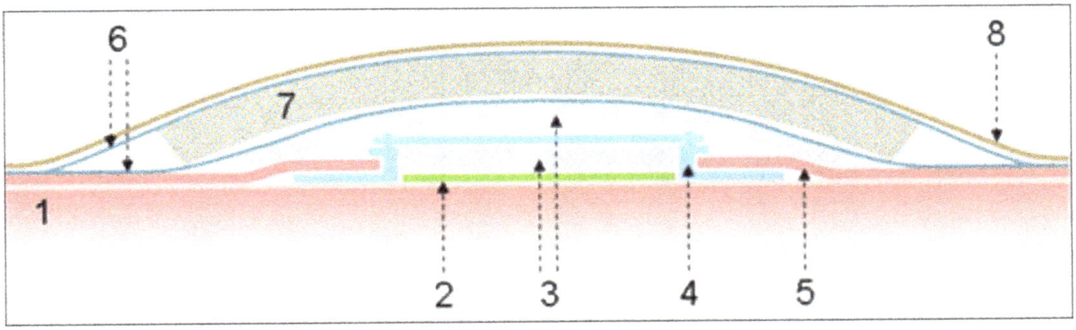

Fig. 3 Multilayered shock absorber device. Schematic drawing showing the architecture of the device: *1* muscle including overlying fascia, *2* skin analogue, *3* soft silver foam, *4* silicone chamber, *5* pig skin, *6* fixation tapes encasing the foam, *7* foam, *8* customized suit. (Adapted from Ref. [5])

with a soft silver foam dressing, fixation tapes encasing the foam, with additional foam and the made-to-measure suit (*see* Fig. 3).

4. After putting on the suits, the pigs should be individually housed with visual and olfactory contact to fellow animals. Group housing was no further possible, as the animals manipulated each other's suits until the suits were ripped apart or completely removed.
5. If possible, pigs should be acclimatized to the made-to-measure protective suits for a few days prior to the skin transplantation procedure. However, it might not be possible to put on the suit stress-free in the conscious animal. In our study, we therefore put them on while the animal was still anesthetized, and carefully observed the animal during the recovery phase.

   Diazepam (0.2 mg/kg BW i.m.) can be administered to support a smooth recovery and to reduce any stress due to the unfamiliarity of the suit. All animals further received diazepam (0.4 mg/kg BW; SID, p.o.) for three days to minimize agitation until the animals accustomed to the suit.
6. One should obtain at least two made-to-measure suits for each animal, taking the growth of the pig into consideration. We exchanged the suits during each inspection of the transplant site or in case of soiling and wetness.
7. Despite the fleece padding, we observed some redness under the arm pits and groin. Therefore, a soothing crème (Vita-Hexin® or Bepanthen® ointment) should be applied on a regular basis.

References

1. Sullivan TP, Eaglstein WH, Davis SC et al (2001) The pig as a model for human wound healing. Wound Repair Regen 9(2):66–76
2. Lindblad WJ (2008) Considerations for selecting the correct animal model for dermal wound-healing studies. J Biomater Sci Polym Ed 19(8): 1087–1096
3. Swindle MM, Makin A, Herron AJ et al (2012) Swine as models in biomedical research and toxicology testing. Vet Pathol 49(2):344–356
4. Bottcher-Haberzeth S, Biedermann T, Reichmann E (2010) Tissue engineering of skin. Burns 36(4):450–460
5. Schiestl C, Biedermann T, Braziulis E et al (2011) Skingineering II: transplantation of large-scale laboratory-grown skin analogues in a new pig model. Pediatr Surg Int 27:249–254
6. Fleischmann T, Clutton E, Haga HA et al (2023) Anesthesia and analgesia in laboratory pigs. In: Anesthesia and analgesia in laboratory animals, vol 17, 3rd edn. Elsevier, pp 411–439

Chapter 22

Ex Vivo Human Skin Model to Study Immediate and Early Local Responses to External Stimuli

Ines Foessl, Thomas Birngruber, Petra Kotzbeck, and Elisabeth Hofmann

Abstract

Local responses of skin tissue to external stimuli are complex, thus exceeding the capability of in vitro methods but do not need the complexity of a whole organism to be simulated. Therefore, we established a human ex vivo skin model in our lab to study the local effects of various stimuli, such as burns and local topic treatments. In this chapter, we describe the experimental setup of the ex vivo model and additional technology of collecting dermal interstitial fluid. Further, we describe how exosome-like vesicles can be enriched from the dermal interstitial fluid, including their proof via nanoparticle tracking.

Key words Ex vivo, Skin, Early response, Human model, Dermal interstitial fluid

1 Introduction

The processes that occur after external stimuli in the skin can be simulated at different scales [1]. Cellular processes can be simulated well in vitro [2] but complex questions such as studying systemic effects as a response to changes in the skin microenvironment and the interplay of the human skin with other organ systems requires whole-organismal models and need to be answered in clinical studies or animal experiments [3, 4]. Local interaction and response, however, can be well modelled at the tissue level [5].

Human skin explants are by-products that are collected during abdominoplasty surgery, circumferential body lift, and other tissue-reducing operations. With the donors' written informed consent and approval by an ethics committee, they are used for various experimental procedures, including ex vivo models. Such models of human skin are valuable assets for studying a wide range of scientific challenges ex vivo [5, 6].

To study the local responses to various stimuli, we established a standardized ex vivo model and an optimized workflow [7]. Burn injuries can be introduced to the model through contact burn

Thomas Biedermann and Sophie Böttcher-Haberzeth (eds.), *Skin Tissue Engineering: Methods and Protocols*, Methods in Molecular Biology, vol. 2922, https://doi.org/10.1007/978-1-0716-4510-9_22,

injury simulation via a heated steel block and temperature sensors placed prior to the burn stimulus can monitor the stimulating temperature. Inflammatory processes can be simulated with the injection of LPS. The model from fresh, non-frozen skin is suitable for short-term observations within approximately 30 h. During this time, the tissue-specific processes in the metabolically active skin can be closely monitored by means of biopsies. With these biopsies, a plethora of further analysis can be performed, such as histology, electron microscopy, protein expression by immunoblot, cytokine analysis, and gene expression analysis. In addition, combined with open flow microperfusion (OFM) [8], the model can provide valuable results with regard to the diffusion of metabolites in the tissue. From the OFM perfusate, which is used to sample interstitial fluid in the skin, vesicles can be fractioned and these fractions can be tested.

In this chapter, we describe the preparation and maintenance of the model, including the stimulation with a burn as well as LPS injection. We also discuss the application of OFM [9] and exosome-like vesicle (ELV) enrichment for further analyses such as nanoparticle tracking that are established in our laboratory [5].

2 Materials

2.1 Consumables

1. Absorbent bed mat.
2. Antiseptic solution Octenisept®.
3. Sterile gauze pads.
4. 1 μg/mL lipopolysaccharide in 0.9% NaCl (*E. coli* O5:B55).
5. 0.9% NaCl.
6. Cyanoacrylate adhesive (OFM).
7. 10% neutrally buffered formalin solution.
8. Liquid nitrogen.
9. Linear dOFM probes for skin and adipose tissue.
10. Sampling unit.
11. Sterile 10 mL perfusate bags.
12. Sterile push-pull tubing kit (OFM-PP2-100-LB, JOANNEUM RESEARCH—HEALTH, Graz, Austria.
13. Physiological saline (electrolyte composition in mM: 140 Na^+, 5 K^+, 2.5 Ca^{2+}, 1.5 Mg^{2+}, and 108 Cl^-) with 1% human serum albumin (20% infusion solution).
14. Filtered deionized water.

2.2 Instruments

1. Heating chamber with adjustable temperature and humidity.
2. 1.9 kg stainless steel block 5 × 5 cm (Zultner, Graz Austria).
3. Intradermal temperature sensors with temperature data logger (Testo 176T4, Testo GmbH, Austria).
4. High-frequency ultrasound (LOGIQ e R6, GE Healthcare, United Kingdom) with 22 MHz transducer.
5. Microperfusion Pump for Preclinical OFM Studies (MPP102 PC, JOANNEUM RESEARCH—HEALTH, Graz, Austria).
6. NanoSight LM10 (Malvern Panalytical Ltd., Kassel, Germany) with a 532 nm (green) laser.

3 Methods

3.1 Preparation of the Tissue Samples

1. During surgery, the integrity of the skin explants including the subcutaneous lipid layer should be kept intact to ensure maximum resemblance of the resulting explant to the in vivo situation.
2. The explants usually come in two sections—cut in half along the abdominal midline—as a result of the operating technique. These two skin explants can be used as technical duplicates. We suggest to conduct all experiments on our ex vivo skin model in technical duplicates and biological triplicates, at least.
3. Transfer the explants to a food-grade sealable storage container immediately after surgery for the transport to the laboratory. Keep the explants at room temperature and do not cool them.
4. Place the explants on an absorbent mat on a Styrofoam board immediately upon arrival in the laboratory with the epidermis facing upwards. Pin down the edges of the skin explants (Fig. 1).
5. Clean the epidermis with antiseptic solution and sterile gauze pads.
6. Place the explants into a climate chamber with standardized conditions of 32 ± 2 °C and 40–60% humidity and acclimatize them for 1 h, until the experiment can proceed.
7. Mark the application sites on the skin explants using a stencil to standardize the stimulation pattern.

3.2 Burn Injury Model

1. Preheat the stainless-steel block to 100 °C.
2. Optional steps for temperature sensor application (*see* **Note 1**).
3. For deep partial thickness burn wounds, apply stainless steel metal block for 10 s without additional pressure.
4. Optional injection of LPS (*see* **Note 2**).

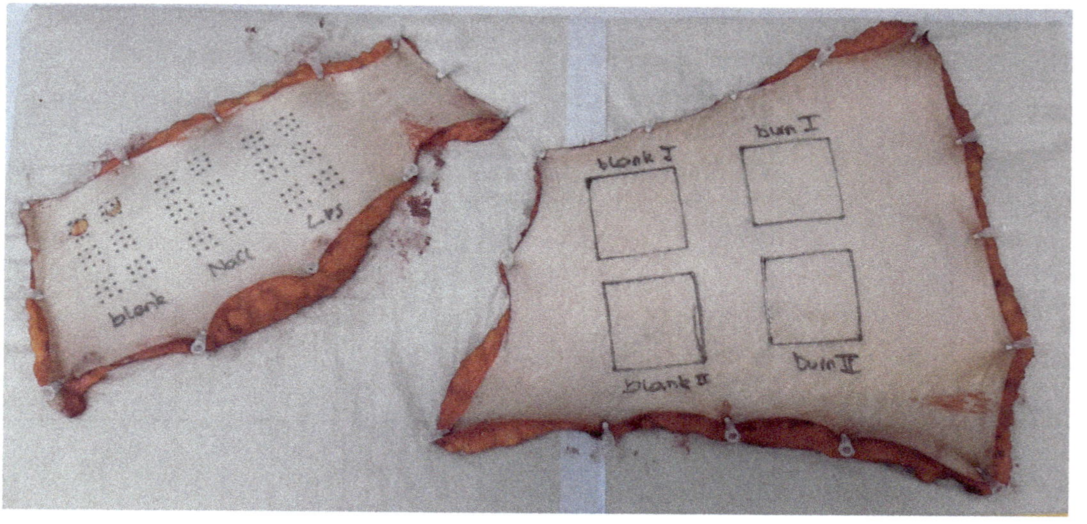

Fig. 1 Skin flaps as prepared for the LPS injection (left) or the burn model (right). Flaps are placed on a Styrofoam board covered with an absorbent mat and secured into place with cannulas. The epidermis is thoroughly cleaned with aseptic solution and marked with the help of a stencil

3.3 Biopsies for Histology and RNA Gene Expression Analysis

1. Take punch biopsies (8 mm) at application sites and untreated control sites 1 h, 4 h, and 24 h after the burn injury or injection.
2. Remove subcutaneous tissue from the biopsies and wash them in sterile PBS.
3. For RNA and protein analyses, snap freeze the unfixed biopsies in liquid nitrogen immediately and store them at −80 °C until further processing.
4. For histology and TEM, fix the biopsies in formalin.

3.4 Dermal Open Flow Microperfusion (dOFM)

1. Mark the dOFM implantation sites by using a stencil and a permanent marker.
2. Insert the dOFM probes with the help of guide cannulas. To monitor the dermal temperature, use a temperature sensor (Fig. 2).
3. Seal the implantation sites with cyanoacrylate adhesive.
4. Connect dOFM probes to tubings, the filled perfusate bags and the peristaltic pumps.
5. Connect a sampling tube between the outlet of the dOFM probe and the pull tubing.
6. Flush the dOFM probes at the run-in phase with a flow rate of 10 μL/min for a maximum of 5 min.
7. Reduce the flow rate to 1 μL/min.

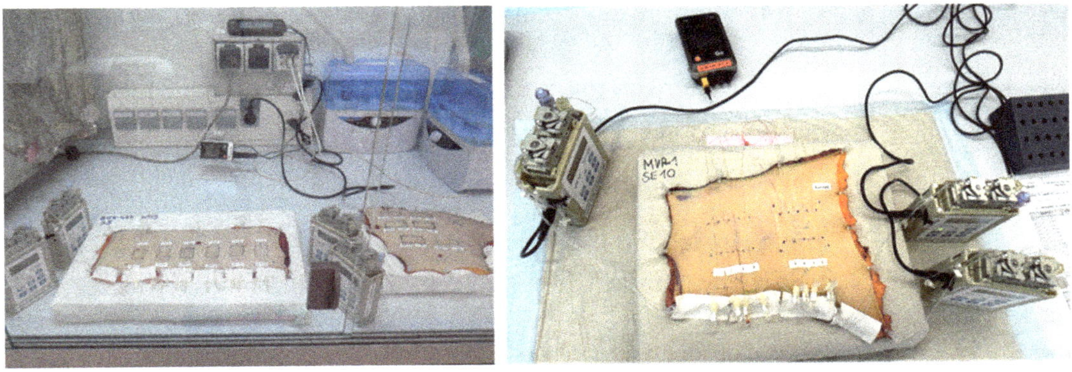

Fig. 2 Experimental setup in the climate chamber for OFM sampling. Left picture: In the climate chamber a heating unit (partly visible top left) and humidifiers (top right) control the conditions of 32 ± 2 °C and 40–60% humidity, that are monitored with digital thermometer and hygroscope. In the foreground, two skin flaps are positioned for the experimental setup of OFM sampling, with catheters positioned in the skin. Right picture: detailed photo of a skin flap with catheters connected to the pumps positioned at the sides. The intradermal temperature sensors with temperature data logger (behind the skin flap) are positioned in the tissue to monitor the tissue temperature during a heat stimulus

8. Discard the run-in samples after 30 min and attach the sampling tubes.
9. Apply the burn stimulus at this point in time (*see* Subheading 3.2).
10. Perform sampling for defined duration (e.g., 4 h).
11. Take biopsies at 1 h, 4 h, and 24 h at burn site and at control sites with no implanted OFM probes.
12. Additional analyses possible (*see* **Note 3**).

3.5 OFM Sample Processing and ELV Enrichment

1. After 4 h of sampling store pooled OFM samples (approximately 60 μL/probe/hour) at 4 °C and process them as soon as possible.
2. Pool all samples from two adjacent probes in order to obtain sufficient sample amount for further processing.
3. Do not aspirate occasional blood smears. Record hemolytic samples by visual assessment.
4. To remove debris, centrifuge perfusate at 500× *g* for 30 min at 4 °C.
5. To remove apoptotic bodies (ApoBD) and cell fragments, transfer supernatant to a fresh tube and centrifuge at 2000× *g* for 20 min at 4 °C.
6. Transfer the supernatant to a fresh tube.
7. Remove 50 μL as mixed particle fraction (MPF).

8. Centrifuge the remaining supernatant at 20,000 × *g* for 70 min at 4 °C.
9. Resuspend the non-visible pellets from each centrifugation step (except for the debris) in 50 μL sterile-filtered PBS and store at 4 °C overnight for nanoparticle tracking. For RNA extraction and immunoblotting, the samples can be frozen at −20 °C for short- term storage.

3.6 Nanoparticle Tracking Analysis

1. For nanoparticle tracking, store OFM samples after ELV enrichment at 4 °C overnight and measure latest on the next day.
2. Dilute the fractions with filtered deionized water (Whatman Anotop® 25 Plus syringe filter, pore size 20 nm) 1:10 or 1:20 (v/v), depending on particle concentration.
3. Measure the samples with the NanoSight with a 532 nm (green) laser at 25 °C with the following settings: camera level 14, slide shutter at 1239, slider gain 366, number of frames 1499, script SOP standard measurement.
4. Take five independent measurements and analyze for average particle properties in terms of size and numbers.

4 Notes

1. Optional steps for temperature sensor application in the burn injury model:
   (a) Intradermal temperature sensors can be implanted in the skin before the contact burn injury thermal treatment by using guiding cannulas.
   (b) Choose the point of implantation just outside the application site with the tip of the temperature sensor positioned in the center of the marked area (*see* also Figs. 1 and 2).
   (c) Confirm the correct intradermal placement of the temperature sensors with high-frequency ultrasound.
   (d) After 1 h of acclimatization, contact burn injuries can be created by using heated metal blocks at the marked application sites.
2. Optional injection of LPS to analyze inflammatory processes independent of tissue disruption by a burn injury:
   (a) Inject of lipopolysaccharide and 0.9% NaCl into the dermis or.
   (b) 450 μL of LPS (E. coli O5:B55, Invivogen, Toulouse, France; 1 μg/ml in 0.9% NaCl, Fresenius Kabi, Bad Homburg, Germany) or 0,9 % NaCl solution injected per 1 cm^2 in 9 single points of 50 μL each (Fig. 1).

3. Samples can be subjected to transmission electron microscopy (TEM) for further analysis.

Acknowledgements

We thank the team of the Division of Plastic, Aesthetic and Reconstructive Surgery, Medical University of Graz, Austria, for the collaboration to obtain the skin explants. We thank the team of Joanneum Research, Graz, Austria, and especially Sonja Kainz, for the thorough handling of the burn wounding and OFM sampling, Selma Mautner for proof reading of the manuscript and all the patients that donate their tissue for research.

References

1. Hofmann E, Nischwitz SP, Holzer JCJ et al (2020) Choice and characterization of preclinical models – towards understanding wound healing. Burns 46(1):251–253. https://doi.org/10.1016/J.BURNS.2019.03.026
2. Hofmann E, Fink J, Pignet AL et al (2023) Human in vitro skin models for wound healing and wound healing disorders. Biomedicines 11(4). https://doi.org/10.3390/BIOMEDICINES11041056
3. Branski LK, Mittermayr R, Herndon DN et al (2008) A porcine model of full-thickness burn, excision and skin autografting. Burns 34(8): 1119–1127. https://doi.org/10.1016/j.burns.2008.03.013
4. Kotzbeck P, Hofmann E, Nischwitz SP et al (2019) Differentiating local and systemic inflammatory responses to burn injuries. Burns 45(8): 1934. https://doi.org/10.1016/j.burns.2018.11.006
5. Foessl I, Haudum CW, Vidakovic I et al (2021) miRNAs as regulators of the early local response to burn injuries. Int J Mol Sci 22(17). https://doi.org/10.3390/IJMS22179209
6. Eberlin S, da Silva MS, Facchini G et al (2020) The ex vivo skin model as an alternative tool for the efficacy and safety evaluation of topical products. Altern Lab Anim 48(1):10–22. https://doi.org/10.1177/0261192920914193
7. Hofmann E, Fink J, Eberl A et al (2021) A novel human ex vivo skin model to study early local responses to burn injuries. Sci Rep 11(1):364. https://doi.org/10.1038/s41598-020-79683-3
8. Bodenlenz M, Aigner B, Dragatin C et al (2013) Clinical applicability of dOFM devices for dermal sampling. Skin Res Technol 19(4):474–483. https://doi.org/10.1111/srt.12071
9. Bodenlenz M, Tiffner KI, Raml R et al (2017) Open flow microperfusion as a dermal pharmacokinetic approach to evaluate topical bioequivalence. Clin Pharmacokinet 56(1):91–98. https://doi.org/10.1007/s40262-016-0442-z

Chapter 23

A Method to Evaluate Skin and Skin Substitute Mechanical Properties Using a Suction Device

Vivienne Woodtli, David Sachs, Alberto Stracuzzi, Serjosha Robmann, Thomas Biedermann, Sophie Böttcher-Haberzeth, and Bettina Thumm

Abstract

Optimizing skin tissue engineering necessitates the precise and objective assessment of scar quality. One approach to achieve this evaluation is through the measurement of the skin's or skin substitutes specific mechanical properties, such as elasticity and firmness. This chapter introduces a measurement protocol for clinical assessment of these properties using a suction device, known as the Nimble. The Nimble device quantitatively evaluates the mechanical characteristics of the skin by applying controlled negative pressure to elevate the skin to a predetermined level, thereby allowing for the measurement of skin elasticity and other related parameters. In this chapter, we provide a detailed description of the Nimble's operation, including the step-by-step process for conducting measurements, and offer guidance on the interpretation of the collected data to assess scar quality effectively.

Key words Mechanical skin response, Suction device, Skin elasticity, Skin substitute mechanical property, Scar quality

1 Introduction

There are various methods to characterize the skin and measure its mechanical response. One innovative and relatively recent technique involves the use of a suction device known as the Nimble [1]. The Nimble was developed at ETH Zurich to assess fibrotic skin tissue. It showed to be suitable to address the unique requirements of pediatric burn care [2]. However, it is also valuable for assessing the mechanical properties of the skin in individuals of other ages or conditions where the skins response may be atypical [3, 4].

The measurement technology is based on a non-invasive suction approach. This technique implies the application of a negative pressure on a portion of the tissue surface by a sealed suction probe (Fig. 1).

Thomas Biedermann and Sophie Böttcher-Haberzeth (eds.), *Skin Tissue Engineering: Methods and Protocols*, Methods in Molecular Biology, vol. 2922, https://doi.org/10.1007/978-1-0716-4510-9_23,

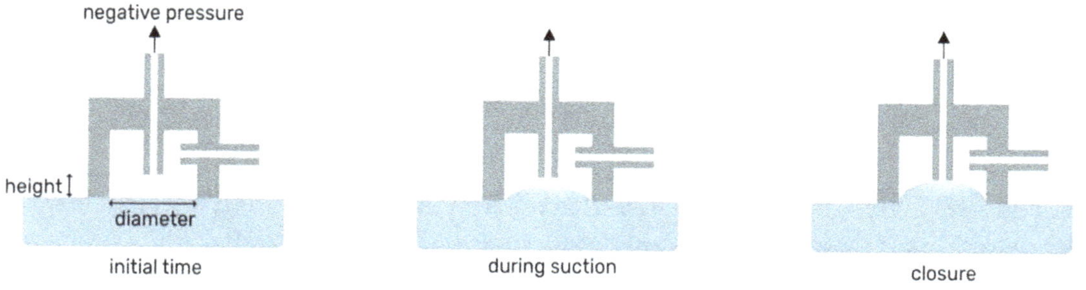

Fig. 1 Schematic representation of a suction experiment on a soft material. The substrate is drawn inside the volume of the cylindrical suction probe up to a fixed height

As a consequence, the tissue is drawn inside the cavity of the suction cup, up to a predetermined height. The internal software algorithm determines the value of pressure needed to reach that level of deformation, which is dependent on the mechanical characteristics of the material under investigation. The suction method has been widely used to characterize the mechanical properties of soft materials and tissues [5–8], since it provides reliable results without damaging the measured material. The Nimble exploits a lightweight design of the suction cup to optimize the sensitivity and precision of the measurement by reducing the disturbance effect of the applied contact force.

With application of various testing protocols, diversified by the specific time history of the aspiration pressure applied, and/or using probes with different geometries, it is possible to determine different mechanical parameters relative to the considered material. Typical materials which can be tested include—but are not limited to—soft elastomers, hydrogels, biological tissues. The device is primarily designed to measure mechanical parameters of human skin in vivo. Additionally, it can be used to assess transplanted skin and tissue-engineered skin both in vivo and in vitro, as well as in animal models.

2 Materials

Nimble device including (Fig. 2):

1. Box.
2. Handheld.
3. Wall adapter for power supply (24 V DC, Mean Well, New Taipei City).
4. Disposable probe set.

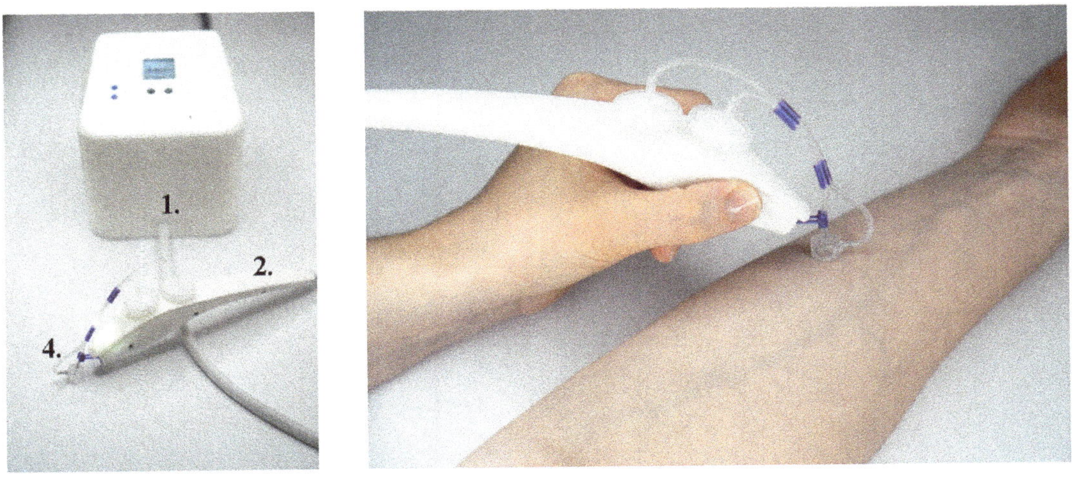

Fig. 2 Composition of the Nimble device, 1. Box, 2. Handheld, 3. Wall adapter for power supply (not shown in this configuration), 4. Disposable probe set. The right image shows a measurement on the volar forearm

3 Methods

3.1 Device Installation

1. To avoid slippage, the box needs to be placed on a rigid and stable surface (*see* **Note 1**).
2. Connect the box to a power supply, using exclusively the provided adapter (24 V DC, Mean Well, New Taipei City). The plug for the power supply can be found on the back side of the box.

3.2 System Start

1. Once the device is connected to the power supply, the device starts up.
2. The user starts the system by pushing the start button (Start Button 1, Fig. 3).
3. Afterwards, any of the two control buttons can be pushed and the screen shows start. The user controls the system with the control buttons on the box.
4. The box is equipped with four push buttons (start and control buttons). When starting the device, the start-button (Start Button 1) has to be pushed.
5. Afterwards pushing any of the two control buttons will activate the device. The start window appears, followed by an animation and finally the operating menu (Fig. 4).
6. The two control buttons are used to navigate through the menu.

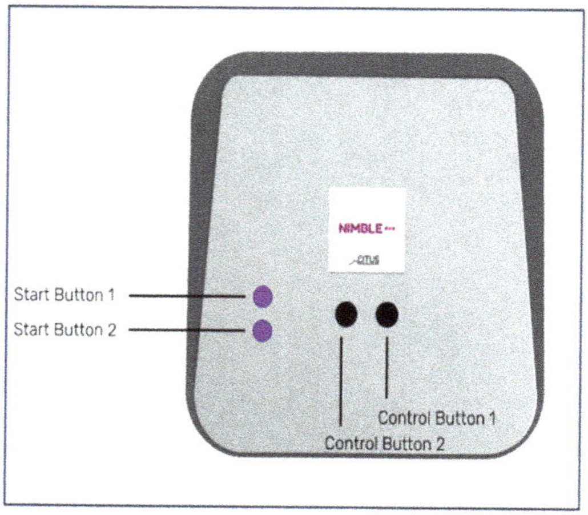

Fig. 3 The box and its associated buttons

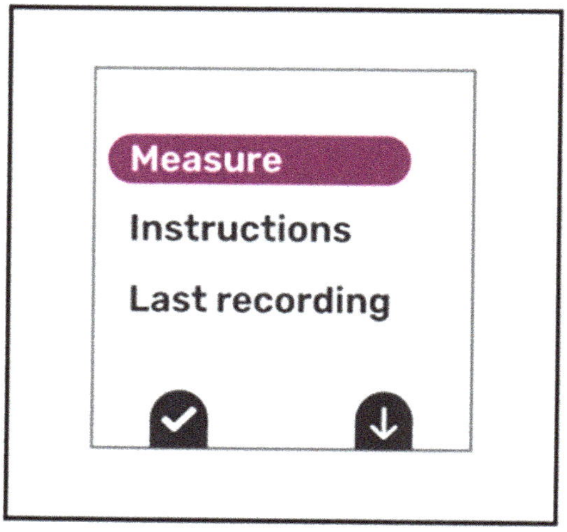

Fig. 4 The operating menu

3.3 Operating Menu

From the operating menu window different operations can be performed:

1. *Measure* = Starting a new measurement.
   (i) During this mode, a low regular noise can be perceived by the user, indicating the correct functioning of the pump.
   (ii) Upon placing the measurement probe on the location of interest, the tissue will be drawn into the measurement probe through the creation of a vacuum. This is indicated on the screen by the "Measuring…" symbol and a colored animation.

(iii) Upon completion of the measurement cycle the output parameters "A" and "B" will be displayed on the screen. The output parameters describe the mechanical behavior of the tested material and are reported in mbar.

- The "A" value describes the *closing pressure* and is related to the stiffness of the material.
- The "B" value describes the *opening pressure* and is related to the elasticity of the material.

(iv) In case of an unsuccessful measurement, the screen will show an error message asking the user to repeat the measurement.

2. *Instructions* = this option of the menu list displays all the steps needed to start a new measurement.

   (i) The user will be guided through the process of mounting a new disposable measurement probe on the handheld.

   (ii) Finally, the user will be asked to prepare the measurement location of interest and by confirming the all-set window the user is led back to the main menu.

3. *Last recording* = the device stores the previous successful measurement.

   (i) To display the parameters "A" and "B" determined in the last measurement, the user can navigate to the operation "Last recording" in the main menu.

   (ii) In case of no previous measurement, the screen will show the information "No last measurement".

4. *Shut down* = the device can be put into stand-by mode during usage.

   (i) This allows turning off the screen when no measurement needs to be performed.

   (ii) To put the device in the stand-by mode, the user needs to navigate to the last operation in the main menu "Shut down" and confirm.

   (iii) To restart the device again, the user can press any of the two control buttons.

3.4 Measurement with Nimble

1. Before a measurement can be performed a disposable measurement probe needs to be mounted on the handheld (*see* **Note 2**).
2. The user then navigates the main menu to the operation "Measure" to start the procedure of a measurement (*see* **Notes 3–5**).
3. The pump will start creating the vacuum (a typical noise of the working pump can be noticed).

4. The measurement probe needs now to be placed gently on the measurement location of interest. When sealing between the probe and the material of interest is achieved, an indication ("Sealed") appears on the screen and the measurement starts.
5. If required, multiple measurements can be performed on one subject with a single probe, minding the indication given above on the maximum number of measurements allowed with a single probe (*see* **Notes 6** and 7).
6. If the same surface location is measured multiple times, a minimum of 30 s should be waited in between measurements.
7. After completing the measurement, the disposable probe needs to be unmounted and trashed according to the local regulations.
8. At the end of a successful measurement, the device shows two values on the screen: the "A" value and the "B" value.
9. The two values need to be written down by hand.

3.5 Determining Measurement Example

Müller et al. [9] provided an example of how to determine the measuring points with the following steps:

1. Identify the specific area of skin to be measured.
2. Divide the scar into longitudinal and transverse axes.
3. Mark the intersection of these axes as Point 1.
4. Designate Point 2 to the left of Point 1 and Point 3 to the right of Point 1.
5. For healthy skin, choose an anatomically comparable site (Fig. 5).

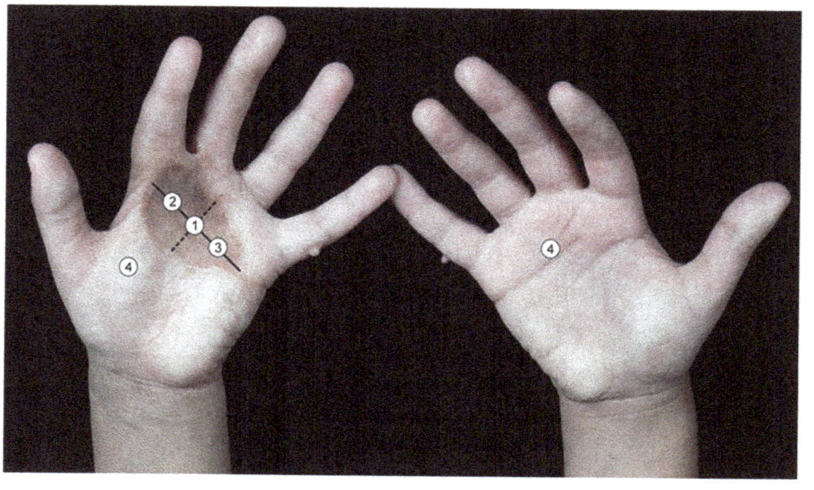

Fig. 5 Examples of possible measuring points as described in [9]: For scars, measure along both the longitudinal and transverse axes. For healthy skin, select an anatomically comparable site

3.6 Interpreting the Data After Measurement

1. The values are calculated during the measurement procedure and depend on the mechanical properties of the tested material.
2. During a measurement, air is drawn from the measurement probe cavity, creating different pressure levels, which lead to a deformation of the tested material.
3. The device records the pressure curves. As soon as a prescribed elevation of the tested material is reached, the pump flow is reversed, and the tested materials is pushed back into its reference configuration.
4. During this cycle, loading and unloading of material, the two output parameters are calculated.
5. The value "A" is related to the material stiffness, and represents the pressure that is needed to elevate the tested material to the prescribed elevation height.
6. The value "B" is related to the time-dependent behavior of the material and represents a measure for the elasticity of the material—the ability to return to its reference configuration.

4 Notes

1. In preparation of a measurement, the material surface under test should be in a stable position and possibly fixed. The user of the device should also be in a stable position. It is advised, when possible, that the elbow of the user is in a stable position on a rigid support, e.g., on a table.
2. A new disposable probe should be used for one single measurement.
3. During the measurement, the user should not move and hold still.
4. Further, the user should not push the measurement probe against the tested material in order not to alter the parameter determination.
5. The force applied during the measurement should be constant and as small as possible.
6. A maximum of 15 measurements can be performed with a single probe on the same material, with negligible influence on the extracted material parameters.
7. A new, clean disposable for every measurement set performed on a different material/substrate under test is suggested, to prevent cross-contamination between different material surfaces. In particular, when the device is used to measure skin of human subjects, it is not allowed to use one probe on different subjects.

Conflict of Interest

Author Bettina Thumm is coinventor of the suction device used in this chapter, further described in the patent: Aspiration device for determining mechanical properties of soft tissues and methods therefore, EP4074248A1.
Authors David Sachs, Alberto Stracuzzi, Serjosha Robmann, and Bettina Thumm are currently working on the commercialization of the suction device used in this chapter.

References

1. Muller B, Elrod J, Pensalfini M et al (2018) A novel ultra-light suction device for mechanical characterization of skin. PLoS One 13(8): e0201440
2. Elrod J, Muller B, Mohr C et al (2019) An effective procedure for skin stiffness measurement to improve Paediatric Burn Care. Burns 45(5):1102–1111
3. Muller B, Ruby L, Jordan S et al (2020) Validation of the suction device Nimble for the assessment of skin fibrosis in systemic sclerosis. Arthritis Res Ther 22(1):128
4. Sachs D, Jakob R, Thumm B et al (2024) Sustained physiological stretch induces abdominal skin growth in pregnancy. Ann Biomed Eng 52(6):1576–1590
5. Mueller B, Elrod J, Distler O et al (2021) On the reliability of suction measurements for skin characterization. J Biomech Eng 143(2):021002
6. Aoki T, Ohashi T, Matsumoto T et al (1997) The pipette aspiration applied to the local stiffness measurement of soft tissues. Ann Biomed Eng 25:581–587
7. Nava A, Mazza E, Kleinermann F et al (2003) Determination of the mechanical properties of soft human tissues through aspiration experiments. In: Ellis RE, Peters TM (eds) Medical image computing and computer-assisted intervention – MICCAI 2003, MICCAI 2003. Lecture notes in computer science, vol 2878. Springer, Berlin/Heidelberg
8. Buffinton CM, Tong KJ, Blaho RA et al (2015) Comparison of mechanical testing methods for biomaterials: pipette aspiration, nanoindentation, and macroscale testing. J Mech Behav Biomed Mater 51:367–379
9. Muller B, Mazza E, Schiestl C et al (2021) Longitudinal monitoring and prediction of long-term outcome of scar stiffness on pediatric patients. Burns. Trauma 9:tkab028

Part IV

Skin Tissue Engineering for Clinical Application

Chapter 24

GMP Compliant Production of Therapeutic Components of Autologous Adipose Tissue

Mustafa Uguten, Joeri van Boxtel, Hieronymus P. Stevens, Martin C. Harmsen, and Joris A. van Dongen

Abstract

Adipose tissue is a popular source of tissue for cellular therapy in the field of regenerative medicine. The regenerative potential is often ascribed to the presence of stromal vascular fraction (SVF) containing extracellular matrix and multipotent stromal cells secreting a plethora of growth factors to create a regenerative environment. SVF can be isolated by means of enzymatic or mechanical isolation procedures and expanded in culture or directly used intraoperatively. Depending on the clinical use of SVF, specific regulatory requirements are demanded and might classify SVF as an advanced therapy medicinal product (ATMP). As an ATMP, SVF must be manufactured, processed, and controlled according to good manufacturing practice (GMP) guidelines to ensure safety and quality. Subsequently, the GMP standards require extensive validation, process control, and characterization of SVF. Here we report a GMP-compliant production of clinical grade tissue (tSVF) by means of fractionation of adipose tissue (FAT) procedure. Previous validation studies demonstrated tSVF to be safe and feasible for clinical use intraoperatively according to GMP standards with the appropriate release criteria. The presented procedures can be used as a template for the development of an investigational medicinal product dossier to be enclosed in future clinical trials (Fig. 1).

Key words GMP, Autologous adipose tissue, SVF, Advanced therapy medicinal product, ATMP

1 Introduction

Human subcutaneous adipose tissue has become a popular source of tissue for cellular therapy in the field of regenerative medicine [1]. Adipose tissue has the advantage to be relatively easy and repeatably accessible with a minimal invasive liposuction procedure yielding a large number of isolated cells. [2] The regenerative potential of adipose tissue can be primarily ascribed to the presence of stromal vascular fraction (SVF) containing adipose tissue-derived stromal cells (ASCs), fibroblasts, vascular cells, immune cells as well as extracellular matrix (ECM) and ECM-bound trophic factors [3–8]. These multipotent stromal cells along with the ECM are able to

Thomas Biedermann and Sophie Böttcher-Haberzeth (eds.), *Skin Tissue Engineering: Methods and Protocols*, Methods in Molecular Biology, vol. 2922, https://doi.org/10.1007/978-1-0716-4510-9_24,

bind and release paracrine factors and cytokines to dictate cell fate, e.g., migration, differentiation, proliferation, or apoptosis and create a regenerative environment [6, 9, 10].

SVF can be isolated by means of enzymatic or mechanical isolation procedures resulting in cellular like SVF (cSVF) or tissue SVF (tSVF) respectively. The classical procedure to generate cSVF involves digestion of lipoaspirate with enzymes, which result in a single cell suspension with disrupted intercellular connections and the destruction of the ECM [2, 11]. Consequently, enzymes like collagenase might alter the natural state of cells. Hence, these cells are extensively manipulated which calls for stricter regulations in many countries. In contrast, mechanical isolation procedures use various shuffling techniques with either centrifugation, filtration, or disruption with blades to isolate tSVF. In tSVF, ECM and most intercellular connections are intact and tissue integrity is preserved. tSVF is therefore classified as minimally manipulated [12, 13].

In clinical practice, mechanical isolation procedures are faster, easier, and less costly than enzymatic isolation procedures making them more suitable for intraoperative use. However, certain clinical indications might require cSVF such as intravascular use of SVF cells to prevent embolus formation. In case of intra-articular use of SVF cells, the ECM should be preserved to maintain the interaction between cells and growth factors to increase the regenerative potential of tSVF [14].

To use cSVF or tSVF as therapy in clinical trials, both the isolation procedure as well as the end product, i.e. cSVF or tSVF, should meet specific regulatory requirements. Two main criteria are used according to the U.S. Food and Drug Administration (FDA) as well as the European Medicines Agency (EMA) [15]. The use of human cells, tissues, and cellular and tissue-based products is defined in two categories: "minimally manipulated" and "extensively manipulated" products [15]. If the tissues original relevant characteristics relating to its utility for reconstruction, repair, or replacement remain unchanged, it is generally considered minimally manipulated. All other products are classified as extensively manipulated. The distinction between minimal and extensive manipulation is pivotal in the regulatory landscape in cell therapy. The second criteria states if donor cells or tissues perform one or more of the same basic functions as the recipients cells, it is considered homologous use and thus a lower risk product. For instance, if tSVF is used for wound healing purposes or as anti-scarring treatment, it is considered homologous use because tSVF is injected into subcutaneous tissue. If tSVF is used to treat osteoarthritis, it is considered non-homologous use because tSVF is injected into an area with no subcutaneous tissue. In case of non-homologues use or extensive manipulation, stricter drug regulations are applied and SVF is classified as an advanced therapy medicinal product (ATMP). As an ATMP, SVF must be manufactured, processed, and controlled according to good manufacturing practice (GMP) guidelines to ensure safety and quality [16, 17]. Consequently, the GMP standards require extensive validation, process control, and characterization of SVF.

Here we report a GMP-compliant production of clinical grade tSVF by means of fractionation of adipose tissue (FAT) procedure [12]. Previous validation studies demonstrated tSVF features (viability, phenotype, and function) to be safe and feasible for clinical use intraoperatively according to GMP standards with appropriate release criteria. The presented procedures can be used as a template for the development of investigational medicinal product dossier to be enclosed in future clinical trials.

2 Materials

2.1 Location

1. Operating room with laminar air flow.

2.2 Liposuction of Adipose Tissue (Fig. 1)

1. Human subcutaneous liposuction derived adipose tissue.
2. 3 mL luer lock syringe.
3. 25 G needle.
4. 2% Lidocaine.
5. Scalpel 11.
6. Spiral multiple-hole infiltration cannula (Arthrex, Inc., Naples, Italy).
7. 50 or 20 mL luer lock syringes to infiltrate.
8. Modified Klein's solution (per 500 mL of saline, 30 mL of xylocaine 2% (epinephrine 1:200000) and 3 mL of natriumbicarbonate 8.4%.
9. 3-port harvesting cannula (Arthrex, Inc., Naples, Italy).
10. VacLok 30 mL syringe (Arthrex, Inc., Naples, Italy).
11. Arthrex Autologous Conditioned Plasma (ACP)® Double-Syringe System (Arthrex, Inc., Naples, Italy).

2.3 Fractionation of Adipose Tissue Procedure (Fig. 2)

1. Arthrex ACP® Double-Syringe System (Arthrex, Inc., Naples, Italy).
2. Centrifuge with the capability to go to 960× *g* (Hettich Rotofix 32, benchtop, swing out rotor).
3. Luer to luer lock connector transparent transfer adapter with one hole of 1.4 mm inside (Arthrex, Inc., Naples, Italy).
4. Luer to luer lock connector transparent transfer adapter with one hole of 2.4 mm inside (Arthrex, Inc., Naples, Italy).
5. Regular luer lock 10 mL syringe.

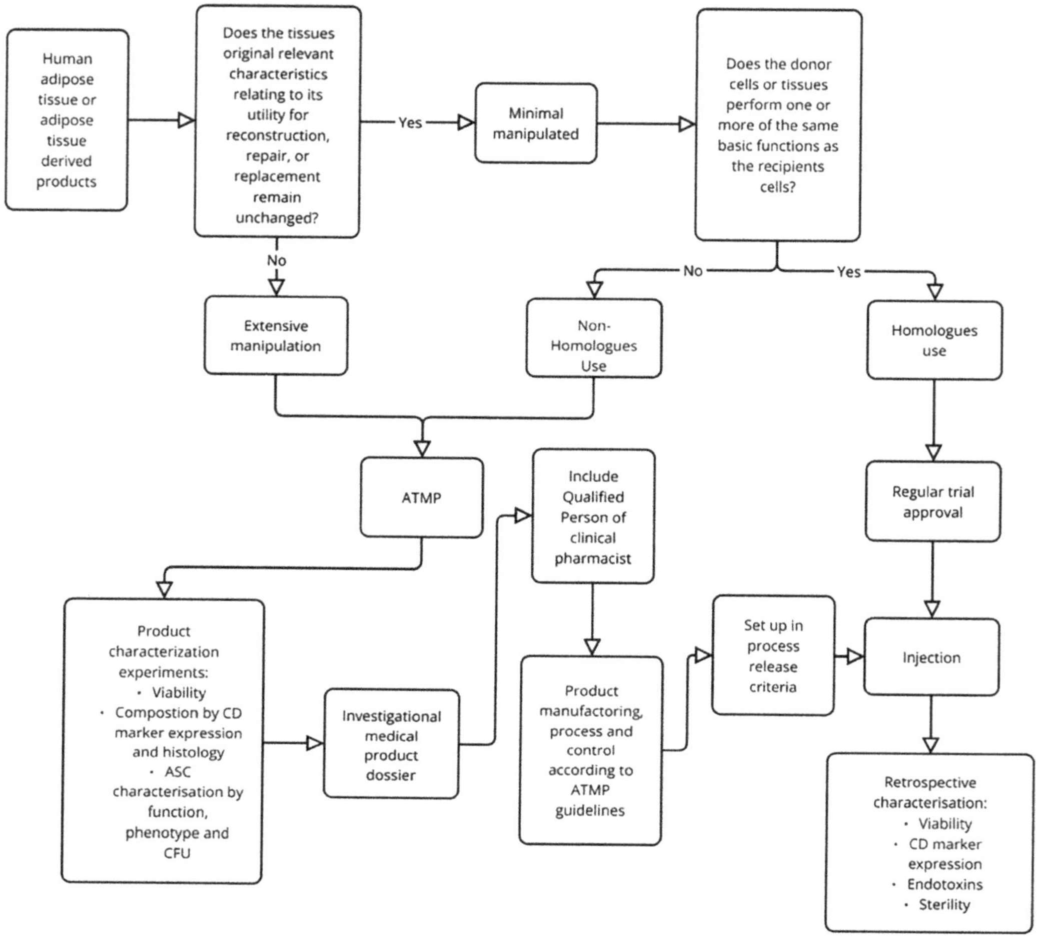

Fig. 1 Flow diagram demonstrating how human adipose products apply as an advanced therapy medicinal product and how the manufacturing and validation process from harvest till injection is controlled

2.4 Immuno-histological Staining of tSVF

1. 10% formalin.
2. 1.8% agarose solution.
3. Xylol.
4. 100%, 96%, 70% alcohol.
5. 0.1 M Tris/hydrochloric acid (HCl) buffer pH 9.0.
6. 10 mM Tris/1 mM ethylenediaminetetraacetic acid (EDTA) buffer pH 9.0.
7. 3% hydrogen peroxidase/phosphate buffered saline (PBS).
8. α-smooth muscle actin (SMA, Abcam #ab7871).
9. Perilipin A (Abcam #ab3526).
10. Von Willebrandfactor (vWF, Dako #A0082).

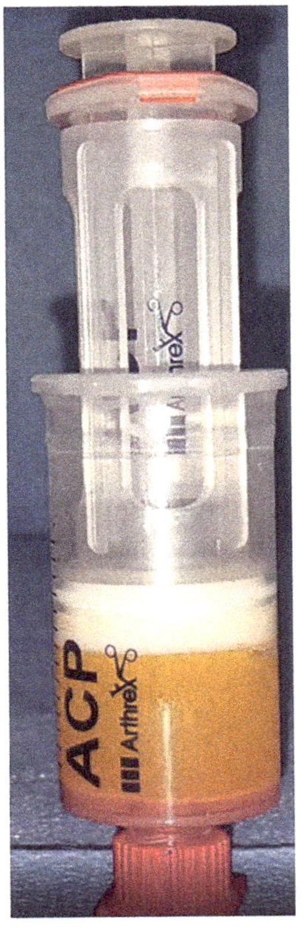

Fig. 2 ACP double syringe closed system (ACP-kit, Arthrex GmbH, USA). The system includes an inner syringe and outer syringe (red plunger)

11. 1% Bovine serum albumin (BSA).
12. 1% Human serum.
13. 1% Swine serum.
14. Polyclonal Rabbit anti-Mouse.
15. Polyclonal Goat anti-Rabbit.
16. Polyclonal Swine anti-Rabbit.
17. 3,3′-Diaminobenzidine.
18. Hematoxylin.
19. Mounting solution.

2.5 Viability of tSVF

1. Caspase-3 (1:100) + 1% BSA + 1% Human serum.
2. 10% formalin.
3. 1.8% agarose solution.
4. Xylol.
5. 100%, 96%, 70% alcohol.

6. 3% hydrogen peroxidase/phosphate buffered saline (PBS).
7. 1% Bovine serum albumin (BSA).
8. 1% Human serum.
9. Polyclonal Goat anti-Rabbit.
10. 3,3′-Diaminobenzidine.
11. Hematoxylin.
12. Mounting solution.

2.6 Masson's Trichrome Staining of tSVF

1. 4% paraformaldehyde (PFA).
2. 1% triton X-100.
3. Bouin fixative:
   (a) 36 mL picric acid (saturated).
   (b) 12 mL 37% formalin.
   (c) 2 mL acetic acid (glacial).
4. Weighert's iron hematoxylin:
   (a) 25 mL stock solution A.
   (b) 25 mL stock solution B.
   (c) Stock solution A:
      (i) 0.5 g hematoxylin.
      (ii) 50 mL 100% alcohol.
   (d) Stock solution B:
      (i) 0.6 g iron chloride.
      (ii) 49.5 mL demiwater.
      (iii) 0.5 mL hydrochloric acid.
5. Biebriech scarlet-acid fuchsine:
   (a) 0.5 g Biebriech scarlet.
   (b) 0.5 g acid fuchsine.
   (c) 49.5 mL demiwater.
   (d) 0.5 mL acetic acid (glacial).
6. Phosphomolybdic-phosphotungstic acid:
   (a) 2.5 g phosphomolybdic acid.
   (b) 2.5 g phosphotungstic acid.
   (c) 50 mL demiwwater.
7. Aniline Blue:
   (a) 1.3 g aniline blue.
   (b) 1 mL acetic acid (glacial).
   (c) 49 mL demiwater.

8. 1% Acetic acid:
   (a) 0.5 mL of acetic acid (glacial).
   (b) 49.5 mL of demiwater.

2.7 Analysis of Histology and Histochemical Stainings

1. Fiji image processing package with ImageJ (public domain).
2. NDP.view Image viewing software.
3. Microsoft Excel spreadsheet program.

2.8 Enzymatic Isolation of Single Cells Derived from tSVF to Characterize Cellular Composition

1. 0.1% bacterial collagenase A solution:
   (a) 50 mg bacterial collagenase A.
   (b) 50 mL PBS/1% BSA.
2. 100 μm filters.
3. Lymphoprep.
4. Lysis buffer.
5. FACS buffer (PBS/0.5% BSA).

2.9 Endotoxin Testing

1. Limulus Amebocyte Lysate (LAL) Reagent (Kinetic-QCL® or PYROGENT® 5000 Reagent).
2. Control Standard Endotoxin (CSE).
3. LAL reconstitution buffer (required for PYROGENT® 5000 Reagent kinetic Turbidimetric LAL assay).
4. LAL reagent water (LRW) (#W50-W640, W50-100, W50-500).
5. Glass dilution tubed (#207).
6. Serological pipettes.
7. 96-well plates (#25-340).
8. Reagent reservoirs (#00190035).
9. Timer.
10. WinKQCL® software.
11. Vortex mixer.

2.10 Sterility Testing

1. BacTec® culture system (BACTEC™ LYTIC Anaerobic/F Culture Vials).
2. Alcohol.

3 Methods

Full development of tSVF-based investigational medical product protocol should contain several product features and acceptable release criteria (Fig. 1). This includes in process release criteria and retrospective analysis of the administered sample (endotoxin determination, sterility testing, cellular composition determination by histology and CD marker expression and indirect viability testing using caspase staining). A Qualified Person (clinical pharmacist) verifies the in process release criteria and determines if a sample can be administrated.

3.1 Liposuction of Adipose Tissue (See Note 1)

1. Infiltrate incision area with 1 mL of lidocaine 2% using the 3 mL luer lock syringe +25 G needle.
2. Make a small stab incision of 5 mm on the donor site (preferably legs or abdomen).
3. Infiltrate the donor site with a modified Klein's solution using the spiral multiple-hole infiltration cannula.
4. Harvest lipoaspirate with the 3-port harvesting cannula and a VacLok 30 mL syringe. Pull the plunger backwards to allow for negative pressure and lock to maintain the negative pressure.
5. Start harvesting by moving the harvesting cannula forward and backwards.
6. Replace the full VacLok 30 mL with a new one and start again from **step 4**.

3.2 Fractionation of Adipose Tissue Procedure

1. Transfer the lipoaspirate to the disposable Arthrex ACP Double-Syringe System from the VacLok 30 mL syringe.
2. Centrifuge at 960× *g* for 4 min. at room temperature (Hettich Rotofix 32, benchtop, swing out rotor, Tuttlingen Germany).
3. After centrifugation, the upper oily fraction is removed by pulling the inner syringe.
4. The lower aqueous fraction is discarded by opening the red closing cap. Remove oil (disrupted adipocytes) by pulling up the small syringe system in an upright position.
5. Connect the small syringes to a 2.4 mm transfer adapter and fill a regular 10 mL syringe with 10 mL of centrifuged adipose tissue.
6. Connect the 1.4 mm transfer adapter between an empty 10 mL syringe and a 10 mL syringe with centrifuged adipose tissue. Transfer the fat between the syringes at least 30×. Push the adipose tissue extensively forward and backwards.
7. Repeat **step 5** and transfer the fractionated fat from the regular 10 mL syringe to the ACP Double-Syringe System.
8. Centrifuge at 960× *g* for 4 min at room temperature.
9. Transfer the oil into the smaller syringe by slowly pushing down on the outer syringe and simultaneously pulling up the plunger of the inner syringe.
10. Remove the cap to remove the small pellet layer at the bottom.

11. tSVF is left in the outer syringe.
12. The abovementioned procedure will take place simultaneously in three separate ACP Double -Syringe System to make three times around 1 ml of tSVF.
13. All three 1 mL samples of tSVF are pooled to one sample of 3 mL of tSVF and then again divided in three samples of 1 mL of tSVF. This process is done in a closed system with 10 mL syringes and a luer to luer connector.
14. 1 mL of tSVF will be subjected to sterility testing for conditional release (Subheadings 3.3, 3.9, and 3.10)
15. 1 mL of tSVF will be formalin fixed for control of the product afterwards (Subheading 3.4)
16. 1 mL of tSVF will be used for treatment.

3.3 In-Process Quality Control (Release Criteria)

1. The volume of tSVF should be visually controlled, which should range between 0.75 mL and 1.75 mL when 10 mL centrifuged adipose tissue is used as starting material. If the product is out of range, the FAT procedure was inadequate and should be repeated (*see* **Note 2**).
2. The color of the end product should be controlled and should be light pink (Fig. 3).

3.4 Immunohistological Staining of tSVF

1. Fix the isolated tSVF in 10% formalin overnight at 4 °C.
2. Embed the tSVF in a pre-heated (60 °C) 1.8% agarose solution (1:2).
3. Place the tSVF/agarose solution at 4 °C for 30 min.
4. Dehydrate samples with the following steps in sequence at room temperature:
   (a) 50% alcohol for 30 min.
   (b) 70% alcohol for 30 min.
   (c) 96% alcohol for 30 min.
   (d) Twice in 100% alcohol for 30 min.
   (e) Twice in Xylol for 30 min.
5. Embed the samples in paraffin.
6. Cut 4 mm sections and deparaffinize them in xylol for 15 min.
7. Refresh the xylol and place the samples for another 10 min in xylol.
8. Move the samples and place them in 100% alcohol for 10 min, then in 96% alcohol for 3 min and finally in 70% alcohol for 3 min at room temperature.
9. Wash the samples in demi water for 3 min.

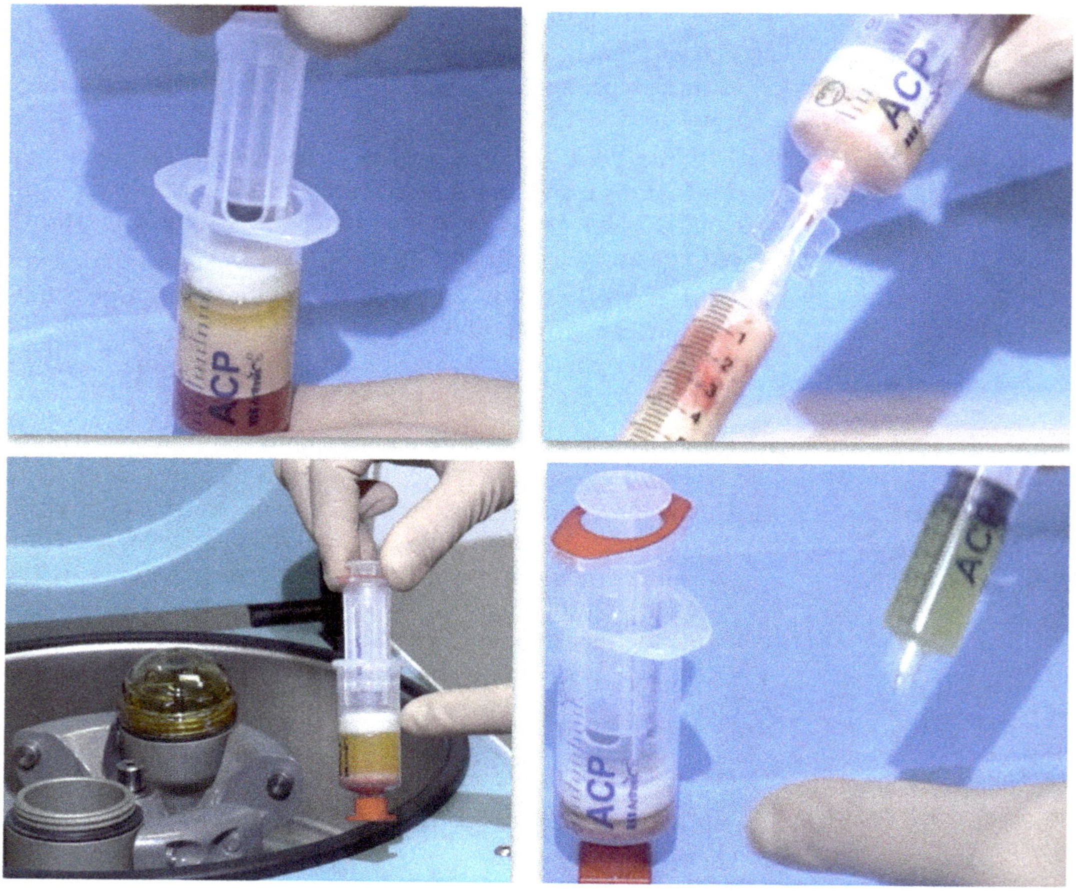

Fig. 3 The pink layer containing stromal vascular fraction with an oily fraction on top

10. Incubate samples overnight with 0.1 M Tris/HCl buffer (pH 9.0) at 80 °C for α-SMA as well as for Perilipin A staining and with 10 mM Tris/1 mM EDTA buffer (pH 9.0) at 80 °C for vWF.
11. Cool down to room temperature for 30 min.
12. Wash samples with PBS three times.
13. Endogenous peroxidase activity is blocked with 3% hydrogen peroxidase in PBS at room temperature for 30 min.
14. Wash samples with PBS three times and incubate samples with primary antibodies at room temperature for 60 min.: for α-SMA (1:200) + 1% BSA + 1% Human serum in PBS, for Perilipin A (1:200) + 1% BSA + 1% Human serum in PBS, for vWF (1:200) + 1%BSA + 1% Swine serum in PBS.
15. Wash samples with PBS three times.

16. Incubate samples with secondary antibodies for 30 min at room temperature: polyclonal Rabbit anti-Mouse (1: 100) + 1% BSA + 1% Human serum in PBS for α-SMA, polyclonal Goat anti-Rabbit (1:100) + 1% BSA + 1% Human serum in PBS for Perilipin A and polyclonal Swine anti-Rabbit (1: 100) + 1% BSA + 1% Human serum in PBS for vWF.
17. Wash samples with PBS three times.
18. Incubate α-SMA sample with a third antibody at room temperature for 30 min with polyclonal Swine anti-Rabbit (1: 100) + 1% BSA + 1% Human serum in PBS.
19. Wash α-SMA samples with PBS three times.
20. Incubate all samples with DAB for 10 min at room temperature in the dark.
21. Wash all samples in demi water three times for 5 min.
22. Incubate all samples with hematoxylin for 1.5 min.
23. Rinse samples in water for 4 min.
24. Mount all the samples and place a coverslip (Subheading 3.7).

3.5 Viability of tSVF

1. Follow **step 1** till **9** of Subheading 3.4.
2. Incubate samples with primary antibodies at room temperature for 60 min, for cleaved Caspase-3 (1:100) + 1% BSA + 1% Human serum in PBS.
3. Wash samples with PBS three times.
4. Incubate samples with secondary antibodies for 30 min at room temperature: polyclonal Goat anti-Rabbit (1:100) + 1% BSA + 1% Human serum in PBS.
5. Wash samples with PBS three times.
6. Incubate all samples with DAB for 10 min at room temperature in the dark.
7. Wash all samples in demi water three times for 5 min.
8. Incubate all samples with hematoxylin for 1.5 min.
9. Rinse samples in water for 4 min.
10. Mount all the samples and place a coverslip.

3.6 Masson's Trichrome Staining of tSVF

1. Follow **step 1** till **9** of Subheading 3.4.
2. Fix samples in 4% PFA for 60 min.
3. Permeabilize samples in 1% triton X-100 for 10 min.
4. Wash samples with PBS three times.
5. Fix samples in Bouins fixative at 51 °C for 5 min.
6. Wash samples in demiwater until color disappears.
7. Incubate samples with Weighert's iron hematoxylin for 20 min.

8. Wash samples in demiwater until color disappears.
9. Incubate samples with Beibrich Scarlet-acid fuchsine for 20 min.
10. Wash samples in demiwater for 2 min.
11. Incubate samples in phosphomolybdic-phosphotungstic acid for 12 min.
12. Wash samples in demiwater until color disappears.
13. Incubate samples in aniline blue for 7 min.
14. Wash samples in demiwater for 2 min.
15. Incubate samples in 1% acetic acid for 5 s.
16. Wash samples in demiwater until color disappears.
17. Dry samples for 20 min.
18. Mount samples in Permount.

3.7 Analysis of Histology and Histochemical Stainings

1. Divide all the different microscopic images of different patients in different folders.
2. Make an excel file for every different staining (e.g., VWF, αSMA).
3. Open a patient in NDP.view (drag the patient file in the viewer).
4. Scroll and zoom around for a representative selection of cells (i.e., a sample) and safe it as a JPEG (three times per patient to rule out a selection bias).
5. Drag the JPEG files to ImageJ.
6. Use "polygon selections" in ImageJ to cut out your sample.
7. Select the RGB stack for the α-SMA, Masson's Trichome, Perilipin A, vWF staining and Caspase-3 staining (Table 1).

3.8 Enzymatic Isolation of Single Cells Derived from tSVF to Characterize Cellular Composition

1. Wash the isolated tSVF with PBS three times.
2. Add 0.1% collagenase A solution 1:1 with tSVF.
3. Stir the collagenase/tSVF mixture in a water bath at 37 °C for 1.5 h.
4. Centrifuge the sample at 600× *g* at room temperature for 10 min.
5. Remove supernatant.
6. Collect cell pellet in PBS/1% BSA.
7. Filter the collagenase/tSVF mixture through filters.
8. Centrifuge the sample at 600× *g* at room temperature for 10 min.
9. Repeat **step 6**, **7**, and **8**.
10. Remove supernatant.

Table 1
Scales of the stained immunohistochemical samples and CD marker analysis. Retrospective composition control of tSVF by means of histology and CD marker expression after administration. Scales of the stained immunohistochemical and CD marker analysis are set by previous extensive characterization and test runs

Test	Method	Exploratory results
Quantification of the number of adipocytes, inflammatory cell counts and blood vessels.	Immuno-histological staining SVF on formalin fixed paraffin sections: perilipin, a-SMA and von Willebrand factor.	Perilipin A: range 0–2000 adipocytes VWF: range 0–10% α-SMA: range 0–25%
Quantification of the relative amount of proteoglycans and extracellular matrix.	Histochemical staining SVF on formalin fixed paraffin sections: Masson's Trichrome.	Masson's trichrome: range 5–50%
Determination of the number of viable cells in SVF.	Histological staining SVF on formalin fixed paraffin sections: caspase-3.	Caspase-3: negative
Determination of composition of cells within SVF	CD marker analysis based on flow-cytometry	Mesenchymal stroma cells range 0–60% Endothelial cells 0–70% Supra-adventitial cells 0–15%

11. Collect cell pellet in 30 mL of PBS/1% BSA.
12. Put 15 mL of lymphoprep in a 50 mL tube.
13. Gently add the 30 mL of PBS/1% BSA with the sample.
14. Centrifuge the sample at 1000× *g* at 4 °C for 20 min. and put the brake on 0.
15. Remove the upper layer.
16. Take cells from the interphase carefully.
17. Add PBS/1% BSA to the cells.
18. Centrifuge the sample at 800× *g* at 8 °C for 10 min.
19. Remove supernatant.
20. Resuspend cell pellet in lysis buffer and place the sample on ice for 5 min.
21. Centrifuge the sample at 800× *g* at 8 °C for 10 min.
22. Remove supernatant.

23. Repeat **step 19** and **23** till all erythrocytes are disrupted and the red color has disappeared.
24. Resuspend cell pellet in FACS buffer and divide cells in multiple tubes. The number of tubes depends on the number of subsets of CD markers used. Additionally, one tube will function as a blank control and for each fluorophore used, an extra tube will be used for the fluorophore specific isotype control (*see* **Note 3**).
25. Centrifuge cells at 300× *g* at 4 °C for 5 min.
26. Resuspend cell pellet in 100 μL FACS buffer.
27. Keep one tube of cells unlabeled and put on ice in the dark for 30 min. This tube functions as a blank control to set up the flow cytometer.
28. Incubate a tube of cells on ice with 5 μL of the preferred fluorophore-conjugated antibodies (1:20) in the dark for 30 min (Table 1 and *see* **Note 4**).
29. Incubate the other tubes of cells on ice with 5 μL of the specific fluorophore isotype control (1:20) in the dark for 30 min.
30. Wash samples with 2 mL FACS buffer.
31. Centrifuge cells at 300× *g* at 4 °C for 5 min.
32. Remove supernatant.
33. Repeat **step 10** till **12** three times.
34. Resuspend cell pellet in 300 μL FACS buffer.
35. Proceed to FACS analysis.

3.9 Endotoxin Testing

1. Reconstitute Control Standard endotoxin (CSE) with LAL reagent water (LWR) to yield a solution containing 50 EU/mL or 100 EU/mL depending on assay method.
2. Vortex for 15 min.
3. Label the tubes with the appropriate endotoxin concentration and add 0.9 mL of LWR to each.
4. Prepare a series of endotoxin standards.
5. Dispense 100 μL of the LRW blank, endotoxin standards, product samples, positive controls into the appropriate wells of the microplate.
6. Pre-incubate the plate for 10 min at 37 °C in the microplate reader.
7. Immediately prior to use, reconstitute LAL and gently swirl.
8. Pour LAL into reagent reservoir and mix gently.
9. Use an eight cannal pipettor to dispense 100 μL of LAL into the appropriate wells of the microplate.
10. Initiate the test by clicking the OK button in the WinKQCL® software (*see* **Note 5**).

3.10 Sterility Testing

1. Remove the flip-off cap from the BD BACTEC vial top and inspect the vial for cracks, contamination, excessive cloudiness in medium and bulging or indented septum.
2. Swab the septum with alcohol.
3. Aseptically inject or draw directly 8–10 mL of specimen per vial.
4. Inoculated aerobic vials should be placed in the BD BACTEC fluorescent series instrument as soon as possible for incubation and monitoring. If placement of an inoculated vial into the instrument has been delayed and visible growth is apparent, it should not be tested in the BD BACTEC fluorescent series instrument, but should rather be subcultured, Gram-stained and treated as a presumptively positive vial.
5. Vials entered into the instrument will be automatically tested every ten minutes for the duration of the testing protocol period of 7 days.
6. Positive vials will be determined by the BD BACTEC fluorescent series instrument and identified by the sensor inside the bottle (*see* **Note 6**).

4 Notes

1. Exclude patients with known articular systemic disease (e.g., rheumatoid arthritis), infectious arthritis of joints, all immunocompromised patients or patients with active prednisone use or immunotherapy. Exclude all patients with known diabetes mellitus type 1 and 2, HIV and hepatitis B or C due to potential differences in the quality of adipose tissue. Perform no liposuction in an area with a possible skin infection.
2. Around 90% of the volume of adipose tissue are adipocytes and 10% is SVF, although this is variable between humans. During the procedure, the volume of the end product should be visually checked, which should range between 0.75 and 1.75 mL when 10 mL of centrifuged adipose tissue is used as starting material. This means a volume reduction of almost ten times the starting volume should be expected. In case the amount of SVF is below or above the abovementioned range, the fractionation is inadequate and the FAT procedure should be performed again with fresh lipoaspirate (**step 1** of the FAT procedure). The range between 0.75 and 1.75 mL is determined after extensive preliminary research and takes interdonor variability into account [12].
3. For a complete characterization of the tSVF, flow cytometry analysis of enzymatic isolated tSVF is advised. Each tube contains all the CD markers used to analyze the preferred cell type

Table 2
Cell composition of tSVF based on CD marker expression. CD45 is used to distinguish between two large populations in tSVF: adipose derived population (CD45^{neg}) and blood derived population (CD45^{pos})

CD markers	Cell population
CD45−; CD90+; CD105+	Mesenchymal stromal cell
CD31+; CD34+	Endothelial cell
CD45+; CD34−	Leukocyte
CD34+/−; CD31−; CD146+	Pericyte
CD45+; CD34+	Hematopoietic stem cell-like cell
CD34bright; CD31−, CD146−	Supra-adventitial cell

present in the tSVF (regardless of a positive or negative expression of the CD marker on the surface of the cell type). It is possible to analyze multiple cell types with the same CD markers in one tube. The maximum number of CD markers used in a single tube depends on the chosen fluorophore (e.g., allophycocyanin (APC) and fluorescein isothiocyanate (FITC)). Each CD marker used in a single tube should contain a different fluorophore.

4. Different types of fluorophore-conjugated antibodies can be combined and used to analyze different cell types in tSVF (Table 2). Table 2 contains a recommended set of CD markers to analyze the most important adipose derived cell types in tSVF [2]. However, there is no consensus regarding the correct subset of CD markers for each cell type [2, 18–23].

5. This test is performed as described in European Pharmacopoeia 2.6.27 (Microbiological control of cellular products) and European Pharmacopoeia 5.1.6 (Alternative methods for control of microbiological quality). We will use BACTECTM LYTIC Anaerobic/F Culture Vials.

6. According to EP 5.1.10, the maximum acceptable endotoxin level is calculated. The dose of tSVF is 1 mL. The endotoxin limit = K/M, where: K = maximum endotoxin dose based on the patient's bodyweight. EP 5.1.10 recommends 5.0 EU/kg bodyweight for intravenous preparations, which results in 5.0×70 (average weight of patients in this study) = 350 EU per SVF Drug Substance administration. M: maximum product dose = 1 mL in this example. The endotoxin limit is thus 350/1 = 350 EU/mL. Based on these considerations, an endotoxin limit of ≤5 EU/mL for the tSVF Drug substance is justified.

References

1. Katz AJ, Llull R, Hedrick MH et al (1999) Emerging approaches to the tissue engineering of fat. Clin Plast Surg 26(4):587–603, viii
2. van Dongen JA, Tuin AJ, Spiekman M et al (2018) Comparison of intraoperative procedures for isolation of clinical grade stromal vascular fraction for regenerative purposes: a systematic review. J Tissue Eng Regen Med 12(1):e261–ee74
3. Ergün S, Tilki D, Klein D (2011) Vascular wall as a reservoir for different types of stem and progenitor cells. Antioxid Redox Signal 15(4): 981–995
4. Zannettino AC, Paton S, Arthur A et al (2008) Multipotential human adipose-derived stromal stem cells exhibit a perivascular phenotype in vitro and in vivo. J Cell Physiol 214(2): 413–421
5. Bourin P, Bunnell BA, Casteilla L et al (2013) Stromal cells from the adipose tissue-derived stromal vascular fraction and culture expanded adipose tissue-derived stromal/stem cells: a joint statement of the International Federation for Adipose Therapeutics and Science (IFATS) and the International Society for Cellular Therapy (ISCT). Cytotherapy 15(6):641–648
6. Eto H, Suga H, Matsumoto D et al (2009) Characterization of structure and cellular components of aspirated and excised adipose tissue. Plast Reconstr Surg 124(4):1087–1097
7. Yue B (2014) Biology of the extracellular matrix: an overview. J Glaucoma 23(8 Suppl 1):S20–S23
8. Briquez PS, Hubbell JA, Martino MM (2015) Extracellular matrix-inspired growth factor delivery systems for skin wound healing. Adv Wound Care (New Rochelle) 4(8):479–489
9. Zuk PA, Zhu M, Ashjian P et al (2002) Human adipose tissue is a source of multipotent stem cells. Mol Biol Cell 13(12):4279–4295
10. Spiekman M, van Dongen JA, Willemsen JC et al (2017) The power of fat and its adipose-derived stromal cells: emerging concepts for fibrotic scar treatment. J Tissue Eng Regen Med 11(11):3220–3235. https://doi.org/10.1002/term.2213
11. van Dongen JA, Stevens HP, Harmsen MC et al (2017) Mechanical micronization of lipoaspirates: squeeze and emulsification techniques. Plast Reconstr Surg 139(6):1369e–1370e
12. van Dongen JA, Stevens HP, Parvizi M et al (2016) The fractionation of adipose tissue procedure to obtain stromal vascular fractions for regenerative purposes. Wound Repair Regen 24(6):994–1003
13. van Dongen JA, Harmsen MC, Stevens HP (2019) Isolation of stromal vascular fraction by fractionation of adipose tissue. Methods Mol Biol 1993:91–103
14. Stevens HP, van Boxtel J, van Dijck R et al (2020) Platelet rich STROMA, the combination of PRP and tSVF and its potential effect on osteoarthritis of the knee. Appl Sci 10(14): Article 4691. https://doi.org/10.3390/app10144691
15. Regulatory Considerations for Human Cells, Tissues, and Cellular and Tissue-Based Products: Minimal Manipulation and Homologous Use. Guidance for Industry and Food and Drug Administration Staff. U.-S. Department of Health and Human Services Food and Drug Administration Center for Biologics Evaluation and Research Center for Devices and Radiological Health Office of Combination Products. July 2020
16. Gimble JM, Guilak F, Bunnell BA (2010) Clinical and preclinical translation of cell-based therapies using adipose tissue-derived cells. Stem Cell Res Ther 1(2):19–27
17. Fekete N, Rojewski MT, Fürst D et al (2012) GMP-compliant isolation and large-scale expansion of bone marrow-derived MSC. PLoS One 7(8):e43255
18. Lin G, Garcia M, Ning H et al (2008) Defining stem and progenitor cells within adipose tissue. Stem Cells Dev 17(6):1053–1063
19. Corselli M, Chen CW, Sun B et al (2012) The tunica adventitia of human arteries and veins as a source of mesenchymal stem cells. Stem Cells Dev 21(8):1299–1308
20. van Boxtel J, Vonk LA, Stevens HP et al (2022) Mechanically derived tissue stromal vascular fraction acts anti-inflammatory on TNF alpha-stimulated chondrocytes in vitro. Bioengineering (Basel) 9(8):345
21. Corselli M, Crisan M, Murray IR et al (2013) Identification of perivascular mesenchymal stromal/stem cells by flow cytometry. Cytometry A 83(8):714–720
22. Zimmerlin L, Donnenberg VS, Pfeifer ME et al (2010) Stromal vascular progenitors in adult human adipose tissue. Cytometry A 77(1): 22–30
23. Traktuev DO, Merfeld-Clauss S, Li J et al (2008) A population of multipotent CD34-positive adipose stromal cells share pericyte and mesenchymal surface markers, reside in a periendothelial location, and stabilize endothelial networks. Circ Res 102(1):77–85

Chapter 25

Quality Controls of Autologous Skin Products According to European GMP Guidelines

Rocio García de la Cruz Valencia, Gloria Carmona Sánchez, Céline Auxenfans, Jean-François Brunet, Ana Fernández González, Vivienne Woodtli, Alexandra Karström, and Marina Trouillas

Abstract

The production of Cultured Epithelial Autografts (CEAs), autologous keratinocyte-sprays, and tissue-engineered skin substitutes in hospital-based laboratories is considered an Advanced Therapy Medicinal Product (ATMP). The production of ATMPs in Europe has to follow GMP (Good Manufacturing Practice) conditions and is regulated by Regulation 1394/2007 and Commission Directive 2009/120/EC. European regulation establishes the need of carrying out a complex process of characterization and quality controls during production, in order to release these products for clinical use. They are carried out at each critical stage of production: reception of the skin biopsy, extraction of keratinocytes and/or fibroblasts, media change, cell growth, cell passaging, and on the finished products. Here, we describe the production for autologous skin products at each critical stage: harvesting and biopsy reception, isolation of keratinocytes and fibroblasts, cell culture, and preparation of the final product.

Key words Advanced therapy medicinal products, Good manufacturing practice, Skin substitutes, Quality controls, EU guidelines

1 Introduction

Hospital-based laboratories in Europe that produce Cultured Epithelial Autografts (CEAs) (Fig. 1), autologous keratinocyte sprays, and tissue-engineered skin substitutes (Fig. 2) must adhere to Good Manufacturing Practice (GMP) conditions. European regulations mandate a rigorous process of characterization and quality control throughout production to ensure these products meet the necessary standards for clinical use. Each batch of these products requires approval from the relevant National Medicines Agency, with most agencies following the European Medicines Agency (EMA) guidelines. These guidelines address both GMP protocols and product-specific requirements, particularly the ones performed

Thomas Biedermann and Sophie Böttcher-Haberzeth (eds.), *Skin Tissue Engineering: Methods and Protocols*, Methods in Molecular Biology, vol. 2922, https://doi.org/10.1007/978-1-0716-4510-9_25,

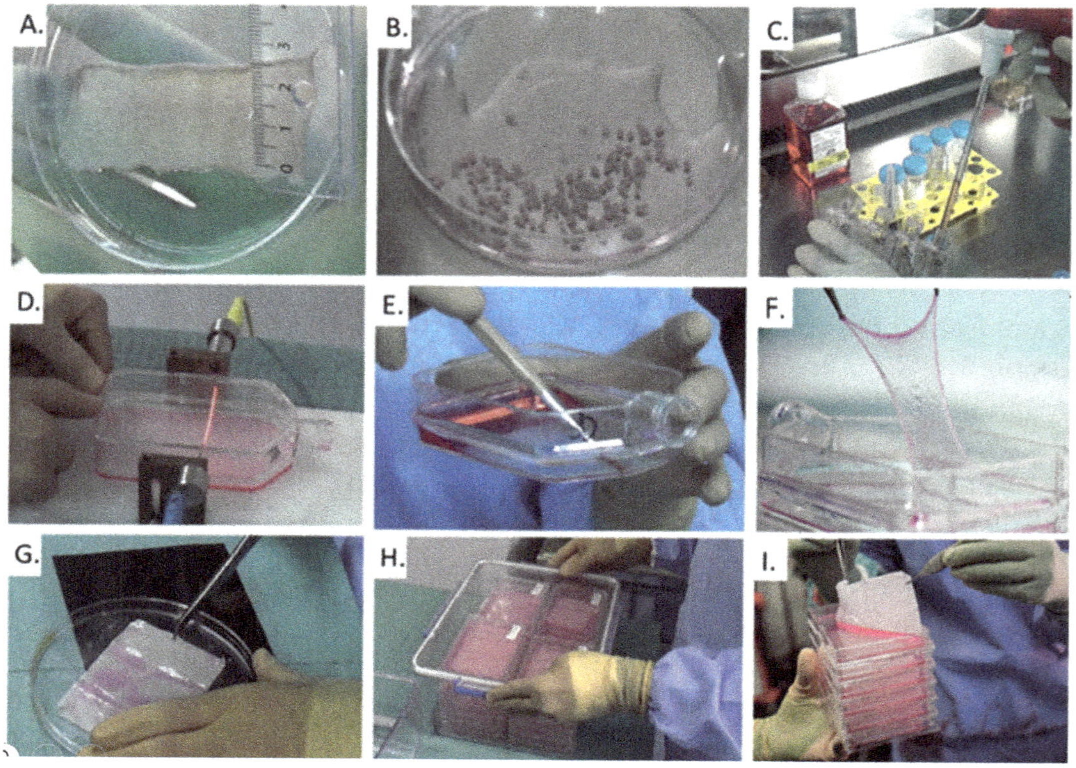

Fig. 1 Graphical illustrations of various steps in manufacturing cultured epithelial autografts (CEA). (**a**) A split-thickness skin biopsy of approx. 10 cm^2 is procured. (**b**) Skin fragments are minced and keratinocytes enzymatically released and subsequently seeded in immortalized 3 T3-mouse fibroblasts in culture flasks. (**c**) Cells are expanded in culture, enzymatically harvested, and re-seeded in new culture flasks to expand cell number further. (**d**) Eventually, keratinocyte culture is driven to stratified sheets that are enzymatically released and harvested from the culture flasks by removing the lid. (**e** and **f**) Cell sheets are carefully removed from the culture vessels. (**g**) Cell sheets are transferred on to vaseline gauze and subsequently put in transport boxes containing a small quantity of culture medium. (**h**) Transport boxes are put in a hermetically sealed box (as secondary container) for shipment to patient. (**i**) CEAs are applied to revised and debrided wounds in the operating theater [17, 18]

during preclinical analyses, and quality controls. In centers across Europe—such as Uppsala, Lausanne, Lyon, Paris, and Granada—the quality control procedures for keratinocyte-based therapies are consistent. These procedures occur at every critical stage of production, including biopsy reception, keratinocyte and fibroblast extraction, media changes, cell growth, and the final product assessment.

Advanced Therapy Medicinal Products (ATMPs), including cell therapies, gene therapies, and tissue-engineered products, represent a novel and complex class of treatments. Their regulatory landscape, governed by Regulation 1394/2007 [1] and Commission Directive 2009/120/EC [2], is continuously evolving,

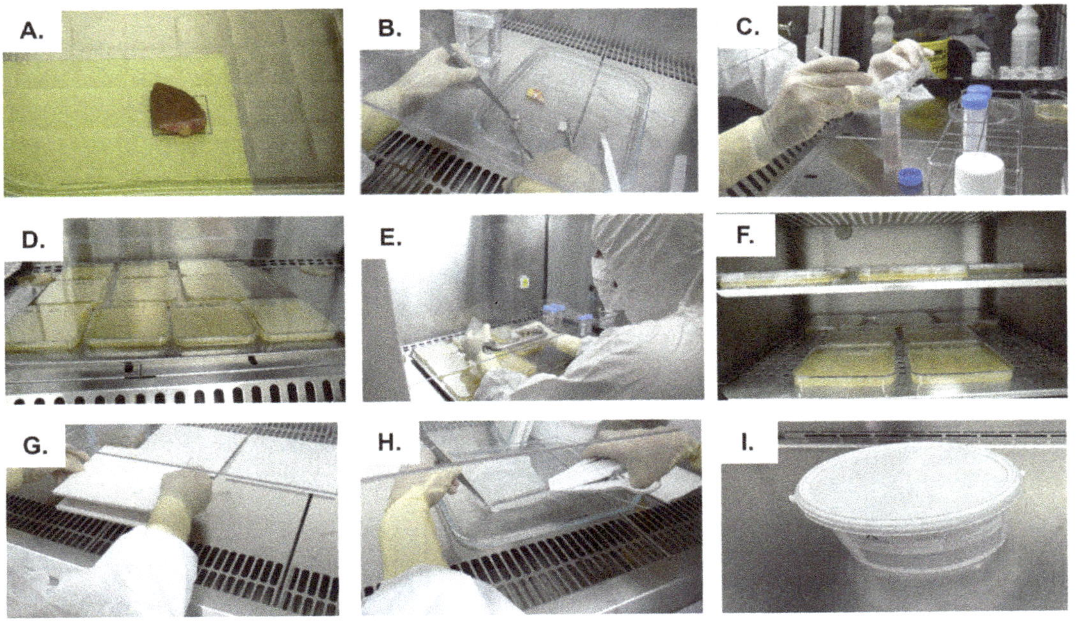

Fig. 2 Graphical illustrations of various steps in manufacturing Bioengineered Autologous Skin Substitutes. (**a**) A skin biopsy of approx. 9 cm^2. (**b**) Mechanical separation of dermis and epidermis. (**c**) Epidermis enzymatic digestion for keratinocytes isolation and subsequently seeded in immortalized 3 T3-mouse fibroblasts in culture flasks. Dermis enzymatic digestion for fibroblasts isolation. Cells are expanded in culture about 3–4 weeks. (**d**) Fibrin-agarose substitutes with human dermal fibroblasts. (**e**) Seeding of keratinocytes onto fibrin-agarose substitutes. (**f**) Maintenance for one week of Bioengineered Autologous Skin Substitutes. (**g**) Pressure dehydration process to reduce the Bioengineered Autologous Skin Substitute thickness and facilitate their implantation and handling in the operating theater. (**h**) Bioengineered Autologous Skin Substitute conditioning. (**i**) Bioengineered Autologous Skin Substitute are put in a hermetically sealed transport container

presenting challenges for both researchers and regulators. For the treatment of burn patients, these products are classified as tissue-engineered products because their mechanism of action is to regenerate, repair, or replace damaged human tissue. In fact, the definition that it is considered according to Regulation 1394/2007 for tissue engineered product is as follows:

Tissue engineered product means a product that:

Contains or consists of engineered cells or tissues, and is presented as having properties for, or is used in or administered to human beings with a view to regenerating, repairing or replacing a human tissue.

A tissue engineered product may contain cells or tissues of human or animal origin, or both. The cells or tissues may be viable or non-viable. It may also contain additional substances, such as cellular products, bio-molecules, biomaterials, chemical substances, scaffolds or matrices.

Products containing or consisting exclusively of non-viable human or animal cells and/or tissues, which do not contain any viable cells or tissues and which do not act principally by pharmacological, immunological or metabolic action, shall be excluded from this definition.

In this chapter, we outline the quality control processes that are specific to these types of products, taking into consideration their classification and intended therapeutic use. However, it is important to notice that in different GMP laboratories in Europe, the strategies producing and controlling autologous artificial skin substitutes are different. All these methods are described in detail in the local SOPs (Standard Operation Procedures) as part of their quality management system, have been designed to comply with EMA regulations and the quality control measures during the production and the release procedure have been authorized by the respective local authorities. It is essential to emphasize that, particularly in critical emergencies, involving autologous skin cells for burn patients, a thorough risk/benefit analysis must be conducted. This assessment will guide the identification of the most critical tests to perform, tailored to the specific conditions and production workflow of each case. Therefore, we show here a collection of equivalent methods that are currently applied for the quality controls of the autologous skin substitutes in Europe, that do not exclude the use of other methods as long as they comply with the EMA regulations and the local guidelines.

2 Materials

2.1 Quality Control of the Biopsy During Harvesting and Reception

1. Disinfectants.
2. Sterile scalpel or dermatome for harvesting skin biopsies.
3. Serologic tests: Closed system for blood sampling.
4. Transport medium.
5. Documentation defining the specifications required for the biopsy.

2.2 Quality Controls During Isolation of Keratinocytes and/or Fibroblasts

1. Local SOPs defining the procedures for cell isolation.
2. Transport medium for biopsy storage and transport as well as cell culture.
3. Cell Culture Material required for the production processes according to local SOPs.
4. Material to test cell viability according to local SOPs: Trypan Blue, Propidium Iodide (PI).
5. Cell counting device.
6. Cell culture microscope.

2.3 Quality Controls During Media Change and Cell Culture

1. Local SOPs defining the procedures for changing cell culture media and splitting cells.
2. Documentation defining cell morphology and staining procedures for flow cytometry.

3. Human CD90 antibody.
4. Human HMB45 antibody.
5. Antifeeder+ (Feeder cells antibody, Clone mEF-SK4).
6. Flow cytometer.

2.4 Quality Controls During Preparation and Release of the Final Product

1. Cell culture microscope.
2. Flow cytometer.
3. Material to test cell viability according to local SOPs: Trypan Blue, Propidium Iodide.
4. Cell counting device.
5. Human CD90 antibody.
6. Human HMB45 antibody.
7. Antifeeder+ (Feeder cells antibody, Clone mEF-SK4).
8. Human CD49f antibody.
9. FDA: fluorescein diacetate.
10. Documentation defining release criteria of final product:
    (a) Material defined in the European Pharmacopoeia to perform the following tests: Sterility (2.6.1), Ph. Eu 04/2011: 20601 corrected 7.7; Microbiological examination of Cell-Based preparations (2.6.27), Ph. Eu 07/2024: 20627; Microbiological examination of non-sterile products: test for specified microorganisms (2.6.13) Ph. Eu. 01/2021: 20613; Microbiological quality of non-sterile pharmaceutical preparations and substances (5.1.4), Ph. Eu. 01/2021: 50104; Bacterial endotoxins (2.6.14), Ph. Eu. 01/2018:20614 corrected 11.0; Guidelines for using the test for bacterial endotoxins (5.1.10), Ph. Eu. 01/2021:50110; Test for bacterial endotoxins using recombinant factor C (2.6.32), Ph. Eu., 01/2021: 20632 corrected 11.0; Mycoplasmas (2.6.7), Ph. Eu. 01/2008:20607 corrected 6.1.; Nucleated cell count and viability (2.7.29), Ph. Eu 01/2024:20729; Flow cytometry (2.7.24), Ph. Eu. 01/2008:20724.
    (b) Or external organizations to perform these tests according to EMA regulations.

3 Methods

The methods described here are equivalent methods that are used in different GMP laboratories in Europe for the purpose of producing autologous artificial skin substitutes to be transplanted to patients. All these methods are described in detail in the local

SOPs (Standard Operation Procedures), have been designed to comply with EMA regulations and the quality control measures during the production and release procedure, and have been authorized by the respective local authorities.

3.1 Quality Control of the Biopsy During Harvesting and Reception (See Note 1)

1. Perform serological tests on blood samples from the donor to identify potential infectious diseases that can be carried by the sample according to European directives.
2. Obtain the biopsy by scalpel excision or a dermatome preferably from a location that corresponds to an unburned area, as described in the local SOPs.
3. Control that the biopsy location is indicated in the documents associated with the biopsy.
4. Control that the time between the harvesting of the biopsy and its processing complies with the specifications defined in the local SOPs.
5. Check that the macroscopic appearance biopsy complies with the specifications defined in the local SOPs.

3.2 Quality Controls During Isolation of Keratinocytes and/or Fibroblasts (See Note 2)

1. Isolate keratinocytes and/or fibroblasts defined in the local SOPs.
2. Check the extraction yield and cell viability after isolation of keratinocytes and/or fibroblasts using the trypan blue [3] or PI method [4].
3. Control that the extraction yield and cell viability comply with the specifications defined in the local SOPs.

3.3 Quality Controls During Media Change and Cell Culture (See Note 3)

1. Check the morphology of squamous cells in the keratinocyte culture with clonal growth at each media change [5].
2. Verify the absence of contamination in the keratinocyte culture by other cell types according to:
   (a) Cell morphology: identifying proliferative spindle cells and melanocytes, seeing as star-shaped dendritic cells.
   (b) Flow cytometry: identify other cell types contaminating the keratinocyte culture by using CD90+ as a marker for dermal fibroblasts [6], Antifeeder+ as marker for feeder cells [3] and HMB45+ as marker for melanocytes.
3. Evaluate cell growth and stemness by calculating the population doubling [7] and/or performing a colony forming unit test [8, 9].

3.4 Quality Controls During Preparation and Release of the Final Product (See Notes 3 and 4)

1. Ensure safety of the final product by:
   - (a) Verify the absence of contamination by other cell types according to:
     - (i) Cell morphology: identifying proliferative spindle cells and melanocytes, as star-shaped dendritic cells.
     - (ii) Flow cytometry: identify other cell types contaminating the keratinocyte culture by using CD90+ as a marker for dermal fibroblasts [6], Antifeeder+ as marker for feeder cells [3] and HMB45+ as marker for melanocytes. Proliferative keratinocyte can be specifically identified using CD49f [3, 10].
   - (b) Test microbiological contamination of the culture media:
     - (i) Monitor the macroscopic appearance of the culture medium, which must remain clear at each media change and at the final stage of culture.
     - (ii) Assess microbiological contamination using methods described in the European Pharmacopoeia: Sterility (2.6.1), Ph. Eu. 04/2021: 20601; Microbiological examination of Cell-Based preparations (2.6.27), Ph. Eu 07/2024: 20627; Microbiological examination of non-sterile products: test for specified micro-organisms (2.6.13) Ph. Eu., 01/2021: 20613 corrected 11.0; Microbiological quality of non-sterile pharmaceutical preparations and substances (5.1.4) Ph. Eu., 01/2021: 50104.
   - (c) Analyze the presence of endotoxins in the culture media using the methods described in the European Pharmacopoeia: Bacterial endotoxins (2.6.14), Ph. Eu., 01/2023: 20614; Guidelines for using the test for bacterial endotoxins (5.1.10), Ph. Eu., 01/2021: 50110 (01/2021); Test for bacterial endotoxins using recombinant factor C (2.6.32), Ph. Eu., 01/2021:20632.
   - (d) Test the presence of mycoplasma in the culture media using PCR or luminescence as described in the European Pharmacopoeia: Mycoplasmas (2.6.7), Ph. Eu., 01/2008: 20607.
   - (e) Cell count and viability according to European Pharmacopoeia: Nucleated cell count and viability (2.7.29), Ph. Eu 01/2024:20729.
   - (f) Discard genome instability of the cultured cells by analyzing the karyotype and DNA fingerprints to ensure the absence of chromosomal alterations and chimerism [11].
   - (g) Follow up patients according to clinical criteria and state of the art practice.

2. Test quality of finished products by:
   (a) Assess cell culture purity by morphologic appearance or phenotype characterization according to Ph. Eu. 01/2008:20724.
   (b) Evaluate epidermal sheet integrity, cell stratification and viability using pheu Nucleated cell count and viability (2.7.29), Ph. Eu 01/2024:20729 and macroscopic appearance of finished products [11, 16].

4 Notes

1. Quality control of the biopsy during harvesting and reception: Upon receipt, the biopsy macroscopic appearance is checked, and the surface specification varies between 1 and 10 cm^2, the preferred location of the sample being an unburned area. This location is indicated in the documents associated with the biopsy. Serological tests are performed on blood samples from the donor. The time between the biopsy harvesting and its processing must be between 2 h (when the biopsy is transported dry) to 48 h (in a nutrient medium). In the majority of cases (Lyon, CTS, Uppsala, Granada), the biopsy is obtained by scalpel excision and in Lausanne, with a dermatome.
2. Quality controls during isolation of keratinocytes and/or fibroblasts: For extraction of keratinocytes/fibroblast, quality controls are the cell yield and the viability of the cells evaluated with trypan blue method [3]. At this stage, the viability obtained is informative for certain centers, or must be greater than 90% for others. When viability is not assessed with trypan blue, it can be assessed with PI. The extraction yield remains informative because it varies greatly from one patient to another.
3. Quality controls during media change and cell culture: Throughout the cell culture process, it is essential to routinely monitor both cell morphology and identity to ensure the purity and quality of the culture. During each media change, squamous cell morphology, including clonal growth patterns, is carefully assessed. Potential contamination by other cell types, such as proliferative spindle-shaped cells and melanocytes (recognized by their star-shaped dendritic morphology), is actively checked. Cell growth and stemness are measured through population doubling (Lausanne) or colony-forming unit (CFU) assays (Lyon, Paris) [8, 9]. Assessing cell identity is critical at various stages of culture, including both intermediate and finished products. Proliferative keratinocyte markers, such as CD49f, are used to verify keratinocyte identity in some GMP facilities like in Lyon, Paris, and Granada [3, 10]. Additionally,

potential contamination by non-keratinocyte cell types is monitored using markers such as CD90+ in Lyon and Paris [6]. (Lyon, Paris) or antifeeder markers in Granada [3]. Dermal fibroblasts and HMB45+ melanocytes can also be identified using flow cytometry and immunohistology, ensuring that these cells do not compromise culture integrity. In Granada, karyotype analysis and DNA fingerprinting are further employed to confirm the absence of chromosomal abnormalities and chimerism [11].

However, ensuring accurate cell identity throughout the entire process requires careful selection of markers. While markers like CD90 and CD49f are used to identify fibroblasts and keratinocytes in some GMP facilities, they may lack the specificity needed to reliably distinguish these cell types. CD90, for instance, can also be expressed by stem cells, and CD49f is not exclusively specific to keratinocytes. Undoubtedly, finding an ideal marker for fibroblasts and keratinocytes is challenging, but it is crucial to explore more specific molecules. Cytokeratin is a specific marker of the epithelial lineage and could improve specificity in keratinocyte identification during identity tests. In contrast, selecting an ideal fibroblast marker is more challenging, but vimentin is a promising candidate that could help enhance the accuracy of fibroblast identification. Ultimately, selecting the most appropriate markers is essential for precise cell identity evaluation and maintaining the quality and purity of the culture.

The absence of microbiological contamination is controlled at all stages by analyzing culture supernatants or transport media. In all centers, the controls are done on blood culture bottles. This absence of microbiological contamination is also checked by monitoring the macroscopic appearance of the culture medium, which must remain clear orange-red, at each change of medium. A control of the absence of mycoplasmas is carried out routinely (by PCR in Lausanne and Granada and by luminescence in Lyon and Paris) [11–13]. This search for contaminants is supplemented by the search for endotoxins (Lausanne, Uppsala, Paris and Granada) [11, 12, 14, 15], eubacteria and hominis pneumoniae (Lausanne). Impurities resulting from the use of certain products during the process are evaluated in Granada unit.

4. Quality controls during preparation and release of the final product: Epidermal sheets integrity, cell stratification of finished products and viability are evaluated in Lausanne, Paris, Uppsala, and Granada [11, 16]. These quality controls make it possible to ensure both the safety of transplanted patients (absence of contamination, absence of cellular contamination, and patient follow-up) and the quality of the finished products

(cellular purity assessed by the microscopic appearance of the cells or the characterization of their phenotype, maintenance of the proliferative potential, integrity of the finished product).

References

1. Regulation (EC) No 1394/2007 of the European Parliament and of the Council of 13 November 2007 on advanced therapy medicinal products and amending Directive 2001/83/EC and Regulation (EC) No 726/2004
2. Commission Directive 2009/120/EC of 14 September 2009 amending Directive 2001/83/EC of the European Parliament and of the Council on the Community code relating to medicinal products for human use as regards advanced therapy medicinal products
3. Fernández González A, Lizana Moreno AM, de Pablos Ramos M et al (2016) Optimization of human keratinocyte culture to develop an artificial human skin model: cell alternatives as feeder layer of advanced therapies. Actual Medica 101:85–94. https://doi.org/10.15568/am.2016.798.or04
4. Crowley LC, Scott AP, Marfell BJ et al (2016) Measuring cell death by propidium iodide uptake and flow cytometry. Cold Spring Harb Protoc 2016(7). https://doi.org/10.1101/pdb.prot087163
5. Barrandon Y, Green H (1985) Cell size as a determinant of the clone-forming ability of human keratinocytes. Proc Natl Acad Sci USA 82(16):5390–5394. https://doi.org/10.1073/pnas.82.16.5390
6. Kisselbach L, Merges M, Bossie A et. al. (2009) CD90 expression on human primary cells and elimination of contaminating fibroblasts from cell cultures. Cytotechnology 59(1): 31–44. doi: https://doi.org/10.1007/s10616-009-9190-3
7. Benyumovich MS (1966) Determination of the time of population doubling in monolayer cultures. Bull Exp Biol Med 62:840–843. https://doi.org/10.1007/BF00805288
8. Auxenfans C, Colloud M, Debard AL et al (2006) Use of allogenic epidermal sheets for difficult wound healing: selection and testing of relevant growth factors. Biomed Mater Eng 16(4 Suppl):S73–S83
9. Alexaline MM, Trouillas M, Nivet M et al (2015) Bioengineering a human plasma-based epidermal substitute with efficient grafting capacity and high content in clonogenic cells. Stem Cells Transl Med 4(6):643–654. https://doi.org/10.5966/sctm.2014-0155
10. Schlüter H, Paquet-Fifield S, Gangatirkar P et al (2011) Functional characterization of quiescent keratinocyte stem cells and their progeny reveals a hierarchical organization in human skin epidermis. Stem Cells 29(8): 1256–1268. https://doi.org/10.1002/stem.675
11. Fernández-González A, Lizana-Moreno A, Fernández S et al (2017) Clinical, histological and homeostasis evaluation of an autologous tissue bio-engineered skin substitute in a patient with 70% of total body surface area (TBSA) burn. Cytotherapy 19:S233. https://doi.org/10.1016/j.jcyt.2017.02.339
12. Egea-Guerrero JJ, Carmona G, Correa E et al (2019) Transplant of tissue-engineered artificial autologous human skin in Andalusia: an example of coordination and institutional collaboration. Transplant Proc 51:3047–3050. https://doi.org/10.1016/j.transproceed.2019.08.014
13. Ph. Eur. (2.6.7.) Mycoplasmas 01/2008: 20614 corrected 6.1
14. Ph. Eur. 11.5 (2.6.14) Bacterial endotoxins 01/2008: 20614 corrected 11.0
15. Ph. Eur. 11.5 (5.1.10) Guidelines for using the test for bacterial endotoxins 01/2021: 50110
16. Martin-Piedra MA, Carmona G, Campos F et al (2023) Histological assessment of nanostructured fibrin-agarose skin substitutes grafted in burnt patients. A time-course study. Bioeng Transl Med 8(6):e10572. https://doi.org/10.1002/btm2.10572
17. Chemali M, Laurent A, Scaletta C et al (2021) Burn center organization and cellular therapy integration: managing risks and costs. J Burn Care Res 42:911–924. https://doi.org/10.1093/jbcr/irab080
18. Chemali M, Laurent A, Scaletta C et al (2021) Corrigendum to: burn center organization and cellular therapy integration: managing risks and costs. J Burn Care Res 42:1064. https://doi.org/10.1093/jbcr/irab130

Chapter 26

Standardized Outcome Measures for the Clinical Application of Tissue Engineered Products

Anna S. van den Bosch, Melinda Farkas, Frederique M. Kemme, Matthea M. Stoop, Paul P. M. van Zuijlen, Annebeth Meij-de Vries, Esther Middelkoop, and Anouk Pijpe

Abstract

The clinical application of tissue engineered products aims to regenerate skin and enhance the appearance, texture, sensation, and functionality of affected skin or scars. To evaluate the efficacy and effectiveness of novel products, level 1 randomized controlled trials are required. However, currently a significant challenge in this field is the clinical and statistical heterogeneity of trial designs and outcome measures for tissue engineered products. To address this challenge and improve clinical outcomes, it is essential to standardize high-quality outcome measures. These measures are clinimetrically developed tools designed to reliably and validly assess specific health outcomes across diverse patient populations, clinical settings, and studies. This chapter proposes a core set of outcome measurements designed to evaluate the most relevant, feasible, reliable, and valid parameters in the clinical application of tissue engineered products. For routine clinical practice, we recommend the inclusion of photography, graft take, re-epithelialization, and scar quality assessment using the Patient and Observer Scar Assessment Scale. For clinical trials, additional measures such as colorimetry and elasticity assessment are advised. Detailed protocols for these methods are provided, resulting in a standardized core outcome set. High-quality outcome assessment also necessitates specialized training and the presence of trained personal at all assessment points. The chapter further addresses specific considerations for outcome assessment in pediatric patients, remote assessment strategies, and additional recommendations for optimizing clinical and research practices in this field. The implementation of standardized outcome measurement is essential for improving the outcomes of patients treated with tissue engineered products.

Key words Tissue engineering, Skin regeneration, Outcome assessment, Scar quality, Clinical trials

1 Introduction

The clinical application of tissue engineered products aims to regenerate the skin and/or improve the affected skin or scar. The goal is for the regenerated skin to function, look, sense and feel as natural as possible, matching the qualities of unaffected skin. Most professionals in the global scientific/medical community recognize

Thomas Biedermann and Sophie Böttcher-Haberzeth (eds.), *Skin Tissue Engineering: Methods and Protocols*, Methods in Molecular Biology, vol. 2922, https://doi.org/10.1007/978-1-0716-4510-9_26,

substantial evidence supporting the effectiveness of tissue engineered products [1]. However, only two-third of the professionals who have experience using these products considered the body of evidence sufficient, and the majority emphasized the absence of clear indications and treatment protocols. An equally important challenge in this field, as in many medical specialties, is the extensive heterogeneity in the nature and design of clinical studies on tissue engineered products. For the ultimate goal of improving outcomes in the clinical application of tissue engineered products, future studies should prioritize intra-patient study designs and greater uniformity in high-quality outcome measures. Intra-patient study designs, where the same patient serves as both the experimental and control subject, are beneficial as it reduces variability, requires smaller sample sizes, and improves statistical efficiency. Particularly in tissue engineering, intra-patient study designs allow for precise comparisons within the same biological environment, leading to more reliable and accurate outcomes.

Standardized outcome measures are systematically developed tools or instruments used to assess and quantify specific health outcomes in a consistent and reliable manner across different patients, clinical settings, and studies. Standard outcome measures are essential for evaluating the effectiveness of healthcare interventions, tracking patient progress, and for comparing results across different studies or patient populations. This comparability is critical for meta-analyses and systematic reviews, which aggregate findings to provide stronger evidence on treatment efficacy and other clinical questions, such as the safety of interventions, the long-term effects of treatments, and patient satisfaction. Clinicians can also use these measures to benchmark their patients' outcomes against established norms or clinical guidelines, helping to identify areas needing improvement or intervention.

We distinguish the following types of outcome measures:

- Patient-Reported Outcome Measures (PROMs): these are questionnaires or surveys completed by patients to assess their own health status, quality of life, or the impact of a disease or treatment.
- Clinician-Reported Outcome Measures (CROMs): these are assessments made by healthcare providers based on their observations and clinical judgment.
- Biometric and Physiological Measures (BPM): these include standardized methods for measuring physical or biological indicators such as blood pressure, cholesterol levels, and imaging results.

The quality and effectiveness of outcome measures rely on a combination of validity, reliability, responsiveness, feasibility, interpretability, relevance, standardization, and generalizability.

It is essential to develop and validate clinical measures that accurately assess health outcomes. We refer interested readers to the COSMIN methodology (https://www.cosmin.nl/) [2]. Also vital for high quality outcome assessment in clinical application of tissue engineered products is specialized training. Training for the assessment of clinical outcomes must be comprehensive, practical, and continuous. It should focus on theoretical knowledge, standardized protocols, hands-on practice, and feedback.

When assessing the outcome of clinically applied tissue engineered products, a large number of relevant parameters exists. Unfortunately, neither in the course of routine clinical treatment, nor in the standardized set-up of a clinical trial is it possible to measure and document all of these many parameters, mostly due to limitations in time and resources. Therefore, in this chapter, we present a recommended core outcome measurement set, which in our opinion, assesses the most relevant, feasible, available, reliable, and valid parameters [3–10]. We acknowledge that some of the presented assessment methods in this core outcome set may not be available to part of the community due to financial reasons. We then advise to select at least the instruments that are available at low or no costs.

2 Materials

2.1 Wound Healing Assessment

1. The expertise of an experienced healthcare professional familiar with skin grafting.
2. The pre-determined time points for the assessment (*see* **Note 1**).

2.2 Photograph

1. Digital camera.
2. Good and uniform lighting conditions.
3. Consistent and suitable background.

2.3 POSAS (Patient and Observer Scar Assessment Scale)

1. Printed version of the Patient Scale.
2. Printed version of the Observer Scale.
3. Pen.

2.4 Colorimeter

1. Colorimeter DSM device (read-out: erythema (E) and melanin (M) indices).
2. Pen.

2.5 Cutometer

1. Cutometer device (weight: 3.9 kg, power supply: external 100–240 VAC, 47–63 Hz, DC 12 V/9A, port: USB).
2. Probe (weight: 165 g incl. air tube, measuring aperture: 6 mm Ø standard, measurement principle: suction (pressure setting

up to 500 mbar), units: millimeter (penetration depth), accuracy: ±3%).

3. Laptop/computer (Windows® XP, Vista or 7 (or newer versions), USB 2.0).
4. Extension cord.
5. Pen.

3 Methods

3.1 Outcome Measures

Recommended non-invasive core outcome set for clinical application of tissue engineered products, in regular clinical setting and/or clinical trial setting (Table 1):

Outcome measures can focus on the patient level (e.g., quality of life) or the scar level (e.g., elasticity). This standardized core outcome set is focused on measurements at the scar level. Outcome measures can be conducted both in clinical trial settings and in clinical/real-world settings. For the daily clinical setting, include at least the following: photography, graft take and re-epithelialization, and scar quality assessment by means of the POSAS. For clinical trials, extend these measures to include colorimetry and elasticity assessment.

3.2 General Measurement Conditions

1. The skin should be cleansed of all products, preferably 2–3 h prior to the measurements.
2. For accurate evaluation of the biophysical skin parameters, an acclimatization time of at least 20–30 min is needed before starting the measurements. Any clothing, dressings or bandages covering the measuring site have to be removed in time, so that there will be no influence of occlusion on the measurements. After physical strain, the patient needs some time to acclimatize, so that blood circulation can return to normal. Emotional stress can lead to more transpiration.
3. Do not perform measurements at high temperatures or at very high ambient humidity (sweating). To get comparable results, temperature and humidity should be approximately the same for all measurements (1–2 °C of difference can be neglected, but differences should not exceed 5 °C). Ideal conditions are a room temperature of 20–22 °C and a relative humidity of 40–60%.
4. The first and follow-up measurements in one patient should ideally be performed at approximately the same time of the day (morning or afternoon).
5. Results are more comparable if the measurements are performed by the same person.

Table 1
Outcome set for clinical application of tissue engineered products

Parameter	Recommended tool/device (including webpage)	Recommended assessment time-points	Time estimate per scar (a) assessment (b) preparation and processing	Costs	Feasible in children?
Scar quality by the patient (PROM)	POSAS 3.0 (www.posas.nl)	3, (6, 9,) 12 months after tissue engineered product placement	a) 5–10 min b) 10 min	Licence is free for clinic/ academia	Yes, but often by proxy; pediatric version under development
Scar quality by the physician's/ researcher (CROM)	POSAS 3.0 (www.posas.nl)	3, (6, 9,) 12 months after tissue engineered product placement	a) 3 min b) 5 min	Licence is free for clinic/ academia	Yes
Visual outcome (CROM)	Photography, any camera will do but use the same within a study and for every visit	Pre-operation, per-operation, post-operation (1 week and 1 month after application), 3, (6, 9,) 12 months after application	a) 2 min b) 1 min	Low costs	Yes
Initial take rate (CROM)	Assessment by an experienced physician/(research) nurse	At 5–7 days	a) 2 min b) 1 min	No/low costs	Yes
Re-epithelialization (CROM)	Assessment by an experienced physician/(research) nurse	At 5–7 days and at 1 month after application	a) 2 min b) 1 min	No/low costs	Yes
Wound healing complications	Assessment by an experienced physician/(research) nurse	Till wound has healed	a) 2 min b) 1 min	No/low costs	Yes
Erythema and melanin (BPM)	Colorimeter DSM (Cortex; https://www.cortex.dk/)	3, (6, 9,) 12 months after tissue engineered product placement	a) 3 min b) 10 min	Additional costs for the device	Yes
Elasticity / Pliability (BPM)	Cutometer (Courage-Khazaka Electronic GmbH (https:// www.courage-khazaka.com/ en/)	3, (6, 9,) 12 months after tissue engineered product placement	a) 5–10 min b) 20 min	Additional costs for the device	Yes, if the child is able to stay still for a certain time (12 s per measurement)

6. All measurements should be performed in a neutral (i.e., most relaxed) position.
7. All assessments are only useful if exact location (body area), time-point, and results are well documented.

3.3 Equipment/ Experience

Vital for high quality outcome assessment is specialized training and having trained personnel available at all time points of assessment (*see* **Note 2**). In addition, implementing standard operating procedures (SOPs) or work instructions (WIs) ensures that measurements and analyses are always done in the same way (*see* **Note 3**). We advise using the same instrument for every assessment in the same patient and within a clinical trial. Different versions of instruments or software can yield different outcome (e.g., POSAS 2.0 vs 3.0, colorimeter DSM II vs III, Cutometer SEM 474 vs MPA 580). Information about outcome assessment in children and remote assessment/home visits is provided in *see* **Notes 4** and **5**.

3.4 Order of Measurements

To minimize any potential influence on the outcome parameters, the following order of measurements is recommended:

1. Wound healing assessment.
2. Photograph: capturing the initial state without any influence.
3. POSAS (Patient and Observer Scar Assessment Scale): perform the subjective assessment of the scar quality; light physical interaction for observer scale assessment.
4. Colorimeter: obtain accurate erythema and melanin measurement; light physical interaction through placement of the probe on the skin.
5. Cutometer: measure the elasticity of the scar, which involves the most profound physical contact of the devices in the core outcome set.

3.5 Wound Healing Assessment

3.5.1 Initial Take Rate

Objective: To assess the initial take rate, the % of split-skin graft which is adherent, after the treatment with tissue engineered products.

1. Divide the wound in imaginary 4 quadrants.
2. Determine viability of the graft (take) based on color and aspect in all 4 quadrants.
3. Determine the mean take by adding the take percentage in the four quadrants and divide by 4.

3.5.2 Re-epithelialization Rate

Objective: To assess the re-epithelialization rate, the % of the wound which is epithelialized, after the treatment with tissue engineered products.

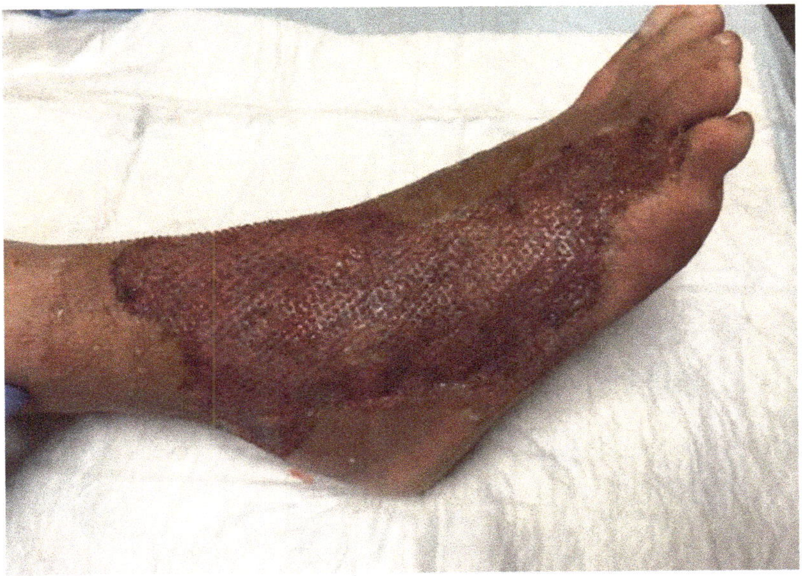

Fig. 1 Photograph of a patient 7 days post tissue engineered product application + split-thickness skin graft placement. Initial take rate: 98%; re-epithelialization rate: 90%

1. Divide the wound in imaginary 4 quadrants.
2. Determine the percentage of the quadrant that is closed with vital epithelium, taking into account the area covered by the transplant, e.g., if a 1:3 expanded meshed skin transplant was used as a transplant and the take of that transplant is 100%, then the minimum area that is epithelialized is 33%. If outgrowth in the meshes of the transplant can be noted, the % epithelialization will be higher.
3. Calculate the total percentage epithelialization by adding the numbers for each quadrant and divide by 4.

Examples are shown in Figs. 1, 2, and 3.

3.5.3 Complications

Objective: To record complications when delayed wound healing is observed, providing valuable insights into potential underlying issues.

1. Document any adverse events noted in the wound area (*see* **Note 6**).

Example is shown in Fig. 4.

3.6 Photographs

Objective: To establish a consistent and reproducible method for taking photographs of scars in patients for the purpose of accurate assessment and monitoring over time.

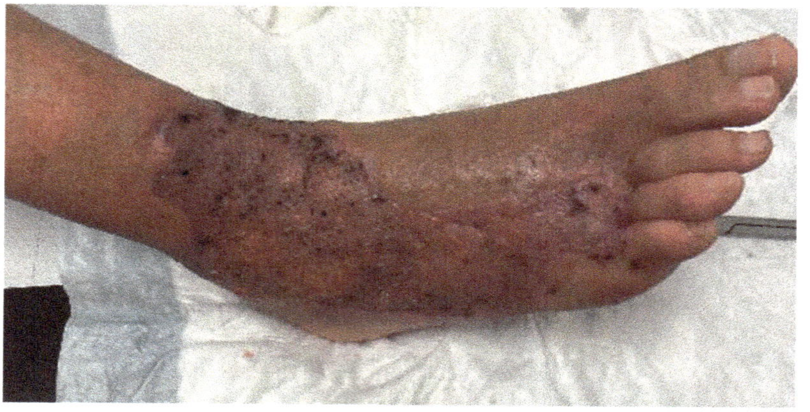

Fig. 2 Photograph of a patient 14 days post tissue engineered product application + split-thickness skin graft placement. Re-epithelialization rate: 99%

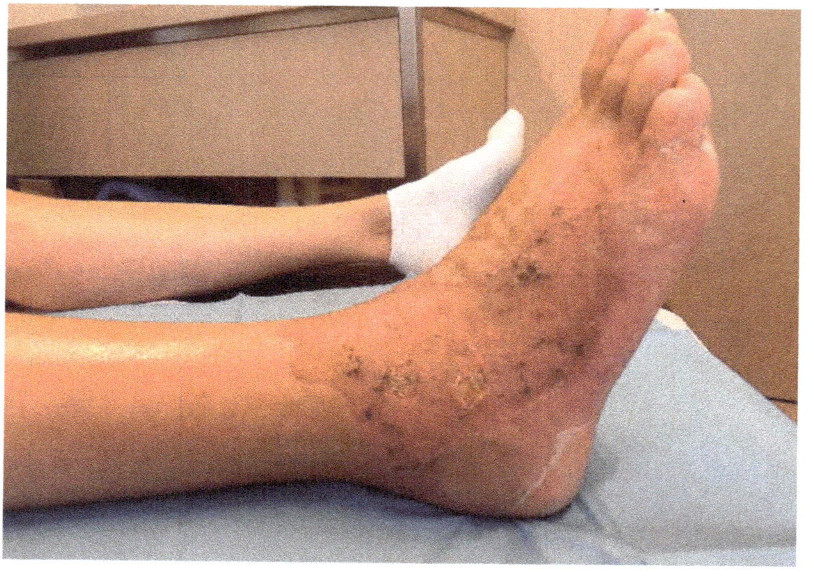

Fig. 3 Photograph of a patient 28 days post tissue engineered product application + split-thickness skin graft placement. Re-epithelialization rate: 100%

1. Follow local hospital protocols for hand hygiene and disinfection.
2. Wear appropriate personal protective equipment (PPE) as per hospital guidelines.
3. Have the patient remove pressure garments from the scar area 30 min before photographing.
4. Use the same digital camera for all follow-ups.
5. Ensure uniform lighting and avoid shadows or glare.

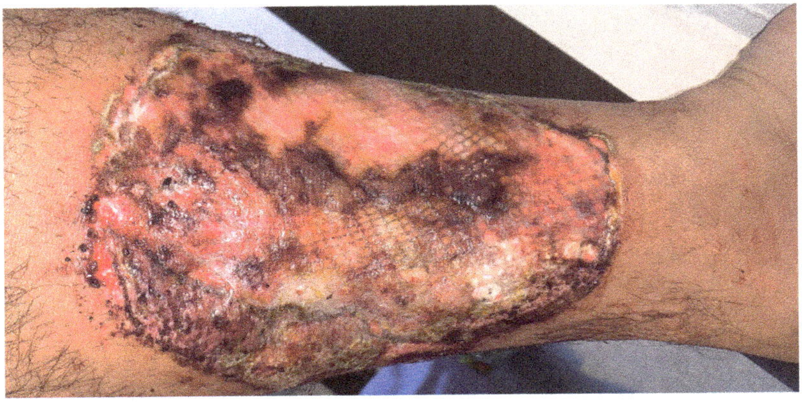

Fig. 4 Photograph of a patient 7 days post tissue engineered product application + Split-thickness skin graft placement. The graft has partially failed, displaying areas of necrosis

6. Use a consistent background.
7. Position the patient consistently and document their posture.
8. Ensure the patient is comfortable with the scar fully exposed.
9. Take one overall and several detailed photos from different angles.
10. Optionally, include a ruler for measurements.
11. Store photos securely, labelling with patient ID/study ID, date, and scar position.
12. Log patient preparation, camera settings, lighting, posture, and other relevant details.
13. For additional analysis methods, *see* **Note 7**.

3.7 POSAS (See Note 8)

Objective: To evaluate scar quality in all types of scars by professionals and patients. Scar quality is a construct referring to the visual, tactile, and sensory characteristics of the scar.

1. Determine which (part of the) scar you want to measure.
2. Complete the Observer Scale by scoring each of the individual items listed in the questionnaire, *see* **Note 9**.
3. Let the patient complete the Patient Scale. Explicitly explain which (part of the) scar the patient has to score. If the patient is unable to fill out the scale themselves, go through the questions of the Patient Scale with the patient, but try not to influence the patient's opinion on the different aspects.
4. For analysis methods, *see* **Note 10**.
5. For an additional PROM to consider for intra-patient clinical trials, *see* **Note 11**.

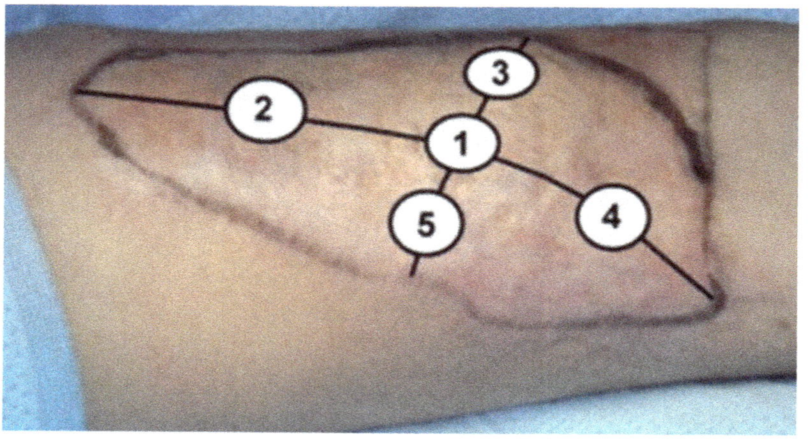

Fig. 5 Example of a scar on the upper arm. The scar is divided by drawing lines through the horizontal and the vertical axis of the scar. Point 1 is measured at the intersection of both lines. Point 2–5 are measured halfway along the lines from the intersection to the edges of the scar. If the scar is too small to allow five measurements, the maximum number of measurements possible should be performed and averaged for statistical analysis

3.8 Colorimeter DSM

Objective: To accurately measure the erythema and melanin levels of a scar using the Colorimeter DSM, compared with unaffected skin.

1. Clean/Calibrate the Colorimeter DSM (*see* **Note 12**).
2. Press the ON/MEASURE button to turn on the instrument. An "auto shut-off" feature automatically turns off power after 60 s of inactivity.
3. Press the MODE/CAL button briefly to switch between display mode: E&M for erythema and melanin indices.
4. Perform measurements on five prestratified locations on the scar (Fig. 5 for location specifics).
5. Place the probe perpendicular to the skin while using minimal pressure to avoid skin or scar bleaching and press the ON/MEASURE button again.
6. Perform one measurement of the healthy skin on the contralateral side. If this is not possible, measure an adjacent location.
7. The white LED light sources in the probe will flash, and the display will show the reading.
8. Record the measurements in the Case Report Form (CRF).
9. For analysis methods, *see* **Note 13**.

3.9 Cutometer

Objective: To measure skin elasticity using negative pressure to mechanically deform the skin (Fig. 6)

1. Clean/Calibrate the Cutometer (*see* **Note 14**).
2. Choose the right settings (*see* **Note 15**).

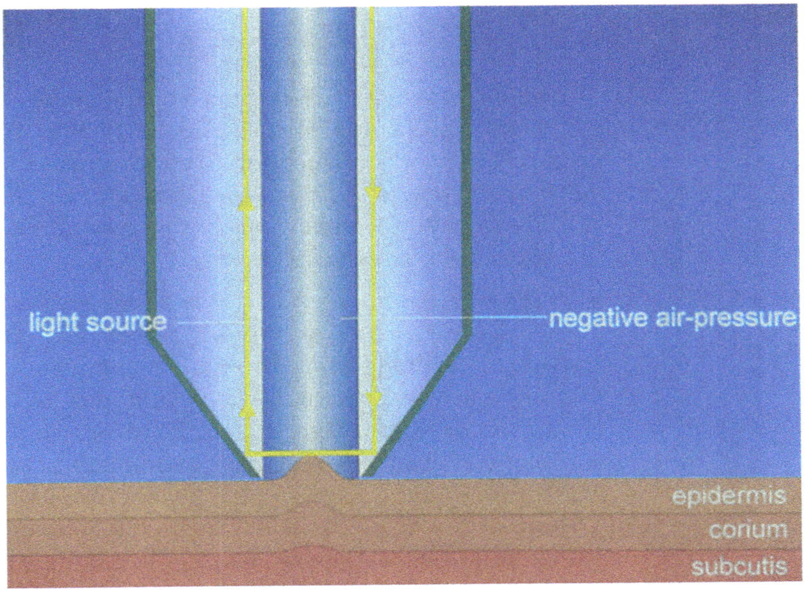

Fig. 6 The Cutometer applies negative pressure and draws the skin into the aperture of the probe, releasing it after a predefined time. Permission to use image from Courage+Khazaka

3. Perform measurements on five prestratified locations on the scar (Fig. 5 for location specifics).
4. Click on the "Measurement"-button in the upper right corner: the device is building up pressure.
5. Check and note the offset parameter (*see* **Note 16**).
6. Place the probe perpendicular on the skin and click "measure" on the computer screen. Held the probe steady during the measurement and ensure constant pressure on the skin.
7. Measuring healthy control skin: two to three measurements on healthy skin are performed on the contralateral side of the body. When this is not possible, these measurements are performed on a location adjacent to scar. This is preferably an anatomical identical location (joints/sharp curvature). Mark the control area with a pencil and take photographs to ensure standardization of location of the measurements over time.
8. When all measurement points are ready, click on "results "in the lower left corner of the screen to see results of all curves simultaneously. Click on "Excel "on the right side to open an Excel®-sheet with all results.
9. Code the data for the patient and save it in the correct folder.
10. Clean the probe after the measurement and cover the probe with the supplied protection cap.
11. For a detailed analysis of the results, *see* **Note 17**.

4 Notes

1. The time points for the assessment of the initial take rate and the re-epithelialization rate should be defined in the study protocols or work instructions. We recommend to assess the initial take rate and the re-epithelialization rate after 5–7 days and 21–28 days. Complications can be monitored at any time throughout the assessment period.
2. Vital for high quality outcome assessment is specialized training and having trained personnel available at all time points of assessment. Training for the above outcome assessments must be comprehensive, practical, and continuous. It should focus on theoretical knowledge, standardized protocols, hands-on practice, and feedback. By incorporating these key features, training programs can ensure that healthcare and research professionals are well-prepared to conduct accurate, reliable, and meaningful outcome assessments. We recommend to train enough personnel to have a backup person for each measurement tool. As a time saver and for feasibility, one may consider to perform the assessments with two persons, one person handling the measurement probe and another person doing the documentation.
3. Implementing standard operating procedures (SOPs) or work instructions (WIs) ensures that measurements and analyses are always done in the same way. To obtain reliable and valid results, it is essential that all trained personnel is trained according to the same SOPs/WIs, and that training is registered on a training log, which is to be coordinated by one or two trained and experienced colleagues (coordinators). In case the SOP/-WIs are updated, all trained personnel are additionally trained. New personnel is to be trained officially and includes doing a brief internship along a trained person/a coordinator, and performing a number of assessments under the eye of a trained person/a coordinator which are thereafter evaluated and discussed. We also recommend to, once a year, as a team, perform the core outcome set in one volunteer/patient, and compare and discuss the obtained scores and data.
4. Outcome assessment in children [11]:
   - Child-friendly environment: create a welcoming environment with lots of fun/toys to reduce anxiety. Depending on the character and age of the child, distract well or explain each step of the procedure to the child in advance, helping them to feel more comfortable.
   - Age-appropriate communication: use language and explanations that are suitable for the child's age. Visual aids can be

helpful for younger children to understand the procedures. Consider age differences in children's understanding of health-related terms.

- Parental involvement: involve parents, caregivers, a nurse, or preferably a medical pedagogical employee in the assessment process. Their presence can comfort the child, and they can assist with understanding. However, try to not let them involve the subjective measurement outcomes.
- Time management: keep the visits as short as possible.
- Scheduling: if children cannot stay immobile for the time of a measurement-series and a procedure is being scheduled with anesthesia, measurements can be performed during this time. Also, a reduced number of outcome assessments or time points can be chosen to shorten the total time (but with loosing statistical power).
- PROMs by proxy: the use of a proxy might lead to different outcomes as opposed to when the patient would fill out a questionnaire themselves. For proxies it can be difficult to answer questions about pain and itch; a child might have several proxies depending on who is accompanying them to the visit; children might answer the questions differently depending whether their parents are in the room or not. We advise to register along with the outcome data who the proxy is, and try to arrange the same proxy for every visit.

5. All recommended outcome measurements can equally be done in an out-patient setting, in rehabilitation or at the patient's home. The Cutometer and the Colorimeter are both transportable although, the Cutometer needs to be handled with special care and calibration has to be checked after each transport.
6. For assessing complications following treatment with tissue engineered products, it is recommended to document any adverse events noted in the wound area. This includes swelling, redness, hypergranulation, necrosis, signs of wound colonization or infection, and bacteriology results. Documenting these complications is crucial as they directly relate to various aspects of wound healing. Figure 4 shows a photograph of a patient 7 days after the application of a tissue engineered product and split-thickness skin graft, with partially failed graft take due to necrosis.
7. If interested in using photographs for more than illustration of visual outcome, it is possible to assess scar surface area or scar volume by calculating length times width while using a ruler, and/or using digital innovations like 2D analysis on the computer or 3D imaging. 3D scar surface area measurement is more valid and reliable than measurements with 2D techniques; however, the devices available are expensive and analysis is

time-consuming making it less feasible for the clinical setting. Several mobile 3D applications exist and these may provide cheaper and quicker data feasible for daily practice but their reliability and validity has not yet been fully established.

8. The first version of the Patient and Observer Scar Assessment Scale (POSAS) has been introduced in 2004 by prof dr. P.P.-M. van Zuijlen. In 2005, after some minor adjustments, the second version followed, POSAS 2.0. This version is available in at least 25 languages. In 2022 the third version of the POSAS has been developed by gaining input from international scar experts (i.e., patients as well as professionals) on its content. The POSAS 3.0 is available in at least ten languages with more translation in progress. Most of the translations are also culturally validated. For more information or to request permission for translation, visit the POSAS website at www.posas.org or contact the POSAS team at info@posas.nl.

   The POSAS includes a Patient Reported Outcome Measurement (PROM), the Patient Scale and a Clinician Reported Outcome Measurement (CROM), the Observer Scale. The POSAS patient and observer scales are cognitively simple and take approximately 5 min to complete. Each item is scored on a 5-point scale (version 3.0). The lowest score is "not" and corresponds to the normal skin or no complaints (normal pigmentation, no itching etc.). The highest score is "extremely" and corresponds to the situation of the worst imaginable scar or sensation.

   The POSAS 3.0 is available for linear scars caused by trauma or surgery and generic scars, caused by all types of traumas, including burns. The patient scale contains an overall satisfaction question and 16 questions (generic) or 17 questions (linear scar) on specific items. The pediatric patient scale of the POSAS is in development. The observer scale contains an overall quality question and seven questions (generic) or nine questions (linear scar) on specific items. Additionally, observer category boxes are available to score nominal items (e.g., hypo/hyper/mixed pigmentation). Ideally, the Patient Scale and the Observer Scale should be completed independently. If this is not possible, fill out the Observer Scale first to avoid potential influence of the patient's opinion. Likewise, do not share the completed Observer Scale with the patient.

9. The characteristics of the scar should be compared to the unaffected skin, e.g., the skin before the injury, using the skin on the contralateral side for the best comparison or the surrounding skin. The directions for use of the Observer Scale POSAS 3.0 are as follows:

- Vascularity: Presence of vessels in scar tissue assessed by the amount of redness, tested by the amount of blood return after applying light pressure with the index finger.
- Pigmentation: The light (hypopigmentation) or dark (hyperpigmentation) color of the scar, which remains after eliminating the influence of the vascularization on skin color by pressing on the scar with your finger.
- Relief/texture: Irregularities of the scar surface (surface roughness or relief) are particularly seen after split skin autografting by using a meshed split skin graft and can be tested by moving your fingers over the scar surface. The irregularities result from secondary healing of the interstices of the meshed skin graft.
- Thickness: Gently move your fingers over the scar to feel if the scar is raised (thicker) or sunken relative to normal skin (thinner).
- Pliability: Pliability is evaluated by picking up a scar, pinch it between thumb and index finger, and rolling it between the fingers.
- Surface area: The surface area of scars can either reduce or enlarge as a result of scar contraction or expansion, respectively. Let the patient move the body part containing the scar to test this.

10. The individual items of the POSAS can be analyzed separately to allow analysis of differential aspects of a scar and/or total scores can be compared between different scar areas. For the interpretation of the POSAS 3.0 results, use the sum score of items 1–16/17 (patient) and items 1–7/9 (observer), excluding item 0 about the general satisfaction/overall quality. Compare these scores within groups of patients and observers or within the same patient/observer over time. Since the POSAS is a subjective measurement instrument, there is no threshold defined for good or bad scars. The minimal important change (MIC) and the minimal clinically important difference (MCID) of the POSAS 3.0 will be determined in the near future.
11. Another PROM to consider for intra-patient clinical trials is patient preference regarding skin grafts with or without the use of tissue engineered products. Patients are asked, "If you had a new comparable (burn) wound requiring skin grafting, which technique would you prefer?" with three options: tissue engineered products plus skin grafting, skin grafting alone, or no preference. This method likely precisely weighs all the advantages and disadvantages from the patient's perspective. This measure helps gauge patient perspectives and preferences, contributing to a more patient-centered approach in evaluating the effectiveness of different skin grafting techniques.

12. For normal cleaning use a dry cloth. For more thorough (antibacterial) cleaning, use isopropanol or Incidin foam. Note that alcoholic cleaning agents can be aggressive to plastic materials. Keep the calibrator out of direct sunlight. Store it in a dark place to prevent color change due to UV-radiation. Regular re-calibration is recommended, e.g., once a week, although it is not necessary to re-calibrate each time the unit is turned on. Calibration is performed on the white surface on the supplied calibrator. Hold the probe against the surface, then press and hold the MODE/CAL button a few seconds. Once calibrated, the display will show 0.
13. Calculate the erythema score by subtracting: erythema index scar—erythema index uninjured skin. The absolute value is taken to represent the deviation from normal skin. This is particularly relevant in analyzing larger group data. However, it should be realised that positive (more red) as well as negative (paler) values can occur. Similarly, the melanin score is calculated.
14. With each measurement of the Cutometer®, sebum and residuals of products on the skin enter the probe. This might eventually result in spoiling of the measurement prisms inside the probe. Therefore, the probe should be cleaned regularly (preferably before/after each measuring session every day) with isopropanol and the small brush that is delivered with the instrument. Clean the probe on a weekly basis: (1) Insert one of the supplied brushes into the liquid and shake off any surplus; (2) Move the brush in the probe opening up and down while turning it. Be careful not to damage the glass prisms inside; (3) Wipe the probe with a soft, dry tissue. Let it dry completely.

    The Cutometer works with highly sensible probe(s). The accurate calibration can be checked anytime. You need the supplied check calibration cap. Please proceed as follows every time before you measure a different patient: (1) Clean the opening of the probe thoroughly as described and let it dry completely before continuing; (2) Call up the command "Check calibration" under "Measurement," a window with instructions appear; (3) Put the probe on the flat side of the check calibration cap and press the "start"-button. Compare the displayed measurement value (Calvalue) with value A on the cap. The value should fall into the tolerance mentioned on the cap; (4) Put the cap tightly on the probe head cap and press the "start"-button. Turn the cap around for a full turn without loosen the pressure on the cap. Compare the displayed measurement value (Calvalue) with value B on the cap. The value should fall into the tolerance mentioned on the cap; (5) In case one or both values are not within tolerance, clean the probe opening again with the special brush and repeat (3) and (4). In case the probe is very dirty, clean the inside of the probe again. The inside must be absolutely dry for the check. Please take care not to damage the glass prisms facing each other at the inside in the probe opening. If the values remain out of tolerance, please contact the service. The check calibration does not replace a regular recalibration in the factory as any measurement tool must have. It is recommended to send the probe in for calibration once per year. Each calibration will be accompanied by a calibration certificate.

15. These settings are recommended for (burn) scars by the Manufacturer when calculating Ua, Ur, Ue, Uv, and Uf altogether:
    - Measurement mode: 1
    - Pressure: 450 mbar
    - Probe aperture: 6 mm Æ
    - On-time: 2 s
    - Off-time: 2 s
    - Repetitions: 3
    - Total time: 12 s
16. Applying pressure to the probe of the Cutometer causes a mechanical deformation of the skin underneath the probe aperture, pressing the skin into the probe [12]. The extent of this deformation is being displayed as the offset value (mm). The more pressure the researcher applies on the probe, the larger the skin deformation and thus the offset value will be. The offset value forms the zero base of the Cutometer measurements. So, the true skin deformation (offset + Uf) is larger than the skin deformation that is displayed in the Cutometer curves (Uf). A high offset value may also be caused by a dirt in the Cutometer probe. Therefore, the observer should always clean the probe before starting the measurements.

    Due to the influence of the offset value on the reliability of the test, the observer should always (1) Try to keep the offset value as low as possible. This can be achieved by minimizing pressure on the Cutometer probe. If there is an outlier, repeat the measurement; (2) When doing multiple measurements on the same scar, attempt to keep the offset values between measurements comparable to each other; (3) Always document the offset value of a measurement.
17. The Cutometer curve displays the viscoelastic properties of the skin. Each phase (suction and relaxation) consists of two parts:
    - Suction Phase: During the first part of the suction phase, skin enters the probe immediately and straight. This is the immediate elastic deformation and described as

Ue. Normally, Ue is reached after 0.1 s of the suction time. In the second part of the suction phase the skin "creeps" into the probe. This part represents the viscoelastic part of the suction phase, Uv. The more elastic a material, the smaller Uv. As age increases, Uv becomes higher. The maximum penetration after the suction time is displayed as Uf.

- Relaxation Phase: Due to viscoelastic properties of the skin, the complete relaxation (Ua) can be divided into two parts: the immediate elastic return Ur and the flat, viscoelastic part Ua—Ur. Young, highly elastic skin will have a very similar Ur and Uf, in older skin Ur becomes visibly smaller. Uf—Ua shows the overall ability of the skin to return to its original shape.
- Relevance of including uninjured skin: skin elasticity varies with anatomical location. Therefore, in comparing different scars on different anatomical locations, comparing elasticity parameters of a scar area relative to its comparative uninjured skin will eliminate these differences (Fig. 7).

The output of the Cutometer is represented by parameters R0 to R9. With the following formulas, it is possible to calculate the parameters Ua to Uv (Table 2).

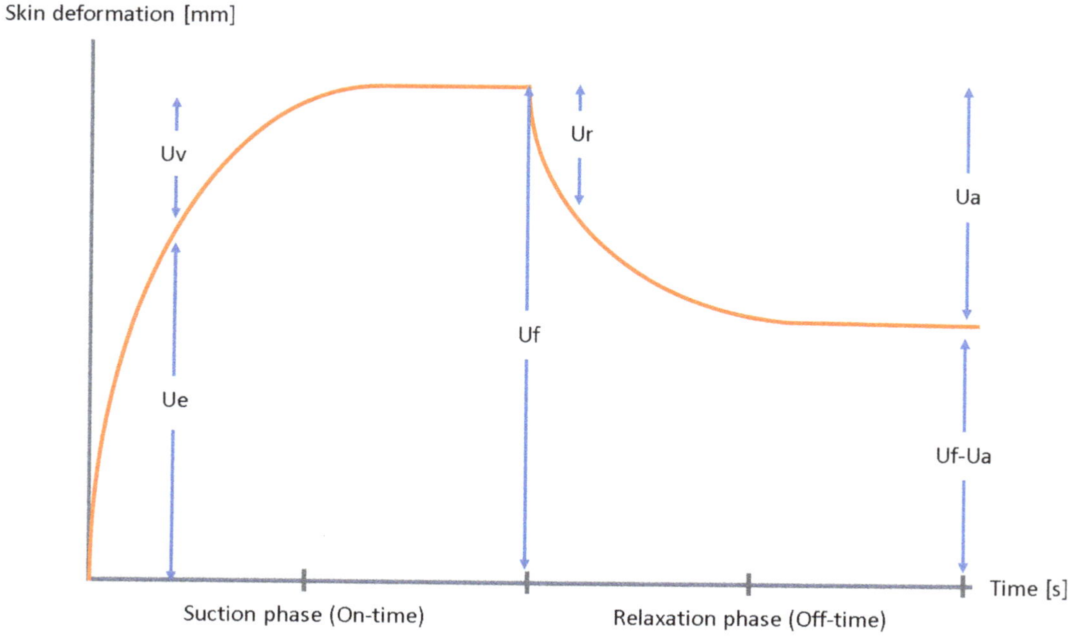

Fig. 7 The Cutometer measurement curve is divided into several parts, which are explained in more detail in this section. Reproduced and slightly modified from Klimitz et al., 2023 [13] under the terms of the Creative Commons CC BY license

Table 2
Calculation parameters Ua to Uv

Pliability	Ua	RO-R1 and approximately the same R3-R4
Retraction	Ur	R0*R7
Elasticity	Ue	(R0*R7)/ R5
Viscoelasticity	Uv	(R0*R7*R6)/ R5
Maximum extension	Uf	RO

Recommendations

We want to encourage all physicians and researchers to publish the results of their studies and evaluations on the clinical application of tissue engineered products, to increase the knowledge in this field.

References

1. van den Bosch AS, Verwilligen RAF, Pijpe A et al (2024) Application of dermal substitutes in the surgical treatment of full-thickness wounds: outcomes of an international survey. Int Wound J 21(7). https://doi.org/10.1111/iwj.14932
2. Mokkink LB, Terwee CB, Patrick DL et al (2010) The COSMIN study reached international consensus on taxonomy, terminology, and definitions of measurement properties for health-related patient-reported outcomes. J Clin Epidemiol 63(7):737–745. https://doi.org/10.1016/j.jclinepi.2010.02.006
3. Tyack Z, Simons M, Spinks A et al (2012) A systematic review of the quality of burn scar rating scales for clinical and research use. Burns 38(1):6–18. https://doi.org/10.1016/j.burns.2011.09.021
4. Mundy LR, Miller HC, Klassen AF et al (2016) Patient-reported outcome instruments for surgical and traumatic scars: a systematic review of their development, content, and psychometric validation. Aesth Plast Surg 40(5):792–800. https://doi.org/10.1007/s00266-016-0642-9
5. Vercelli S, Ferriero G, Sartorio F et al (2015) Clinimetric properties and clinical utility in rehabilitation of postsurgical scar rating scales. Int J Rehabil Res 38(4):279–286. https://doi.org/10.1097/MRR.0000000000000134
6. Durani P, McGrouther DA, Ferguson MWJ (2009) Current scales for assessing human scarring: a review. J Plast Reconstr Aesthet Surg 62(6):713–720. https://doi.org/10.1016/j.bjps.2009.01.080
7. Rennekampff HO, Rabbels J, Reinhard V et al (2006) Comparing the Vancouver scar scale with the Cutometer in the assessment of donor site wounds treated with various dressings in a randomized trial. J Burn Care Res 27(3):345–351. https://doi.org/10.1097/01.BCR.0000216311.61266.00
8. Draaijers LJ, Botman YAM, Tempelman FRH (2004) Skin elasticity meter or subjective evaluation in scars: a reliability assessment. Burns 30(2):109–114. https://doi.org/10.1016/j.burns.2003.09.003
9. Busche MN, Thraen ACJ, Gohritz A et al (2018) Burn Scar Evaluation Using the Cutometer® MPA 580 in Comparison to "Patient and Observer Scar Assessment Scale" and "Vancouver Scar Scale". J Burn Care Res 39(4):516–526. https://doi.org/10.1093/jbcr/irx009
10. van der Wal M, Bloemen M, Verhaegen P et al (2013) Objective color measurements. J Burn Care Res 34(3):e187–e194. https://doi.org/10.1097/BCR.0b013e318264bf7d
11. Matza LS, Patrick DL, Riley AW et al (2013) Pediatric patient-reported outcome instruments for research to support medical product labeling: report of the ISPOR PRO good research practices for the assessment of children and adolescents task force. Value Health 16(4):461–479. https://doi.org/10.1016/j.jval.2013.04.004
12. Bonaparte JP, Ellis D, Chung J (2013) The effect of probe to skin contact force on Cutometer MPA 580 measurements. J Med Eng

Technol 37(3):208–212. https://doi.org/10.3109/03091902.2013.779325

13. Klimitz FJ, Neubauer H, Stolle A, Ripper S, Daeschler SC, Aman M, Boecker A, Thomas B, Kneser U, Harhaus L (2023) Objective burn scar assessment in clinical practice using the Cutometer©: introduction and validation of a standardized measurement protocol. J Burn Care Res 44(1):95–105. https://doi.org/10.1093/jbcr/irac154

Index

Thomas Biedermann and Sophie Böttcher-Haberzeth (eds.), *Skin Tissue Engineering: Methods and Protocols*, Methods in Molecular Biology, vol. 2922, https://doi.org/10.1007/978-1-0716-4510-9,

V

W

GPSR Compliance
The European Union's (EU) General Product Safety Regulation (GPSR) is a set of rules that requires consumer products to be safe and our obligations to ensure this.

If you have any concerns about our products, you can contact us on

ProductSafety@springernature.com

In case Publisher is established outside the EU, the EU authorized representative is:

Springer Nature Customer Service Center GmbH
Europaplatz 3
69115 Heidelberg, Germany

www.ingramcontent.com/pod-product-compliance
Lightning Source LLC
Chambersburg PA
CBHW052317250425
25725CB00005BA/84
* 9 7 8 1 0 7 1 6 4 5 0 9 3 *